*B*iogeography
an ecological and evolutionary approach

Biogeography
an ecological and evolutionary approach

by C. BARRY COX PhD, DSc
and PETER D. MOORE PhD
Division of Life Sciences
King's College London
Franklin-Wilkins Building
Stamford Street
London

SIXTH EDITION

Blackwell
Science

© 1973, 1976, 1980, 1985, 1993, 2000
by Blackwell Science Ltd
Editorial Offices:
Osney Mead, Oxford OX2 0EL
25 John Street, London WC1N 2BL
23 Ainslie Place, Edinburgh EH3 6AJ
350 Main Street, Malden
   MA 02148 5018, USA
54 University Street, Carlton
   Victoria 3053, Australia
10, rue Casimir Delavigne
   75006 Paris, France

Other Editorial Offices:
Blackwell Wissenschafts-Verlag GmbH
Kurfürstendamm 57
10707 Berlin, Germany

Blackwell Science KK
MG Kodenmacho Building
7–10 Kodenmacho Nihombashi
Chuo-ku, Tokyo 104, Japan

First published 1973

Second edition 1976
Third edition 1980
Fourth edition 1985
Reprinted 1988, 1989, 1991
Fifth edition 1993
Reprinted (three times) 1995, 1996
Sixth edition 2000

Set by Graphicraft Limited, Hong Kong
Printed and bound in Great Britain
at the Alden Press, Oxford and
Northampton

The Blackwell Science logo is a
trade mark of Blackwell Science Ltd,
registered at the United Kingdom
Trade Marks Registry

A catalogue record for this title
is available from the British Library

ISBN 0-86542-778-X

Library of Congress
Cataloging-in-publication Data

Cox, C. Barry (Christopher Barry),
1931–
    Biogeography : an ecological
and evolutionary approach /
by C. Barry Cox and
Peter D. Moore. — 6th ed.
     p.   cm.
    Includes bibliographical references.
    ISBN 0-86542-778-X
1. Biogeography.   I. Moore, Peter D.
II. Title.
QH84 .C65   1999
578'.09—dc21   99–14649
                           CIP

DISTRIBUTORS

Marston Book Services Ltd
PO Box 269
Abingdon, Oxon OX14 4YN
(Orders: Tel: 01235 465500
         Fax: 01235 465555)

USA
Blackwell Science, Inc.
Commerce Place
350 Main Street
Malden, MA 02148 5018
(Orders: Tel: 800 759 6102
            781 388 8250
         Fax: 781 388 8255)

Canada
Login Brothers Book Company
324 Saulteaux Crescent
Winnipeg, Manitoba R3J 3T2
(Orders: Tel: 204 837-2987)

Australia
Blackwell Science Pty Ltd
54 University Street
Carlton, Victoria 3053
(Orders: Tel: 3 9347 0300
         Fax: 3 9347 5001)

For further information on
Blackwell Science, visit our website:
www.blackwell-science.com

Contents

Acknowledgements, ix

1 **Introduction to biogeography, 1**
References, 9

2 **Biodiversity, 10**
How many species are there?, 10
Gradients of diversity, 14
Biodiversity hotspots, 24
Diversity in time, 26
Marine biodiversity, 31
Summary, 32
Further reading, 32
References, 32

3 **Patterns of distribution, 34**
Limits of distribution, 34
Overcoming the barriers, 36
A successful family: the daisies (Asteraceae), 38
Patterns of dragonflies, 41
Magnolias: evolutionary relicts, 44
Climatic relicts, 45
Endemic organisms, 49
Physical limitations, 51
Environmental gradients, 52
Interaction of factors, 56
Species interaction, 59
Reducing competition, 63
Predators and prey, 67
Summary, 70
Further reading, 70
References, 70

4 **Communities and ecosystems, 72**
The community, 72
The ecosystem, 75
Ecosystems and biodiversity, 77
Biotic assemblages on a global scale, 80
Patterns of climate, 84
Climate diagrams, 88

Modelling biomes and climate, 88
Biomes in a changing world, 92
Summary, 93
Further reading, 93
References, 93

5   The source of novelty, 95
Natural selection, 95
Darwin's theory and Darwin's finches, 96
The controlling force within the organism, 100
From populations to species, 101
Barriers to interbreeding, 103
Polyploids, 104
Clines and 'rules', 105
Competition for life, 106
Controversies and evolutionary theory, 107
Evolution and the human race(s), 109
Summary, 110
Further reading, 110
References, 110

6   Patterns in the past, 112
Plate tectonics, 113
The evidence for past geographies, 115
Changing patterns of continents, 118
Early land life on the moving continents, 119
One world—for a while, 120
The early spread of mammals, 124
The great Cretaceous extinction event, 125
The rise of the flowering plants, 127
Late Cretaceous and Cenozoic climate changes, 128
Late Cretaceous and Cenozoic floral changes, 131
Summary, 133
Further reading, 134
References, 134

7   Patterns of life today, 136
Mammals: the final patterns, 137
The distribution of flowering plants today, 140
Mammalian vs. flowering plant geography: comparisons and contrasts, 142
Floral patterns in the Southern Hemisphere, 144
The Old World tropics: Africa, India and South-East Asia, 144
Madagascar, 148
The Cape flora, 148
Australia, 149
Wallacea, 151
New Zealand, 152
South America, 152

The Northern Hemisphere: Holarctic mammals and Boreal plants, 155
Summary, 158
Further reading, 159
References, 159

8 Interpreting the past, 161

Phylogenetic biogeography, 162
Cladistic biogeography, 163
Panbiogeography, 164
Phyletic tracks and patterns, 167
Endemicity and history, 169
Pleistocene problems, 170
Centres of dispersal, 172
Fossils and historical biogeography, 172
Palaeobiogeography, 173
Summary, 173
Further reading, 174
References, 174

9 Ice and change, 175

Climatic wiggles, 176
Interglacials and interstadials, 177
Biological changes in the Pleistocene, 182
The last glacial, 183
Causes of glaciation, 192
Summary, 195
Further reading, 195
References, 195

10 The making of today, 197

The current interglacial: a false start, 197
Forests on the move, 200
The emergence of humans, 203
Modern humans and the megafaunal extinctions, 206
Domestication and agriculture, 207
The dry lands, 214
Changing sea levels, 215
Time of warmth, 217
Climatic deterioration, 218
The environmental impact of early human cultures, 220
Recorded history, 221
Summary, 221
Further reading, 222
References, 222

11 Projecting into the future, 224

The changing climate, 224
Nitrogen and sulphur overload, 227

Biogeographical consequences of global change, 229
Changing communities and biomes, 234
Summary, 236
Further reading, 236
References, 236

12   Drawing lines in the water, 238
Introduction, 238
Zones in the ocean and upon the sea floor, 239
The basic biogeography of the seas, 241
The open-sea realm, 241
   The history of the ocean basins, 241
   The dynamics of the ocean basins, 243
   The biogeography of hydrothermal vent faunas, 251
The shallow-sea realm, 251
   Faunal breaks within the shelf faunas, 252
   Trans-oceanic links and barriers between shelf faunas, 253
   Latitudinal patterns in the shelf faunas, 254
   Coral reefs, 255
Summary, 259
Further reading, 259
References, 260

13   Life (and death) on islands, 262
Types of island, 262
Getting there: problems of access, 263
Dying there: problems of survival, 266
Integrating the data: The Theory of Island Biogeography, 267
Second thoughts about the theory, 269
The Theory of Island Biogeography and the design of nature reserves, 271
Starting afresh: the story of Rakata, 273
Evolving there: opportunities for adaptive radiation, 279
The Hawiian Islands, 280
Summary, 288
Further reading, 288
References, 288

Index, 291

Acknowledgements

Barry Cox wishes to express his grateful thanks to Dr Martin Angel of the Southampton Oceanography Centre for his great help in advising on Chapter 12. He would also like to thank Dr Robert Whittaker of the Department of Geography, University of Oxford, for his comments on the section on the island of Rakata, and Professor Robert Hall of the South-East Asia Research Group, Department of Geology, Royal Holloway College, University of London, for providing the basemaps for Figures 6.3 and 6.4.

Peter Moore would like to thank Professor Martin Kemp of the Department of Geographical Sciences, University of Plymouth, for providing an extensive and detailed review of the previous edition of this book. He would also like to express his thanks to Colin Prentice of Lund University, Sweden, for his willingness to provide the data for the two biome images reproduced in Plate 4.2.

CHAPTER 1: *Introduction to biogeography*

There is one thing that we all have in common; we all share the same planet. For all of us this is home. For this reason, and also because rising human population and declining resources are placing the earth under greater strain, we are now looking to the scientists who study the earth and its living creatures to advise us how best to manage the planet to ensure its future, and with it our own.

Among the sciences involved in this difficult but vital task is biogeography, the study of living things in space and time. Biogeographers seek to answer such basic questions as why are there so many living things? Why are they distributed in the way that they are? Have they always occupied their current distribution patterns? Is the present activity of human beings affecting these patterns and if so, what are the prospects for the future?

One of the most impressive features of the living world is the sheer diversity of organisms it contains, and one of the main problems facing a biogeographer is how to explain this diversity, and also the reasons for the varying patterns of occurrence of different species over the surface of the planet. Why, for example, is there more than one species of seagull? And why do different species of gull have different patterns of distribution, some being widespread and others very local? Why are there so many different types of grass growing in the same field, all apparently doing precisely the same job? Why are there more species of butterfly in Austria than in Norway? It is the task of the biogeographer not only to answer such specific questions, but also to seek general rules that can account for many such observations, and which will provide a general framework of understanding that can subsequently be used for predictions about the consequences of tampering with the natural world.

One of the world's foremost experts on biological diversity, Edward O. Wilson of Harvard University [1], has claimed that the diversity of life on earth was greater at the time of the origin of the human species than it had ever been before in the course of the earth's history. The arrival and cultural development of our species has evidently had, and continues to have, a profound impact on the world's biogeography, modifying species' ranges and bringing some to extinction. From the point of view of biological diversity (the subject of Chapter 2), the evolution of humankind was something of a catastrophe, and it would be quite unrealistic to attempt any synthesis of biogeography without taking the human impact into account. Very few species escape the effects of human activity in some aspect of their ecology and distribution. For this reason our species will play an important role in this book, not only in terms of our influence on other species of plant and animal, but also because we too are one species among many, perhaps as many as 30 million other species, and we obey essentially the same rules as the others. The more we can understand about the Hawaiian goose, the oak tree and the dodo, the more we shall appreciate our own position in the order of things.

Because it faces such wide-ranging questions, biogeography must draw upon an extensive range of other disciplines. Explaining biodiversity, for example, involves the understanding of climate patterns over the face of the earth, and the way in which the productivity of photosynthetic plants differs with climate and latitude. We must also understand what makes particular habitats desirable to animals and plants; why locations of particular soil chemistry, or moisture levels, or temperature range, or spatial structure, should be especially attractive. Hence, climatology, geology, soil science, physiology,

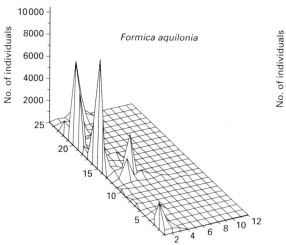
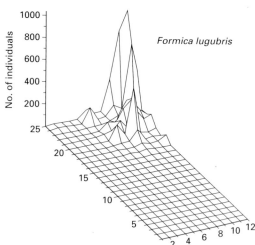

Fig. 1.1 The abundance of two ant species caught in a grid of pitfall traps in the boreal coniferous forest of southern Finland. The numbers on the grid refer to 5-m intervals of trap spacing, so the entire grid is 125 × 60 m, with 300 traps. The two ant species can be seen to exclude one another spatially. Data from Niemela *et al.* [2].

ecology and behavioural sciences must all be invoked to answer such questions.

But the answers to this kind of question are very dependent on the spatial scale at which we approach them. Two species of ant, for example, may share the same continent, even the same area of woodland, but they are unlikely to share the same nest space. When we examine their spatial patterns at a scale of square metres rather than square kilometres, the two species may never be found together. Biogeography concerns itself with all levels of scale, as will be discussed in Chapter 3.

Let us look at an example of the effects of scale at a local level. In a very detailed examination of the small-scale biogeography of an area of boreal coniferous forest in southern Finland, some researchers from the University of Helsinki [2] set a grid of traps, sunk into the ground ('pitfall traps'), in which they gathered those hapless, ground-dwelling insects and invertebrates that happened to stumble into them. In this way they could detect exactly where in their

study area these small, rambling creatures were foraging. They laid out a grid measuring 125 m on one axis and 60 m on the other, with traps placed at 5-m intervals, 300 in total. They recorded all the ants and the ground beetles (carabids) that fell into these traps, pooling their records for an entire summer, and some of their results are shown in Figs 1.1 and 1.2.

In Fig. 1.1 we see the pattern of abundance of two ant species that occupy the area. They have very sharp peaks in abundance relating to their nest sites, with *Formica aquilonia* having a number of nest centres and *F. lugubris* having just one main nest. Although the two species are found together when examined at a sample size of hundreds of square metres, they are quite strictly separated when we look in greater detail. Competitive interaction between the colonies is evidently leading to strict spatial segregation.

In Fig. 1.2 we have displayed the results for some of the ground beetles occupying the same area. The first two shown, *Leistus terminatus* (a) and *Cychrus caraboides* (b) have quite similar patterns of distribution. Both are commoner towards the bottom right part of the sample plot, and when we compare their patterns with those of the two ants it is tempting to conclude that they avoid areas where the ants are densest. Such negative relationships between

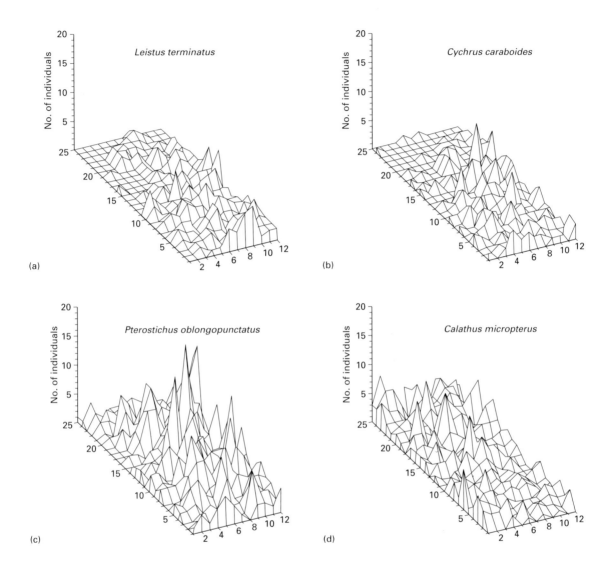

Fig. 1.2 The trap grid shown in Fig. 1.1 but with the records for four species of ground beetle. The species shown in (a), (b) and (c) decline in abundance in the area rich in ants, whereas species (d) is more abundant in the ant regions. It probably preys upon ants. Data from Niemela *et al.* [2].

ground beetles and ants are often the case in nature. The fact that the two species of beetle are found together suggests that they have no such scruples about sharing space with one another, so they probably do not consume one another, nor, in all likelihood, do they compete for food, otherwise one or the other would probably be excluded by the better competitor. *L. terminatus* preys upon tiny animals (the springtails and mites) that feed in leaf litter on decomposing plant material. *C. caraboides*, on the other hand, is a specialist feeder that preys on molluscs—slugs and snails. So the two have quite different food requirements and can cohabit in an area without any negative interactions.

A third beetle, *Pterostichus oblongopunctatus* (c), is more abundant than either of the other two, and also overlaps with them in its range. It is a generalized predator so, although it may well compete with the other two, it has a wider range of food resources to choose from, allowing it to coexist with them. As in the case of the former two species, however, it also seems to avoid the ants. The fourth beetle, *Calathus micropterus* (d), has a rather different pattern. Unlike species a, b and c, it appears to be more abundant at the top end of the sample area. At present it is not known precisely what this species eats, but many of its close relations prey upon ants, so this is the most likely explanation for its abundance in the ant-infested parts of the study area.

This example of a detailed local study of animals reveals how complex the patterns of organisms can be when examined at this scale. This set of data can mainly be explained by feeding relationships, but similar patterns of other organisms often relate to such factors as temperature differences, light microclimates, humidity, etc., or some animals may be active at different times of day (or night), so adding a time dimension to biogeography even at this scale of study.

The interaction between organisms (eating one another, competing with one another for food, etc.) is clearly very important in biogeography, and may determine whether or not different species can be found together in communities (a concept examined further in Chapter 4). These interactions can become so complicated that an alteration in the abundance of one animal or plant can often have very unexpected consequences for the rest of the community, and it is extremely important that ecologists and biogeographers should be aware of these relationships if they are to be capable of predicting the outcome of environmental change, or of the adoption of certain land-use or management practices.

A particularly intricate example of the potential complexity of these interactions has been described by Clive Jones of the Institute of Ecosystem Studies at Millbrook, New York, and his colleagues [3]. They examined the relationship between oak trees, mice, moths, deer, ticks and people in the eastern United States and found a complicated web of interactions.

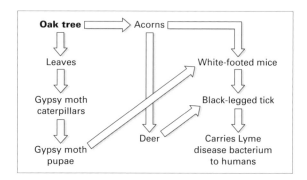

Fig. 1.3 A complex web of interactions relating acorn production by oak trees to mouse and deer abundance and hence the risks of humans contracting Lyme disease from the black-legged tick. Changes in the crop of acorns can affect all other parts of the system.

Oak trees produce a crop of acorns during the autumn ('masting'), but in some years the crop is much bigger than others. Roughly every 2–5 years comes a good 'mast year'. The acorns are eaten by white-footed mice, but mast years also attract deer, which spend 40% of their time in the oak woods in these mast years compared with only 5% in non-mast years. A third consumer of oak is the gypsy moth, whose caterpillars feed on oak leaves and may, in plague years, when its population rises to unusually high levels, defoliate the trees to such an extent that the acorn yield is reduced. Just to add to the complexity, mice eat the soil-dwelling pupae of gypsy moths and can control moth populations in that way (Fig. 1.3).

White-footed mice carry a resident spirochaete bacterium called *Borrelia burgdorferi* and they pass it on to parasitic ticks that initially feed on the mice and then move on to deer, thus spreading the disease. A tick carrying this bacterium may also attach itself to a human wanderer in the woods, who will then contract Lyme disease. The web is given a dynamic element by the occurrence of mast years, because when acorns are abundant mice populations rise, deer spend more time in the woods and the ticks consequently have a particularly successful time. This was demonstrated experimentally by the researchers by artificially supplying extra acorns and then watching the mouse population expand, the deer move into the oak forest, and the ticks proliferate. In a separate experiment, the mice population was artificially reduced by trapping, and the gypsy moth population then rose with the potential to reduce first oak foliage and, subsequently, acorn supply to the system. In the natural situation, tick abundance, and hence risk to human health, is greatest two seasons after a mast year because of the time taken to build up a population of mice. It is therefore possible to predict health risks to wood-walking humans on the basis of an understanding of this complex set of interrelationships within a forest community. This is just one example that has been particularly well studied, but biogeographers need to be aware of the existence of such complexities in the systems they study.

The time factor is also clearly important in biogeography and, like space, it needs to be considered at different scales. Bats and barn swallows hunt flying insects at different times in the diurnal cycle, so they can coexist. Fluctuations in climate from year to year may influence masting in oak trees and plagues in herbivorous insects. Climate changes over centuries may have an impact on the distribution patterns of organisms at a greater spatial scale, and even more so when millenia pass and major climatic cycles are involved. Millenia develop into millions of years, and the process of evolution becomes ever more important in the consideration of biogeographers.

The study of fossils has long made it clear that the diversity of living organisms on earth has not always been the same. New species have arisen and old species have become extinct. Appreciating biodiversity therefore involves understanding the mechanisms by which new species arise, and it is an essential part of the biogeographer's work to study the source of novelty, the means by which new species can be generated. We worry, very reasonably, about the current rate of extinction on the planet. But how fast can we expect new species to evolve to replace them? How fast can species adapt to cope with the modifications which humankind is making to the climate and the living conditions

of the world? This aspect of biogeography will be considered in Chapter 5.

We can observe species changing in their form, behaviour and physiology even as conditions change in the present day, and we can see from the fossil record how they have become modified and adjusted to changing conditions in the past. Take the woodrat, for example. There is now strong evidence that this animal has changed in its body size according to climatic conditions in the geologically recent past. There is an observation in biogeography, often referred to as Bergmann's Rule, that animals in warm climates tend to be smaller than their close relatives in cold climates. Large animals, it seems, are more liable to be killed by very high temperature than small ones because of their failure to lose body heat fast enough. Small animals, on the other hand, may radiate too much heat under cold conditions when they cannot afford to do so. In the case of the North American bushy-tailed woodrat, it has proved possible, on the basis of fossil records, to study changes in size over the past 14 000 years. The study, by Felisa Smith of the University of New Mexico and her colleagues [4], does not actually use fossils of the animals themselves, however, but is based on their faecal pellets found in waste heaps, or middens, that the woodrats have deposited at various times in the past.

It is easy to demonstrate experimentally that large woodrats produce large faecal pellets, and a precise relationship has been established between the two, so that we can determine from a population of faeces just how large the woodrats were that produced them. Figure 1.4 shows the results from fossil middens of different ages dated by radiocarbon techniques (see Chapter 10). At 14 000 years ago, the Ice Age was just terminating, and evidence from the middens shows that large faecal pellets were being deposited, presumably by large woodrats. Pellet size falls off to a minimum around 6000 years ago when the climate is thought to have been warmer than at present, and then rises again as the climate cooled. In this way biogeographers can follow microevolutionary developments, in this case in animal size, and can test such hypotheses as

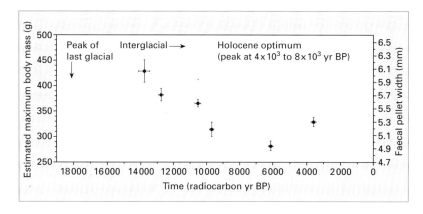

Fig. 1.4 Graph showing the changing size of the woodrat in North America over the past 14 000 years. The body size of the animals has been estimated on the basis of fossil faecal pellets deposited in middens at various times in the past. Data from Smith *et al.* [4] with permission.

Bergmann's Rule. Here the hypothesis seems to hold up well.

The fossil record also enables us to look into the past and observe the changing distribution patterns of organisms during the passage of time. Our current experience of nature represents just one point in a constantly changing mosaic of animals and plants that are responding to an endless sequence of environmental and climatic change. Biogeography therefore can never be a static discipline; it must always be aware of, and take into account, the modifications that organisms are making to their distribution patterns as conditions favour them or render their life more difficult in a particular region. When conditions change too fast to permit genetic adjustment and evolution, populations, species and even whole groups of species may pass into extinction. Extinction is not an exception to the norm in evolutionary terms, it is a regular feature of the constantly changing pattern of life. Novel forms prosper at the expense of old-fashioned, poorly adapted groups, which consequently disappear.

Understanding biogeography must therefore involve the time perspective, not simply encompassing the decades or centuries by which we measure human history, but covering the millions of years, or even hundreds of millions of years, during which whole groups of plants and animals may rise to prominence and progress to extinction. Within this time scale even the continents have not remained static. Convection forces in the earth's mantle have

carried land masses over the surface of the globe, resulting in the fragmentation of supercontinents and occasional continental collisions of great violence, and these movements have had significant repercussions in the distribution patterns of plants and animals. Land-mass fragmentation and movement have led to related groups of organisms being widely separated, and collisions have brought together unexpected groups of quite alien origins. Therefore, no account of biogeography can be complete without a consideration of these long-term geological movements. Chapters 6 and 7 provide an overview of these processes and their impact on biogeographical thinking.

Biogeography, then, is concerned with the analysis and explanation of patterns of distribution, and with the understanding of changes in distribution that have taken place in the past and are taking place today. It concerns itself with what units of life, or 'taxa', are found where, and what are the geographical definitions of that 'where': is it North America, or land between 2000 and 5000 m altitude, or land between 17°N and 23°S latitude? Once the pattern has been established, two sets of questions arise. There is the internal question: how is the organism adapted to the conditions of life in this area? This in turn generates the opposite, externally directed questions: why does the organism not exist in adjacent areas, and what are the factors (biological or environmental) that prevent it from doing so? These factors can be thought of as forming a barrier that prevents the further spread of the organism. Islands are a particularly distinct example of such isolation (Chapter 13).

Sometimes the organism's pattern of distribution is discontinuous or 'disjunct', the species being found in several separate areas. Two different explanations may be offered for this. Firstly, the organism may originally have been present in only one area, and been able to cross the intervening barrier regions to colonize the other areas. Alternatively, the barrier may have appeared later, dividing a once-continuous, simple pattern of distribution into separate units. These two explanations are, respectively, termed 'dispersalist' and 'vicariance' (see Chapter 8).

In the case of living organisms, all of these questions can be investigated in detail, examining all aspects of their ecology and adaptations to life and dispersal. These investigations can extend over several decades and analyse, for example, the effects of a series of dry summers that may cause local ecological changes affecting the distribution of individual species over a few thousand square metres. The wide range of phenomena available for analysis under these circumstances has led to this type of biogeographical investigation becoming known as 'ecological biogeography'.

Biogeographical changes that have taken place over the last few centuries are well documented, and their causes are often relatively well understood. They are also restricted to observable, much-studied phenomena, such as climatic change or minor geographical changes in the distribution of land and water. Furthermore, they affected taxa that are comparatively well defined and that can be studied as living organisms. The distributional and climatic data available from cores of lake or deep-sea sediments, or analysis of tree rings, together with radiocarbon dating, have made it possible to extend this ecological biogeography back many thousands of years before the present, which often provides a rich supply of information and explanation for modern distribution patterns of species.

At a slightly greater remove in time, the biogeographical changes of the Ice Ages can be seen as extrapolations of ecological biogeography. They affected entire biota rather than merely individual species, involved changes in the patterns of distribution and biological zonation over several thousand kilometres, and were caused by systematic changes in the earth's orbit or by minor changes in sea levels or currents. Nevertheless, the processes, phenomena and organisms involved are similar in nature to those of today, and the evidence is comparatively abundant.

Understanding even longer-term changes and more ancient patterns of life requires a different approach. These changes took place tens or even hundreds of millions of year, ago, and involved the splitting, moving or fusion of whole continents, raising new mountain chains or causing the appearance or disappearance of major oceans and seas, with accompanying changes in climate. The biota affected were transported over thousands of kilometres and include groups that are now wholly extinct and are therefore not available for ecological study. The data involved in such long-term historical biogeography are therefore restricted and incomplete, and cannot be approached merely as an extrapolation of the methods used to analyse the phenomena of the more recent past. The different types of data, and the methods that have been proposed for analysing them, form the basis of Chapter 8.

The last two million years of the earth's history, the subject of Chapters 9 and 10, have been of particular significance to biogeography for a number of reasons. Within this time period the climate, which has been getting colder for the last 70 million years or so, has become particularly unstable and has entered a series of oscillations in which there is an alternate expansion and contraction of the earth's ice caps. Within these episodes, new ice caps expanded and contracted over the continental land masses of North America and Eurasia. This glacial/interglacial series of cycles has had a particularly important part to play in affecting the current distribution patterns of plants and animals, not only in the direct effect of changing climates, but also in the alterations these have caused to global sea levels. Rising and falling sea level has periodically resulted in the formation and disruption of land bridges linking islands to land masses and even continents to one another, as in the case of eastern Asia and Alaska. Modern distribution patterns of organisms often provide examples of stranded populations and islands from which a particular organism is unexpectedly absent. Since most of the species that now exist have been present for much of the last million years or so, it has proved easier to reconstruct the detailed ecological changes of this ultimate stage in the earth's geological history than has been possible for earlier stages.

One reason why this recent period has a particular interest for the biogeographer is the fact that our own species has evolved and come to prominence within these last two million years or so. The emergence of *Homo sapiens* from the position of a social ape of the savannas to its present position of global dominance has an important message for modern humans. The picture that develops from a biogeographical approach to human development is one of clever manipulation of the environment, deflecting energy from other parts of the natural food webs into the support of one species. This is not unknown in the animal kingdom (some ants herd greenflies and grow fungi in gardens), but in the case of our own species the development of agriculture and, later, industrial processes have created a new set of conditions in the modern world that we find hard to comprehend. It is difficult to grasp that one species has achieved such an impact that it now affects the massive geochemical and climatic processes of the entire planet.

We have also created new barriers for organisms by fragmenting habitats into ever more isolated units. An example of this shrinkage and fragmentation is provided by the rain forests of Madagascar (Fig. 1.5), a habitat that is rich in species that are limited to this particular region (endemic) [5]. This type of development has raised many questions for conservationists, such as how species will cope when reduced to small populations, genetically out of touch with other populations of the same species [6]. Are such populations more liable to extinction?

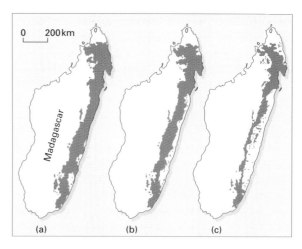

Fig. 1.5 Distribution of rainforest in the island of Madagascar off East Africa. (a) The probable original extent of the forest; (b) the situation in 1950; and (c) in 1985. Data from Green & Sussman [5]. Copyright 1990 by the AAAS.

Will they evolve separately from one another and form new races? How will they respond if the earth's climate changes? Will they be able to migrate to suitable new areas, or will they be stranded in their isolation and simply decline into extinction?

The study of biogeography has particular relevance to some of these problems, because an understanding of the processes and the factors that have influenced the successes and the failures of other species may well assist in the amelioration of our present condition. Biogeography will hopefully develop, in time, into a predictive science, as explained in Chapter 11.

All of the preceding chapters have dealt with different aspects of biogeographical distribution where they are most obvious and have been more thoroughly investigated and understood—on the continental land masses. The oceans have long seemed to be both inexhaustible in their contribution to our food supply, and so vast as to be beyond our powers to change or damage. But it is now clear that our increasing abilities to gather food from the oceans have affected adversely the populations of marine organisms on which we rely to replenish those resources. So, despite its very different environment, and despite the fact that our understanding of its faunas and patterns is still at a much more elementary stage when compared with those on land, it is now imperative that we develop the discipline of marine biogeography as rapidly as possible. Chapter 12 outlines where we stand today in that enterprise.

Islands, too, are quite different from the great continental land masses. Their comparatively simple ecosystems, and individual variety, have contributed unique perceptions to our studies of the evolutionary process. The simplicity of their ecosystems allows us to analyse the composition and interaction of the different elements of an island's biota much more easily than in the case of the complex, extensive biotas of the continents. In the case of the East Indies island of Rakata, stripped of its life in a great volcanic explosion in 1883, we also have the opportunity to study the ways in which such an ecosystem is rebuilt by colonists arriving by air and by sea.

The variety of islands and their biota also provides us with the challenge of trying to understand the factors that contribute to these patterns, and the ways in which these factors interrelate. Ideally, we would be able to integrate these, and construct a formula that might enable us to predict how an island biota might behave in given circumstances. Such a development would not only provide a synthesis satisfying to the academic scientist, but it might also provide a useful tool to ecologists trying to design nature reserves that would be both conservation-effective and cost-effective. All these topics are dealt with in our final chapter (Chapter 13).

In conclusion, biogeographical understanding may help to underline the necessity for conserving those resources that yet remain to us, so that we may use them wisely. Biogeography may also supply the knowledge we need to develop that wisdom. We have much to learn from the history of the Hawaiian goose—a bird that approached the very brink of extinction, yet recovered, and also from the oak trees—a group of species that remains resilient and adaptable after a long and successful evolutionary history; but we must not forget the dodo. Historical biogeography has its warnings as well as its encouragements.

References

1 Wilson EO. *Biodiversity*. New York: National Academic Press, 1988.
2 Niemala J, Haila Y, Halme E, Pajunen T, Punttila P. Small-scale heterogeneity in the spatial distribution of carabid beetles in the southern Finnish taiga. *J Biogeogr* 1992; 19: 173–81.
3 Jones CG, Ostfeld RS, Richard MP, Schauber EM, Wolff JO. Chain reactions linking acorns to gypsy moth outbreaks and Lyme disease risk. *Science* 1998; 279: 1023–6.
4 Smith FA, Betancourt JL, Brown JH. Evolution of body size in the woodrat over the past 25,000 years of climate change. *Science* 1995; 270: 2012–14.
5 Green GM, Susmann RW. Deforestation history of the eastern rain forests of Madagascar from satellite images. *Science* 1990; 248: 212–15.
6 Turner IM, Corlett RT. The conservation value of small, isolated fragments of lowland tropical rain forest. *Trends Ecol Evol* 1996; 11: 330–3.

CHAPTER 2: *Biodiversity*

Biodiversity is a term that encompasses all of the living things that currently exist on earth. It includes all animals and plants that zoologists and botanists have discovered and described, and also all those that yet remain undiscovered and still await a scientific description. In addition, biodiversity includes all fungi, bacteria, protozoa and viruses which, on the whole, are even less well known than the animals and plants. But biodiversity goes even beyond this. When we take a species and analyse its composition, we find that it consists of a series of populations, sometimes adjacent to one another and sometimes fragmented and isolated, and within these populations there is often great variation between individuals. Biodiversity includes the whole range of populations, together with all the genetic variations found within each species.

Just as we can regard a species as a collection of component populations, so we can interpret communities of organisms as assemblages of many populations of a variety of different species, all interacting together. When these communities are placed in the setting of their non-living environment, they make up ecosystems. Our concept of biodiversity should therefore include the rich variety of ecosystems that occupy the earth, many of which have an important human component. If we wish to conserve biodiversity, and most ecologists believe that this is basically a sensible aim, then the conservation of whole ecosystems and their habitats is the most appropriate starting point.

Why should the conservation of biodiversity be regarded as important? Why do we need, or why do we want, to maintain the biotic richness of the earth? Needing and wanting are very different experiences. We need something when it is useful to us, and one could argue that the other living organisms of the earth are useful. Some provide food, or materials for building homes and making fabrics; others are a source of pharmaceuticals. Very many organisms are part of our general support system on earth, being involved in the maintenance of the gaseous balance of the atmosphere, the healthy structure of the soil, and even the modification of our climate. So there is a strong utilitarian argument for the maintenance of the earth's biodiversity. Even those species that are not currently regarded as useful may one day be found to be so.

There is a further argument, however, that may be associated with 'wanting' rather than 'needing'. We may want to have birds and flowers in the world simply because we enjoy having them there—a kind of aesthetic argument. We may regret that we shall never have the opportunity to see a passenger pigeon, a dodo or a great auk. Or we may take a view that is even less human-centred and claim that the organisms have rights of their own to exist, perhaps even as great a right as ours. It is our responsibility therefore as a species with a high impact on the earth's ecosystems, to ensure that extinction is minimized. Conservation then becomes a matter of ethics.

If we accept any of these arguments it leads us to a position where we need to know whether the earth is actually losing species and, if so, how fast? We also need to be concerned about whether human beings are contributing substantially to the rate of loss and whether there is anything that we can do about it. But to understand extinction rates and their causes we have to go back further still and ask just how many species there are on earth so that we can calculate how rapidly we may be losing them.

How many species are there?

No-one likes to lose things, but the severity of the

loss can best be measured on the basis of what is left. The loss of a dollar will be felt more severely by a pauper than by a millionaire. The importance to humanity of the loss of species from the earth can only be judged therefore if we can view it in proportional terms and view the loss from the perspective of what remains. We need to appreciate how many species occupy the earth in order to evaluate the importance of current extinction rates.

One of the most surprising things about science is how little we know. You might assume, for example, that biologists would have a reasonably good idea of how many species of living organisms exist on earth, but in fact this is not so. The question is still hotly debated, and the estimates that biologists have made range between 12 and 30 million species. What they are all agreed upon is that only a very small proportion of the total is currently known to science and has been adequately described. This confusion may seem surprising to non-biologists, but the sheer wealth of species makes it difficult to be sure that all of those that have been described are valid and are not duplicates, or that those described as a single species do not, in fact, consist of a number of species that we have ignorantly lumped together. The herring gull (*Larus argentatus*), for example, is found in coastal areas right around the world in the temperate zone of the Northern Hemisphere (termed a 'circumboreal' distribution), but it occurs in a variety of different forms in different places and is regarded by some as several different species.

On the other hand, of the approximately 1.8 million species of organism that have been described, many may have been named twice, or even more times! Among beetles, for example, 40% of the species described have only ever been recorded at the site of their first description. This is very unlikely to be a true reflection of their distributions, and it is quite likely that many of the species are in fact duplicates with different names in different localities. This so-called 'alias problem' will inflate the numbers of species described [1], but the likelihood is that there are far more species remaining to be found, so that the true number of species

still living on the earth must greatly exceed the number currently described.

A conservative estimate for the possible number of species is 12.5 million [2], but the tropical ecologist Terry L. Erwin [3] has proposed that the total is far greater than this, perhaps as high as 30 million for tropical insects alone. He came to this conclusion as a result of the study of beetles on a single tree species, *Luehea seemannii*, in Panama, which he sampled by 'fogging'. This is an efficient technique for stunning the insects in a canopy by smoking them with an insecticide. The dazed insects fall from the tree and are collected in trays placed beneath the canopy. Erwin examined just 19 individual trees of *L. seemannii* species in the Panamanian forests and managed to obtain 1200 species of beetles alone from this analysis. This large number is not entirely surprising, since beetles are extraordinarily successful insects and may comprise as much as 25% of the total number of species of living organisms. But this study does illustrate the remarkable richness of beetles in the tropical forest.

From these data Erwin made a number of assumptions about the numbers of beetle found specifically on particular tree species, the numbers of tree species found, and the proportions of different organisms in the forest in relation to one another. He extrapolated from the information gathered and came to the conclusion that, if this number of beetles is truly representative of the forest richness, then one might predict a total of 30 million species of insect on earth. The uncertainty of many of his assumptions, however, should make us very cautious in accepting this figure uncritically. Other entomologists, such as Nigel Stork and Kevin Gaston from the University of Sheffield, England [4], have checked Erwin's estimates using data from studies in the tropical forests of Borneo. Stork has generated estimates ranging from 10 million to 80 million for the arthropods (a group of invertebrate animals including the insects). Another independent estimate [5] supports the lower end of this scale, placing tropical arthropods at six to nine million. The range of error in all estimates is still so wide that there is bound to be a great deal of discrepancy in the figures arrived at, but the world's wealth of species is likely to exceed 10 million.

The diversity of microbes, the bacteria, fungi, viruses, etc., are particularly difficult to estimate, because these groups are not nearly as well known as, say, the mammals or the flowering plants. Of the bacteria, for example, only about 4000 species have so far been described, and this may well represent only about one-tenth of one per cent of the total, so much remains to be done in this area. The study of biodiversity among microbes is complicated by the range of genetic and biochemical variation found among wild populations [6,7]. It is also made more difficult by the fact that bacteria can survive deep in geological deposits, far beyond those surface layers of the earth which were once supposed to represent the limits of the biosphere [8]. Their capacity to live in extreme environments and their great potential in the service of humankind makes them particularly interesting and important to us, so it is in our interests to improve our knowledge of microbial diversity. But even the concept of the species has to be reconsidered when dealing with microbes [9].

Among the fungi, some 70 000 species have been described so far, but David Hawksworth of the International Mycological Institute, England, believes that the true total could be around 1.6 million species [10]. In areas of the earth where the fungi have been thoroughly investigated, each higher plant species supports about five or six fungal species. Therefore, if the total flowering plants (once all have been described) is assumed to be about 300 000, the total number of fungi must be in the region of one-and-a-half million or more.

From this example it can be seen that the estimation of possible numbers of organisms is derived by a process of extrapolation. If we have certain facts and make further assumptions, then we can begin to project from what we know into the misty realms of uncertainty. The outcome is not satisfactory, but it is the best we can do so far. Table 2.1 provides

Table 2.1 The numbers of described species in selected groups of organisms, together with the likely total numbers on earth, and the percentage of the group that is currently known. Data from Groombridge [11].

Group	Described species	Likely total	%
Insects	950 000	8 000 000	12
Fungi	70 000	1 000 000	7
Arachnids	75 000	750 000	10
Viruses	5 000	500 000	5
Nematodes	15 000	500 000	3
Bacteria	4 000	400 000	1
Vascular plants	250 000	300 000	83
Protozoans	40 000	200 000	20
Algae	40 000	200 000	20
Molluscs	70 000	200 000	35
Crustaceans	40 000	150 000	27
Vertebrates	45 000	50 000	90

some idea of what we presently know about the richness of the earth's diversity of species for a few groups of organisms and also approximately what proportion of each group is currently thought to have been described [11]. The vertebrates (animals with backbones) are reasonably well known, and relatively few new species may be expected in this area, and the same is true for flowering plants. In recent surveys, 4327 species of mammal were listed, and 9672 species of bird, and it is unlikely that these totals will grow very substantially even with further survey work and research into their classification, although new species do continue to be described at a rate of about 100 species per decade in the case of mammals [12]. But the spiders and mites (arachnids), the algae and the nematode worms, among others, are still very poorly understood and many more species can be expected in these groups. There is the additional frustration, from the biologist's point of view, that smaller organisms tend to be both more abundant and more diverse than are larger ones [13], thus adding to the workload at the most difficult end of the spectrum.

All these difficulties in estimating the biodiversity of the earth apply equally to individual sites

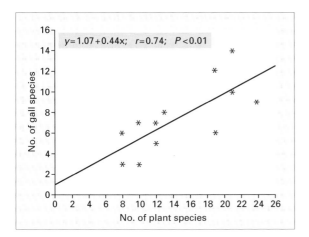

Fig. 2.1 Graph showing the relationship between the richness of gall-forming species and the number of woody shrub species in the Aynsberg Nature Reserve, South Africa. Data from Wright & Samways [14].

and habitats. Even listing all the species present at a site, apart from the genetic and habitat variation that contribute to biodiversity, would be a long, costly, time-consuming and finally inaccurate process. An alternative approach is to assess the richness of certain groups of organisms that are easily observed and identified (such as higher plants, mammals, birds or butterflies) and assuming them to be representative of the less easily observed and identified groups. This method should work well where species are closely dependent on one another (as host and parasite, or as food and feeder), as in the case of gall-forming organisms and plant species in the scrub grasslands of South Africa (Fig. 2.1) [14]. Here, there is a strong linear relationship between the two groups. But the extensive survey work of John Lawton and colleagues in the Mbalmayo Forest Reserve in Cameroon [15], covering birds, butterflies, beetles, ants, termites and nematodes, has shown little overall relationship between one group and another. This demonstrates that it is unwise to regard a single group of organisms as representative of biodiversity in general, but the practical problems involved will inevitably mean that biodiversity

estimates can at best be based on only a limited number of biological groups.

It is clear from this account of the problems in assessing species numbers at a single site that rates of extinction can only be vaguely estimated, since they will be dependent on how many species there are in the first case. Although many known extinctions in the recent past have resulted from human hunting or persecution, the organisms involved have been large and easily targeted. Nothing is known of the smaller creatures, especially the microbes. It is likely, however, that extinction of small organisms is mainly due to a loss of habitat, and some habitats are richer than others, so that calculating possible extinction rates depends on what kind of habitats are being destroyed. Assessing rates of extinction is also made difficult because we can rarely be sure that a species is actually lost, that no isolated members remain. There is still a possibility that the ivory-billed woodpecker survives somewhere in the southern United States, even though it has long been regarded as extinct. Just occasionally animals that were considered extinct have been rediscovered, as in the case of the tamar wallaby, an Australian mammal that disappeared from the mainland almost 100 years ago. It has been found alive and well with a population of over 2000 on an island off New Zealand, where a former Governor of that country had very fortunately introduced it in 1862.

In the case of plants, extinction is even more difficult to record. Many of the plant species of the Atlantic forests of Brazil, now occupying only about 10% of their original cover, were last recorded back in the 1850s and have not been seen since. But it is difficult to be sure that they have finally gone and that not even dormant seeds survive. There are occasions when plants are discovered that have formerly been known only in a fossil state, as in the case of the maidenhair tree (*Ginkgo biloba*) found in east China and previously known only as a 200-million-year-old fossil. Even as recently as 1994 a new tree was discovered in a deep gorge near Sydney, Australia [16], which has been given the name *Wollemia nobilis* and, like *Ginkgo*, closely

Fig. 2.2 Numbers of breeding birds and mammal species in different parts of Central and North America.

resembles a fossil plant of the Cretaceous that was presumed extinct.

But, despite the problems in recording the process accurately, extinction is undoubtedly occurring all around us. The biologist Edward O. Wilson [2] has calculated that the loss of species from the tropical forest area alone could currently be as high as 6000 species per year. This amounts to 17 species each day, and the tropical forests cover only 6% of the land surface area of the earth, so the rate of extinction will be yet higher if we include other vegetation types. Although extinction has always been an inevitable element in the evolutionary process, it is calculated that recent extinction rates may be 100–1000 times greater than those before the emergence of our species. It is also feared that they might accelerate by a further 10 times in the next century [17].

Gradients of diversity

When we look at the way in which biodiversity is distributed over the land surface of the planet, we find that it is far from even. The tropics contain many more species overall than an equivalent area of the higher latitudes. This seems to be true for many different groups of animals and plants, as can be seen from Fig. 2.2, which illustrates the number of breeding birds and mammals found in various Central and North American countries and states. The tropical country Panama, only 800 km (500 miles) north of the Equator and a close neighbour of Costa Rica, has 667 species of breeding birds, three times the number found in Alaska, despite the much greater area of Alaska.

A similar pattern is seen in the number of mammal species at different latitudes in North America

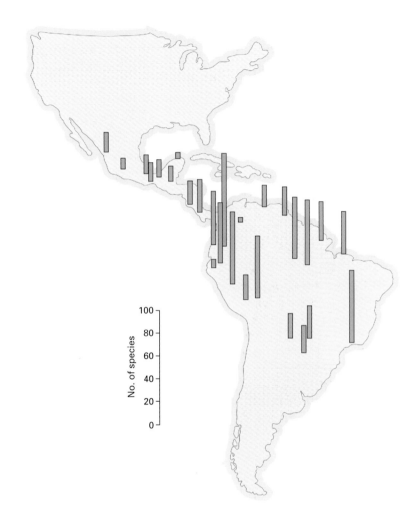

Fig. 2.3 Numbers of frogs in different parts of Central and South America. Data from Groombridge [11].

(Fig. 2.2). If we consider only forest areas from southern Alaska (65°N) in the north, through Michigan (42°N) into the tropical forest of Panama (90°N), these have 15, 35 and 70 mammalian species, respectively. Breaking the mammals down into their component groups by diet and by taxonomy, we find that the bats account for a large part of the difference between the three locations. Moving from the north, one-third of the increase in species from Alaska to Michigan is due to the larger number of bats, and so also is two-thirds of the increase between Michigan and Panama. Yet more informa-

tion becomes available if we consider diet among the mammals. Much of the tropical diversity among mammals is due to the greater predominance of a fruit-eating way of life and to the greater number of insectivores, many of which are eating insects that in turn feed on the fruits of the forest. This may also account for the diversity of frogs in the lower latitudes (Fig. 2.3). Therefore diet is evidently an important aspect of the diversity gradient found among animals.

As we have seen, insect diversity is difficult to measure, because the insect groups are generally

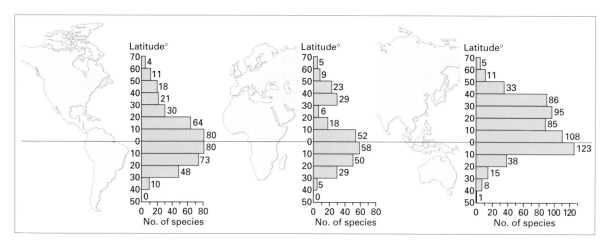

Fig. 2.4 Latitudinal gradients of species richness for swallowtail butterflies in three different parts of the world. Data from Collins & Morris [19].

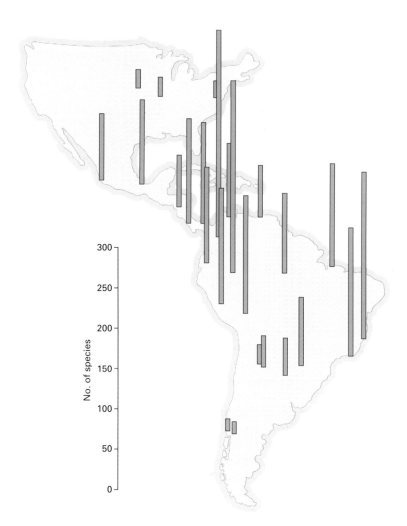

Fig. 2.5 Latitiudinal gradients of tree species richness in the Americas. Data from Duellman [18].

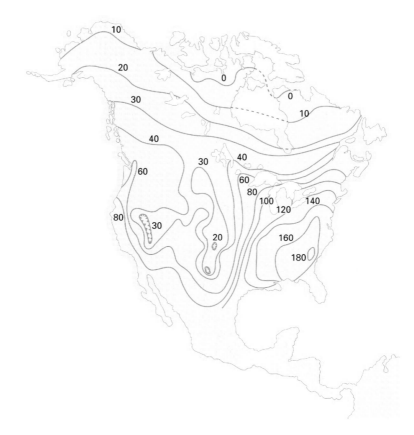

Fig. 2.6 Number of tree species
(i.e. any woody plant over 3 m in
height) found in different parts of
North America. The contours
indicate areas where particular
numbers of tree species were
recorded within large-scale quadrats
(mean area of 70 000 km²). Data
from Currie & Paquin [20].

not well described. Butterflies are among the best
recorded of the insects and Fig. 2.4 shows the latitudinal gradients of richness in just one group of
butterflies, the swallowtails. The high numbers of
species found in the tropics is again apparent, and
here it can be seen that this applies to all tropical
areas of the world. One easily explained anomaly
in the gradients can be seen in the African/Euro-
pean section, where there is a dip in species richness that coincides with the North African desert
region.

Diet may be an important factor in determining
animal diversity, but what of plants? They also
show a general trend towards increasing diversity
in the tropics (Fig. 2.5) but they do not vary in their
diet since they all need sunlight energy for their
photosynthesis.

The relationship between plant species richness
and latitude, however, is not at all a simple one.
David Currie and Viviane Paquin of the University
of Ottawa have constructed a map of the richness of
tree species across North America [20], and this is
shown in Fig. 2.6. From this it can be seen that the
contours of richness do not simply follow the lines
of latitude, especially in the areas south of Canada.
Patches of low diversity occur in the mid-west, and
exceptionally high diversity of trees in the south-
east. When these workers examined the possible
environmental factors that may be associated with
this pattern, the one which correlated most closely
was the sum of evaporation (directly from the ground)
and transpiration (from the surface of vegetation)
combined to give a value for the loss of water from
the land surface (evapotranspiration). Evidently,

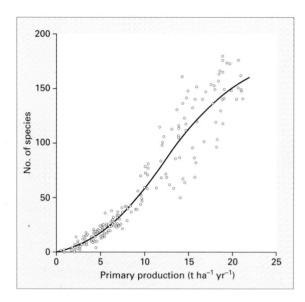

Fig. 2.7 Number of tree species in North American sites (see Fig. 2.6) plotted against the primary productivity of those sites. A distinct positive relationship can be observed. Data from Currie & Paquin [20].

those regions with the highest evapotranspiration are able to support the highest diversity of tree species. But evapotranspiration itself also correlates closely with the potential productivity of a region (the amount of plant material that accumulates by photosynthesis in a given area in a given time), so perhaps plant diversity is essentially determined by how much photosynthesis can be carried out in a given site. Figure 2.7 shows this relationship between primary production and tree species richness, and it can be seen that there is indeed a good correlation between the two.

In general, the equatorial regions are the areas in which highest productivity is possible because of the prevailing climate, which is hot, wet and relatively free from seasonal variation. Figure 2.8 illustrates this by displaying the world distribution of primary productivity. From this map it can be seen that very high productivity is concentrated in the equatorial belt and that this drops off as one moves towards higher latitudes. The picture

is complicated by the arid belt in northern Africa and central Asia, of which more will be said in due course, but the general trend is decreasing productivity at higher latitudes. An examination of the North American part of this map shows a good correlation with the tree-richness map (see Fig. 2.6), especially with regard to the high diversity of tree species and high productivity in the south-eastern United States. One theory suggests therefore that the richness of plant and animal species in an area is dependent on how much energy is captured by the vegetation.

In a sense, we may be asking the wrong question when we try to analyse the factors that contribute to the high species diversity of the tropics, for this is really the perspective of biologists viewing the problem from the temperate regions, which happens to be where most ecologists live and work. From a tropical angle, the question should be why the higher latitudes have lower diversities than the tropics [21]. Perhaps it is simply a matter of land area? The tropics contain a larger surface area of land than higher latitudes—a fact that is not always evident when we examine commonly used projections of the earth's curved surface since this tends to exaggerate the areas of land in the higher latitudes —and some biogeographers regard the latitudinal gradients of diversity as a reflection of this effect [22]. But an analysis of the data by Klaus Rohde [23] does not support this explanation. Although area may contribute to biodiversity it is certainly not the whole story, otherwise large land masses would always be richer.

Productivity seems to be involved, but perhaps its influence is indirect. Where conditions are most suitable for plant growth, i.e. where temperatures are relatively high and uniform and where there is an ample supply of water, one usually finds large masses of vegetation. This leads to a complex structure in the layers of plant material. In a tropical rainforest, for example, a very large quantity of plant material builds up above the surface of the ground. There is also a large mass of material developed below ground as root tissues, but this is less apparent. Careful analysis of the above-ground

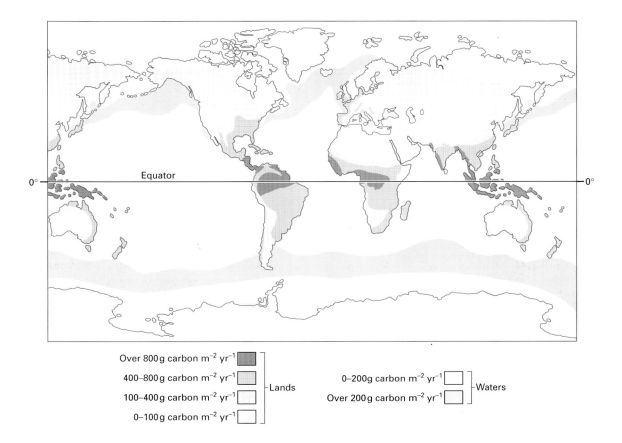

Over 800 g carbon m^{-2} yr^{-1} ▓
400–800 g carbon m^{-2} yr^{-1} ▒ ⎤
100–400 g carbon m^{-2} yr^{-1} □ ⎬ Lands
0–100 g carbon m^{-2} yr^{-1} □ ⎦

0–200 g carbon m^{-2} yr^{-1} □ ⎤
Over 200 g carbon m^{-2} yr^{-1} □ ⎬ Waters
⎦

Fig. 2.8 World distribution of plant productivity. The data displayed here are simply estimates of the amount of organic dry matter that accumulates during a single growing season. Full adjustments for the losses due to animal consumption and the gains due to root production have not been made. Map compiled by H. Leith.

material reveals that it is arranged in a series of layers, the precise number of layers varying with the age and the nature of the forest. The arrangement of the 'biomass' of the vegetation into layered forms is termed its 'structure' (as opposed to its 'composition', which refers to the species of organisms forming the community). Structure is essentially the architecture of vegetation and, as in the case of some tropical forests, can be extremely complicated. Figure 2.9 shows a profile of a mature flood plain tropical forest in Amazonia [24] expressed in terms of the percentage cover of leaves at different heights above the ground. There are three clear peaks in leaf cover at heights of approximately 3, 6 and 30 m above the ground, and the very highest layer, at 50 m, corresponds to the very tall, emergent trees that stand clear of the main canopy and form an open layer of their own. So this site contains essentially four layers of canopy.

Forests in temperate lands are simpler, often with just two canopy layers, so they have much less complex architecture. Structure, however, has a strong influence on the animal life inhabiting a site. It forms the spatial environment within which an animal feeds, moves around, shelters, lives and breeds. It even affects the climate on a very local level (the 'microclimate') by influencing light intensity, humidity and both the range and extremes of

20   *Chapter 2*

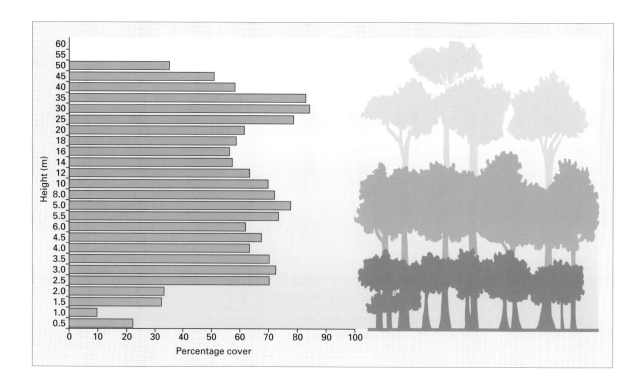

Fig. 2.9 Profile of a tropical rain forest with the percentage leaf canopy cover recorded at different heights above the ground. Note the stratification of the leaf cover into distinct layers. See Fig. 2.18 for further details of canopy structure development in rain forest. From Terborgh & Petren [24].

temperature. Figure 2.10 shows a profile through an area of grassland vegetation that has a very simple structure, and it can be seen that the ground level has a very different microclimate from that experienced in the upper canopy. Wind speeds are lower, temperatures are lower during the day (but warmer at night) and the relative humidity is much greater near the ground. The complexity of microclimate is closely related to the complexity of structure in vegetation and, generally speaking, the more complex the structure of vegetation, the more species of animal are able to make a living there. This is illustrated in Fig. 2.11, which relates the number of bird species found in woodland habitats to the number of leaf canopy layers that can be detected [25]. The high plant biomass of the tropics leads to a greater

spatial complexity in the environment, and this will lead to a higher potential for diversity in the living things that can occupy the region. The climates of the higher latitudes are generally less favourable for the accumulation of large quantities of biomass, hence the structure of vegetation is simpler and the animal diversity is consequently lower.

There is one extension to this line of argument that is worth pursuing. Complexity, or the conception of complexity, depends upon the size of the observer. It was stated above that grassland has a relatively simple structure, but this is only the case if one views it from a human perspective. From an ant's point of view, on the other hand, a grassland environment may be highly complex. For this reason, an area of grassland offers a home to far more ants than it does to humans, cows or bison. The earth as a whole can support far more small creatures than it can large ones. This is illustrated by Fig. 2.12, assembled by Robert May of Oxford University [26], showing the relationship between the number of species of terrestrial (land) animals

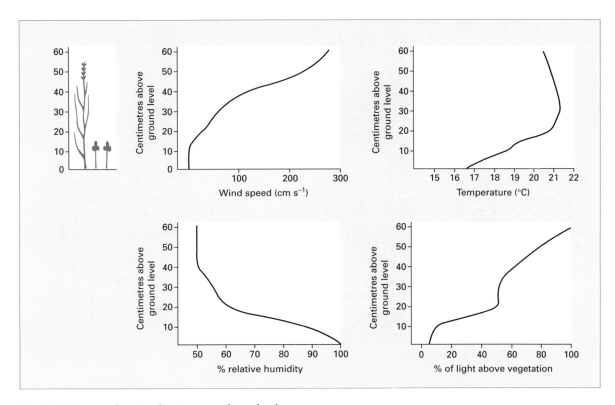

Fig. 2.10 Diagram showing the structure of grassland vegetation and the effect this has upon the microclimate of the habitat.

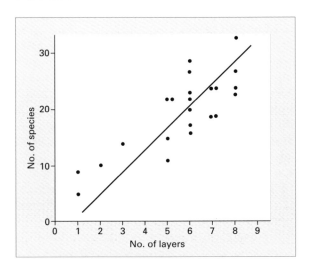

Fig. 2.11 Graph showing the relationship between the number of bird species and the number of layers in the vegetation stratification. Data from Blondell [25].

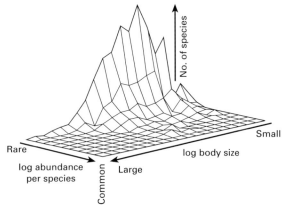

Fig. 2.12 Relationship between body size and abundance among beetle species. The vertical axis represents the number of species of beetle having the body size and abundance indicated on the horizontal axes. 'Small' for a beetle is 0.5 mm and 'large' is 30 mm. From May [26]. Copyright 1988 by the AAAS.

and their respective body sizes. The general picture is one of a greater abundance of small species than large ones. The fact that there appears to be a drop in species richness when dealing with extremely small organisms probably reflects our ignorance of these creatures rather than any real decrease in numbers.

There is one group of very small animals, however, that goes against the rules. The aphids (such as plant-feeding greenflies), of which about 4000 species are known, are less diverse in the tropics than they are in the temperate regions [27]. Most aphid species feed on only one type of plant and are rather poor at locating that plant from a distance, relying on the sheer chance of air-flow patterns to carry them from one suitable host to another. They do best therefore where populations of particular plant species are dense, as is the case in agricultural crops. Unfortunately for the aphids, tropical plant assemblages consist of many species each of which is present only at low density, so aphids are not well suited to such conditions. Aphid diversity is therefore inversely related to plant diversity, so they transgress the general pattern of latitudinal gradients.

But, generally speaking, the richness of the tropics as far as animal life is concerned may be a consequence not simply of the high productivity of these latitudes, but also of their great structural complexity resulting from their high biomass, which can support many species of small animal.

There are, however, alternative explanations concerning the abundance of species in the tropics. Some of these relate to the evolutionary history of certain major groups of organisms.

If we examine the distribution patterns of the various families of flowering plants, we find that they also have a tendency to centre on the tropics. Very roughly, about 30% of flowering plant families are widespread in distribution, about 20% mainly temperate and about 50% mainly tropical. This has suggested to some research workers that the tropics have been a centre for the evolution of many of the flowering plant (angiosperm) groups. This proposal can be examined by looking at the fossil record to see whether the tropics have always been richer

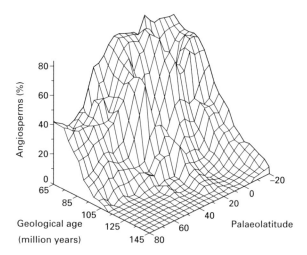

Fig. 2.13 Estimated percentage representation of flowering plants (angiosperms) at different times in geological history and at different latitudes. Angiosperms have always been most abundant in the low-latitude (tropical) regions. From Crane & Lidgard [28].

than the temperate latitudes as far as flowering plants are concerned. This approach has been attempted by Peter Crane and Scott Lidgard of the Field Museum of Natural History in Chicago [28], and some of their results are represented in Fig. 2.13. This shows an analysis of fossil plant material covering the period 145 million years ago to 65 million years ago, and depicts the importance of the angiosperms in different latitudes during this period of time. From this it is clear that the flowering plants first rose to some prominence in the tropics, and that their predominance in the tropics was maintained throughout this period of time as they gained importance in the plant kingdom. The latitudinal gradient in diversity for flowering plants goes back a very long way—right back, in fact, to the evolutionary origins of the group. Tropical biodiversity is clearly an ancient phenomenon and may be related, in this case, to tropical origins and enhanced tropical rates of evolution and diversification.

Arguing from past history to account for the diversity of the tropics with respect to higher latitudes raises the question of changing global climate with time. It is generally accepted that the earth's climate has been constantly changing and that, over the past two million years, it has been considerably colder than was the case for the previous 300 million years. As a consequence, the high latitudes have been disrupted by the development of glaciers over the land surface. The effects of these changes on the biogeographical patterns of plants and animals will be considered later, but it is evident that the most severe disruption, in the form of ice masses that have spread and destroyed all vegetation over major areas, has occurred largely in the high latitudes, and the tropics have consequently been subjected to less obvious climatic stress. This idea of a climatically stable tropical belt, if it is indeed true, could account for some of the diversity still found in the tropics—the plants and animals could be a relic feature from a former age. But has the tropical region been climatically more stable than the temperate region? This subject will be examined in greater detail later in the book, but the general conclusion that has emerged, particularly from the studies of Paul Colinvaux of the Smithsonian Institute, Panama, and his coworkers [29], is that the tropical lowlands of Amazonia have also been considerably colder (perhaps 5 or 6°C colder than at present) during recent times (the last million years or so) and that the tropical forests could not have remained intact over the whole of their current range. This climatic shift has meant that the tropical rainforest has undoubtedly been restricted in the altitudinal range it was able to cover, and has also been at least partially fragmented as a result of cold and drought during 'glacial' periods.

The equatorial forests, however, have still had to endure less disturbance than their temperate counterparts, and some areas have probably maintained themselves in a forested form throughout the period of stress, even though their species composition and architectural structure may well have changed. It has been suggested that, if they were fragmented, this could have actually assisted in the progress of evolution and diversification, for isolated populations, as we shall see in a subsequent chapter, may diverge in their evolution and form separate species that fail to interbreed when brought into contact

once more. In this case, the impact of climatic change in fragmenting the tropical forests may have added to, rather than subtracted from, their diversity. Evidence from molecular studies of some American songbirds obtained by John Klicka and Robert M. Zink of the University of Minnesota, however, does not support this idea [30]. The time of evolutionary divergence can be estimated from the similarity/dissimilarity of their mitochondrial DNA, and most seem to have diverged from common ancestors up to five million years ago, considerably further back in time than can be accounted for by the ice ages of the last two million years.

A further approach to the question of latitudinal gradients of diversity has recently emerged and is based on the observation that high-latitude organisms have broader geographical ranges than those from the low latitudes. This general feature of biogeography was first pointed out by E.H. Rapoport in the 1970s, but has risen to prominence as a result of the work of George C. Stevens [31], who coined for it the term 'Rapoport's Rule'. Much work has now been carried out to test the generalization, and the species of high latitudes do, on the whole, display wide geographical ranges, great altitudinal ranges, and broad ecological tolerances in comparison with tropical species. But there remains doubt as to whether this is a local effect that only makes itself felt in more northerly latitudes (above about 40–50°N), or whether it continues to be operative in the equatorial regions. Klaus Rohde [32], of Armidale, Australia, considers that Rapoport's Rule is of local application only, and cannot be applied in the tropical regions. Certainly, there is very little evidence for the effect within the tropics. The occurrence of broad-range species in the high latitudes could itself be a consequence of the impact of successive glaciations, leaving only the most adaptable species behind. It can also be argued that the greater seasonal fluctuations of the high latitudes will select for wide-tolerance organisms.

The use of Rapoport's Rule (perhaps better termed 'the Rapoport Effect' [33]) as an explanation for the cause of latitudinal gradients of diversity is now challenged. Given lower species richness in the

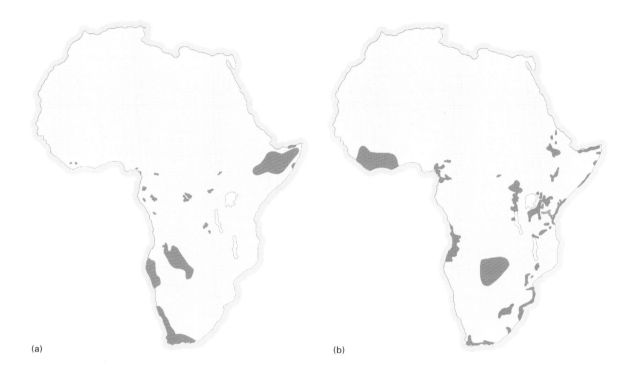

(a)

(b)

Fig. 2.14 The areas of Africa that are particularly rich in (a) plant species compared with those areas that are rich in (b) endemic birds. As can be seen, the two do not always correspond.

high latitudes, it is only to be expected that there will be less competition for resources and that ecological and geographical ranges of species will therefore be more extensive than in the species-dense tropics. The Rapoport Effect is likely to be a consequence of latitudinal gradients in species richness rather than its cause.

It is reasonable to conclude therefore that many factors contribute to the species richness of the tropical regions and the lower richness of higher latitudes; no single explanation accounts for all the observed data.

Biodiversity hotspots

Although species richness does generally follow a latitudinal gradient, it often proves more complex when we examine the picture in detail. For example, the Amazon basin contains approximately 90 000 flowering plant species, whereas equivalent areas in Africa and in South-East Asia contain only about 40 000 each. There appear to be certain areas that are exceptionally rich in species, termed 'biodiversity hotspots' by the conservationist Norman Myers [34]. He originally proposed 10 hotspots, largely identified on the basis of plant diversity, for his argument was based on the idea that if vegetation is diverse all else will follow. But this is not always entirely true, as we have already seen in the case of aphids. Figure 2.14 shows the areas of high plant diversity in Africa compared with areas of high bird diversity and it can be seen that there is relatively little overlap, so we cannot assume parallelism in trends of biodiversity between groups of organisms.

The original work of Myers has been developed and expanded, particularly by the organization Conservation International, based in Washington DC, and Fig. 2.15 shows the location of some of the

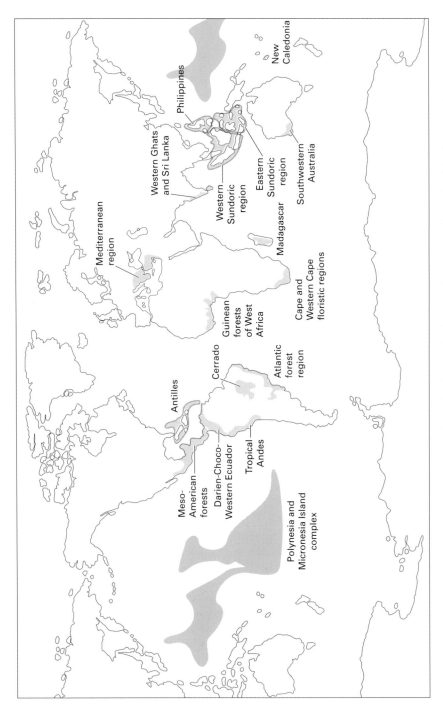

Fig. 2.15 The location of areas of exceptionally high biological richness—biodiversity hotspots, as determined by Conservation International.

hottest spots for biodiversity on earth based on a consideration of many of the better known groups of organisms (plants, mammals, birds, reptiles, amphibians, etc.). When we add all the land surface area of these 17 hotspots together, they comprise only about 1.32% of the earth's terrestrial total, yet they contain about 50% of the world's land-based species. Obviously these areas must form a focus for global conservation activity, but it would be a mistake to confine our attention to them because half of the world's species are located elsewhere.

Human beings are often blamed for the loss of biodiversity in many parts of the world, and frequently we are indeed responsible. The destruction of habitats, such as rainforests, is undoubtedly resulting in species loss, but this is difficult to quantify. Indeed, although 90% of the coastal rainforest of eastern South America has been cleared, no bird species are known to have become extinct as a result [35]. This does not mean to say that other groups of organisms have been as fortunate, nor that extinctions will not occur in the future as populations settle in to new equilibria. Sometimes the human impact on biodiversity can be subtle, as in the case of the cichlid fishes of Lake Victoria in East Africa. At least 500 species of cichlid fish live in this lake and, although many are capable of hybridization, they are kept separate by their brightly coloured patterns; they refuse to accept mates who do not precisely match their colour requirements. But the increasing human activity around the lake is resulting in soil erosion and sewage pollution, which leaves the waters turbid and reduces the visibility for the fish. Lost in this aquatic fog, the breeding barriers may break down and the great diversity of fish in Lake Victoria may soon be lost [36].

Explaining why hotspots are so rich is even more difficult than solving the latitudinal gradient question; indeed, one can detect a general low-latitude concentration in the hotspots, so this whole problem of gradients is clearly involved here. But further mechanisms are also at work. Perhaps these hotspots are centres of evolution; or perhaps they are relict fragments of former diverse communities.

Hotspots are not necessarily those areas that have suffered least from human impact. In the case of the Mediterranean region of Europe, for example, there is high diversity accompanied by a very long history of human activity—sometimes very destructive activity. The island of Crete is only 245 km long by 50 km wide and has been isolated as an island for about 5.5 million years. It has supported human populations since at least the arrival of Neolithic peoples about 8000 years ago. Since then climatic changes have resulted in the development of very dry conditions in summer, and disturbance by earthquake and volcano activity have been experienced. The increase in human populations, their need for agricultural land and their intensive pastoralism has resulted in the stripping of much of the original vegetation [37]. Yet, despite all this, Crete has 1650 species of plant, 10% of which are endemic to the island. The fossils of Crete tell us that many species have become extinct during its recent history, yet it still remains a remarkably species-rich island. Indeed, the Mediterranean-climate areas generally are very rich in plant species, perhaps as a result of the high intensity of habitat patterns and the severity of the local impacts of drought and fire [38].

We may have to delve deeply into geological and ecological history to explain the rich and distinctive biotas of these hotspot areas, and this will be investigated later in this book.

Diversity in time

Some aspects of the spatial gradient in diversity that we find over the land areas of the earth may be explicable in terms of their persistence, even if only as fragments. The study of changing species diversity in time is clearly an important aspect of understanding why there are so many different types of plant and animal, and why they may be concentrated in certain parts of the world. Studies over long periods of time (tens or hundreds of millions of years) can be very informative, but such studies suffer from the disadvantage that it is often difficult to determine how many species there were within a fossil group. One can, however, examine changes in the species composition of a habitat over a short period (decades or centuries). Changes over periods of time of this order are termed successions, especially if they follow a predictable and directional course of development.

A simple example is the invasion of vegetation following the retreat of glaciers in Alaska [39]. Warmer conditions cause the melting of ice, and the ice front gradually recedes, leaving bare rock surfaces and crushed rock fragments in sheltered pockets and crevices. Such primitive soils may be rich in some of the elements needed for plant growth such as potassium and calcium, but are poor in organic matter, so they have a very limited capacity for water retention, poor microbial populations, little structure and low levels of nitrogen. A plant that can grow even under these stressed conditions is the Sitka alder (*Alnus sinuata*). This is a low-growing bushy tree that owes its success in part to its association with a bacterium that grows in association with its roots. This microbe forms colonies in swollen nodules on the alder's roots and is able to take nitrogen from the atmosphere and convert it to ammonium compounds that can subsequently be used (together with materials derived from alder photosynthesis) to build up proteins. Therefore the alder manages perfectly well despite the low levels of nitrate in the soil. In fact, because of the gradual death of roots and the return of litter to the soil from the alder, the growth of the tree increases the amount of nitrate in the soil and thus fertilizes it. But this very process of modifying the soil environment eventually proves the downfall of the alder, because it permits the invasion of other, less highly adapted plants, among them the Sitka spruce (*Picea sitchensis*). After about 80 years, the Sitka spruce trees, which are more robust and faster-growing than the alder shrubs, assume dominance in the vegetation and begin to shade out the pioneer alders. Thus, by their very existence at the site, the alders have effectively sealed their own fate and made the next step in the succession inevitable.

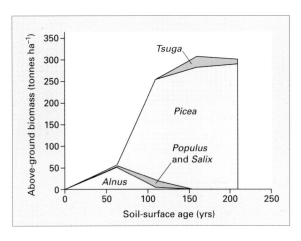

Fig. 2.16 Change in major species composition (expressed in terms of biomass, i.e. above-ground dry weight) during the development of forest following ice retreat in Alaska. The dominance of one species, spruce (*Picea*), is established as the biomass increases with successional development of the plant community. From Bormann & Sidle [39].

This driving mechanism that underlies the successional process is termed 'facilitation', and it ensures a progressive development within the vegetation.

The course of succession also leads to an accumulation of biomass during the course of time. This is shown in Fig. 2.16, in which the biomass of the major tree species can be seen increasing over the course of 200 years of a succession in Glacier Bay, Alaska. Alder, poplar (*Populus*) and willow (*Salix*) are replaced by Sitka spruce and hemlock (*Tsuga*), and, while alder only achieved a maximum biomass of about 50 t/ha, the spruce/hemlock forest grows to a biomass of over 300 t/ha. Such an increase in biomass naturally involves the development of a more complex canopy structure and, as we have seen, the diversity of animal species often follows the increase in structural complexity in vegetation. A gap in a tropical rain forest created by the death of an old tree or by wind damage, for example, gradually becomes filled by the invasion and growth of new vegetation [24]. This operates like a succession in miniature, as is shown in

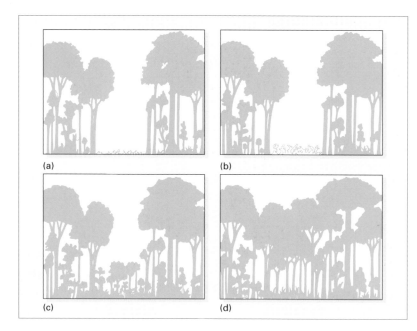

(a)

(b)

(c)

(d)

Fig. 2.17 Gaps in forests, created by the death of old trees, or minor catastrophes such as wind-blow or fire, become filled by the regrowth of young trees, often passing through a succession of different species. Habitat heterogeneity and canopy complexity is an outcome of this process.

Fig. 2.17. The development in structural complexity can be seen in Fig. 2.18, in which plants gradually replace one another and produce successively more complex patterns of canopy cover. This in turn leads to an increasing diversity in the course of succession.

The word 'diversity' should be used with caution, for it can represent different things in different people's minds. It conveys the idea of the number of species per unit area of ground, but this is not an adequate definition of diversity. One should really call this 'species richness'. So the latitudinal gradients that we have been examining earlier in this chapter are actually gradients of richness rather than diversity.

The term 'diversity' should be used only to describe the way in which the number of animals or plants (or their biomass) is allocated among the species present. A community, for example, that contains 100 individuals belonging to 10 species could have 91 individuals belonging to one species and only one each for all the others. This is a less diverse community than one in which there are 10 individuals belonging to each species. Yet both communities have the same species richness.

In the case of successions, both richness and diversity tend to increase with age, especially for the animal component, but this may not always be the case with plants. Later stages in succession, as in the case of the alder/spruce sequence, may become dominated by a few large-bodied species, which effectively reduces plant diversity. In the early stages of succession, however, increasing diversity seems to hold generally true for both plants and animals. This can be illustrated by reference to Fig. 2.19, which shows the number of plant species and their area of cover during the course of an 'old field' succession. This is the development of vegetation following the abandonment of agricultural land and its reversion to woodland. The data shown in the diagram are derived from the work of F.A. Bazzaz [40], who studied an old field succession in Illinois. It covers a period of 40 years and illustrates the increasing number of plant species present (richness) and also a general flattening of the bars, showing that fewer species are dominating the

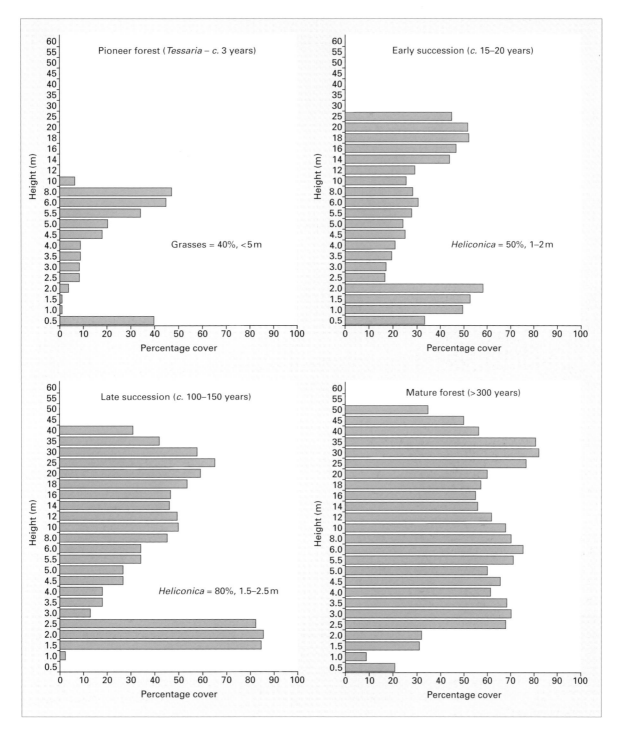

Fig. 2.18 Canopy profiles of a rainforest during the course of successional development over a period of about 300 years. The canopy structure is expressed in terms of leaf percentage cover at different heights above the ground. From Terborgh & Petren [24].

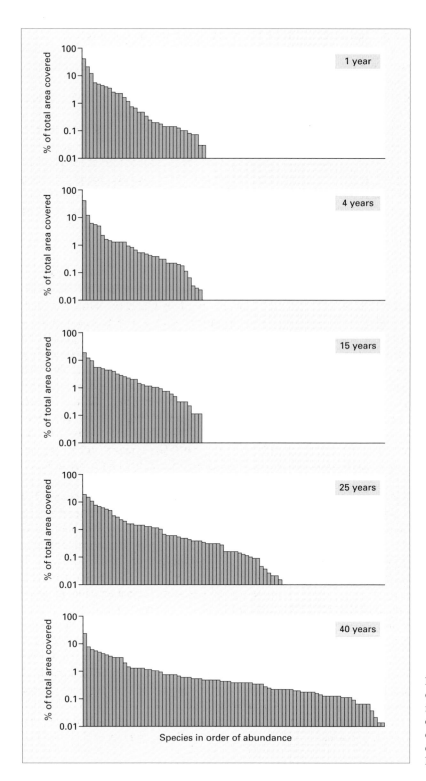

Fig. 2.19 Increasing species diversity in an old field succession in Illinois. Species are ranked in order of abundance, the latter being expressed as percentage area covered on a log scale. Data from Bazzaz [40]. After May [42].

community and more species are occupying a fairer share of the available space. This is a graphic way of expressing the concept of diversity, which can clearly be seen to be increasing through successional time in this instance. There is no strong indication of dominance here, although in later stages one species is beginning to account for a large proportion of the total vegetation cover. (Note that the cover values are expressed as logarithms, which tends to make such dominance less evident.)

The final, mature stage in a succession is termed the 'climax', and represents an equilibrium state. But the achievement of overall equilibrium does not mean that the community is static. Individual trees will become senile and die, minor catastrophes such as wind-blow and fire may cause openings to develop within the canopy, and these gaps, as we have seen, become occupied by small-scale successional developments. Some members of pioneer species groups will find new opportunities to survive for a short while and re-establish themselves within the gap, eventually to be replaced by more persistent species. The final state of vegetation is thus best conceived as an assemblage of individuals of various ages in a constant state of flux, the composition changing in response to the chance arrival of different species and the local extinction of others, all modified by any more gradual environmental change, such as climate.

In the north-eastern United States, a typical sequence in the hardwood forest, following the fall of a mature beech (*Fagus grandifolia*) tree, is yellow birch (*Betula alleghaniensis*) invasion, followed by sugar maple (*Acer saccharum*) and eventually the regrowth of beech [41]. But since this cyclic process of forest healing is taking place wherever a gap has resulted, the 'climax' forest actually consists of a mosaic of patches all in different stages of recovery, together with some patches of mature beech. Therefore the climax vegetation is actually a collection of different-aged patches. This, in fact, adds to the diversity of the whole system, for the vegetation is not uniform but extremely heterogeneous, and many of the species that would be lost from an area, if successional development had effectively ceased,

are still present in some of the forest openings. Thus, the complexity of the time element in vegetation development allows even more species to be packed into a given area.

If we return to the question of why the tropics, and tropical forests in particular, are so rich in species, then we have now established a further means by which high diversity can be maintained. The forest is constantly undergoing disturbance from storms, local fire and the meandering and flooding of rivers. All of these leave the forest in a state of turbulence and active regeneration that contributes to the diversity of the whole.

Marine biodiversity

Patterns of biodiversity in the oceans are even more poorly understood than those on land. For some groups of organisms, such as the bivalve molluscs, gastropods, foraminifera (microscopic, planktonic organisms—see Chapter 10), and corals, there is good evidence for gradients in species number with latitude, similar to that found in many terrestrial groups. Evidence is also increasing that a latitudinal gradient exists in deep-sea animals, especially in the northern hemisphere. This seems to be due largely to a very poor Arctic fauna and a very rich fauna in the equatorial regions of the west Pacific Ocean [43].

The marine realm, however, has many biogeographic problems associated with it. Why, for example, is the Antarctic so much richer in species than the Arctic? Much research remains to be carried out before such questions can be answered.

Explaining the global patterns of biodiversity is a process that requires a consideration of many factors. Some of these have been discussed here, but further understanding of the subject demands a knowledge of many other aspects of biogeography. How is it that many species manage to occupy the same habitat? What factors limit the geographical range of individual species? How have such ranges changed during the course of the earth's history? How do new species evolve and why do they evolve in particular ways? These are some of the questions that must be faced if the complex issue of biodiversity is to be further understood.

Summary

1 Biodiversity means the full range of life on earth, including all the different species found, together with the genetic variation between populations and individuals, and the variety of ecosystems, communities and habitats present on our planet.
2 We are losing species at an unknown, but accelerating rate. We need to know more about the variety of life on earth before we can even appreciate how fast we are losing it.
3 There are probably between 12 and 30 million species on earth, but only about 1.8 million have been described.
4 The tropics are generally richer in species than the high latitudes, possibly as a result of high productivity and food availability, high biomass and hence complex structure, past patterns of evolution, survival of fragments of habitats through the cold episodes of the last two million years, and also the degree of small-scale disturbance resulting in a mosaic of successional processes.
5 The term 'diversity' involves both species number (richness) and the pattern of allocation of numbers or biomass between the different species (evenness). It generally increases during the course of succession.

Further reading

Gaston KJ. *Biodiversity: a Biology of Numbers and Difference.* Oxford: Blackwell Science, 1996.
Groombridge B (ed.) *Global Biodiversity: Status of the Earth's Living Resources.* London: Chapman & Hall, 1992.
Hawksworth DL (ed.) *Biodiversity: Measurement and Estimation.* London: Chapman & Hall, 1995.
Huston MA. *Biological Diversity: The Coexistence of Species on Changing Landscapes.* Cambridge: Cambridge University Press, 1994.
Perlman DL, Adelson G. *Biodiversity: Exploring Values and Priorities in Conservation.* Oxford: Blackwell Science, 1997.
Ricklefs RE, Schluter D. *Species Diversity in Ecological Communities: Historical and Geographical Perspectives.* Chicago: Chicago University Press, 1993.

References

1 May RM, Nee S. The species alias problem. *Nature* 1995; 378: 447–8.
2 Wilson EO. *Biodiversity.* New York: National Academic Press, 1988.
3 Erwin TL. Beetles and other insects of tropical forest canopies at Manaus, Brazil, sampled by insecticidal fogging. In: Sutton SL, Whitmore TC, Chadwick AC, eds. *Tropical Rain Forest: Ecology and Management.* Oxford: Blackwell Scientific Publications, 1983: 59–75.
4 Stork N, Gaston K. Counting species one by one. *New Scientist* 1990; 127 (1729): 43–7.
5 Thomas CD. Fewer species. *Nature* 1990; 347: 237.
6 Pace NR. A molecular view of microbial diversity and the biosphere. *Science* 1997; 276: 734–40.
7 Holms B. Life unlimited. *New Scientist* 1996: 148, 10th February, 26–9.
8 Fyfe WS. The biosphere is going deep. *Science* 1996; 273: 448.
9 O'Donnell AG, Goodfellow M, Hawksworth DL. Theoretical and practical aspects of the quantification of biodiversity among microorganisms. In: Hawksworth DL, ed. *Biodiversity: Measurement and Estimation.* London: Chapman & Hall, 1995: 65–73.
10 May RM. A fondness for fungi. *Nature* 1991; 352: 475–6.
11 Groombridge B (ed.) *Global Biodiversity: Status of the Earth's Living Resources.* London: Chapman & Hall, 1992.
12 Morell V. New mammals discovered by biology's new explorers. *Science* 1996; 273: 1491.
13 Siemann E, Tilman D, Haarstad J. Insect species diversity, abundance and body size relationships. *Nature* 1996; 380: 704–6.
14 Wright MG, Samways MJ. Gall-insect species richness in African fynbos and karoo vegetation: the importance of plant species richness. *Biodiversity Lett* 1996; 3: 151–5.
15 Lawton JH, Bignell DE, Bolton B, Bloemers GF, Eggleton P, Hammond PM, Hodda M, Holt RD, Larsen TB, Mawdsley NA, Stork NE, Srivastava DS, Watt AD. Biodiversity inventories, indicator taxa and effects of habitat modification in tropical forest. *Nature* 1998; 391: 72–6.
16 da Silva W. On the trail of the lonesome pine. *New Scientist* 1997: 155, 6 December, 36–9.
17 Pimm SL, Russell GJ, Gittleman JL, Brooks TM. The future of biodiversity. *Science* 1995; 269: 347–50.
18 Duellman WE. Patterns of species diversity in anuran amphibians in the American tropics. *Ann Missouri Bot Gdn* 1988; 75: 70–104.

19 Collins NM, Morris MG. *Threatened Swallowtail Butterflies of the World. IUCN Red Data Book.* Cambridge: IUCN, 1985.

20 Currie DJ, Paquin V. Large-scale biogeographical patterns of species richness of trees. *Nature* 1987; 329: 326–7.

21 Blackburn TM, Gaston KJ. A sideways look at patterns in species richness, or why there are so few species outside the tropics. *Biodiversity Lett* 1996; 3: 44–53.

22 Rosenzweig ML. *Species Diversity in Space and Time.* Cambridge: Cambridge University Press, 1995.

23 Rohde K. The larger area of the tropics does not explain latitudinal gradients in species diversity. *Oikos* 1997; 79: 169–72.

24 Terborgh J, Petren K. Development of habitat structure through succession in an Amazonian floodplain forest. In: Bell SS, McCoy ED, Mushinsky HR, eds. *Habitat Structure: the Physical Arrangement of Objects in Space.* London: Chapman & Hall, 1991: 28–46.

25 Blondell J. *Biogeographie et Ecologie.* Paris: Masson, 1979.

26 May RM. How many species are there on earth? *Science* 1988; 241: 1441–9.

27 Dixon AFG. *Aphid Ecology: An Optimization Approach*, 2nd edn. London: Chapman & Hall, 1998.

28 Crane PR, Lidgard S. Angiosperm diversification and paleolatitudinal gradiants in Cretaceous floristic diversity. *Science* 1989; 246: 675–8.

29 Colinvaux PA, De Oliveira PE, Moreno JE, Miller MC, Bush MB. A long pollen record from lowland Amazonia: forest and cooling in glacial times. *Science* 1996; 274: 85–8.

30 Klicka J, Zink RM. The importance of recent ice ages in speciation: a failed paradigm. *Science* 1997; 277: 1666–9.

31 Stevens GC. The latitudinal gradient in geographical range: how so many species coexist in the tropics. *Am Naturalist* 1989; 133: 240–56.

32 Rhode K. Rapoport's rule is a local phenomenon and cannot explain latitudinal gradients in species diversity. *Biodiversity Lett* 1996; 3: 10–3.

33 Gaston KJ, Blackburn TM, Spicer JI. Rapoport's rule: time for an epitaph? *Trends Ecol Evol* 1998; 13: 70–4.

34 Myers N. The biodiversity challenge: expanded hot-spots analysis. *The Environment* 1990; 10: 243–56.

35 Brooks T, Balmford A. Atlantic forest extinctions. *Nature* 1996; 380: 115.

36 Seehausen O, van Alpen JJM, Witte F. Cichlid fish diversity threatened by eutrophication that curbs sexual selection. *Science* 1997; 277: 1808–11.

37 Rackham O, Moody J. *The Making of the Cretan Landscape.* Manchester: Manchester University Press, 1996.

38 Cowling RM, Rundel PW, Lamont BB, Arroyo MK, Arlanoutsou M. Plant diversity in mediterranean-climate regions. *Trends Ecol Evol* 1996; 11: 362–6.

39 Bormann BT, Sidle RC. Changes in productivity and distribution of nutrients in a chronosequence at Glacier Bay National Park. *Alaska J Ecol* 1990; 78: 561–78.

40 Bazzaz FA. Plant species diversity in old field successional ecosystems in southern Illinois. *Ecology* 1975; 56: 485–8.

41 Forcier LK. Reproductive strategies in the co-occurrence of climax tree species. *Science* 1975; 189: 808–10.

42 May RM. The evolution of ecological systems. *Sci Am* 1978; 238 (3): 118–33.

43 Clarke A, Crame A. Diversity, latitude and time: patterns in the shallow sea. In: Ormond RFG, Gage JD, Angel MV, eds. *Marine Biodiversity: Patterns and Processes.* Cambridge: Cambridge University Press, 1997: 122–47.

CHAPTER 3: *Patterns of distribution*

The basic units with which biologists have to operate are individual organisms, whether animals, plants or microbes. In the majority of cases these individuals can be sorted into like groups which we call species. Species are thus reasonably distinguishable groups of organisms within which interbreeding can occur and to which it is normally confined (see Chapter 5). Having said this, however, we must bear in mind that there may be considerable variation within any given species, both in visual features, such as form, size and colour, and in physiological and biochemical features that may affect the preferred environment, food or climatic tolerance [1]. Such variations may become so distinctive that they justify the description of subspecies or races as convenient systems for the classification of individuals.

Species are classified into higher units, such as genera and families, again according to the degree of similarity encountered, but such divisions are not always clear and they may not always reflect adequately the degrees of difference or similarity that actually occur between organisms. Now that it is possible to analyse the genetic constitution of plants and animals, some old arguments are being settled, but other new ones are arising. In the case of the chimpanzee, for example, it has long been believed that there is just one species, which can be divided into three subspecies, one from West Africa, one from Central/West Africa from the Congo to the Niger River, and the third from East Africa. But genetic analyses are showing that the picture is more complicated, with a possible fourth subspecies in western Nigeria, and the possibility that the West African populations are a different species altogether [2]. It is arguments of this type that can make it difficult to determine just how many species there are on earth, as discussed in Chapter 2. Some

biologists even consider that the genetic constitution of the chimpanzee is sufficiently similar to that of our human species to warrant our both being placed in the same genus. Such controversy serves to illustrate that the classification of living organisms is by no means cut and dried, and is essentially a convenient system that taxonomists have constructed and are constantly modifying as new information becomes available.

When we examine the geographical distributions of species of organisms we find that there are no two species that are identical in their ranges. Some correspond fairly closely, but others differ totally. When we use terms such as 'distribution' and 'range', we must also be careful about the spatial scale we are considering. Two species may be widespread within a given geographical area, such as the British Isles, and yet occupy different types of habitats (such as woodland or grassland). Even within a habitat, species may occupy different microhabitats. In a New Zealand forest, for example, one may find both the brown kiwi (*Apteryx australis*) and the fantail (*Rhipidura fuliginosa*), a kind of flycatcher. But they occupy different microhabitats, for the kiwi is confined to the forest floor whereas the fantail nests in canopy branches. Therefore, scale, both horizontal and vertical, is an important consideration when studying distribution patterns [3].

Limits of distribution

Whether the species' distribution is considered on a geographical, habitat or microhabitat scale, it is surrounded by areas where the species cannot maintain a population because different physical conditions or lack of food resources will not permit survival. These areas can be viewed as barriers that must be crossed by the species if it is to disperse to

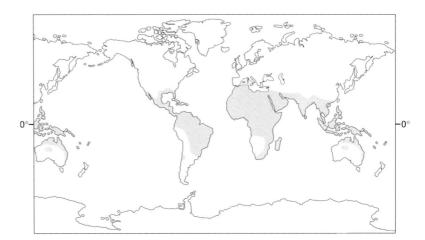

Fig. 3.1 World distribution map of the palm family (Palmae), a pantropical family of plants.

other favourable, but as yet uncolonized, places—much as the European settlers had to cross ocean barriers to colonize North America or Australia. Any climatic or topographic factor, or combination of factors, may provide a barrier to the distribution of an organism. For example, the problems of loco-motion or of obtaining oxygen and food are quite different in water and air. As a result, organisms that are adapted for life on land are unable to cross oceans: their eventual death will be due, in varying proportions, to drowning, starvation, exhaustion and lack of fresh water to drink. Similarly, land is a barrier to organisms that are adapted to life in sea or fresh water, because they require supplies of oxygen dissolved in water rather than as an atmo-spheric gas, and because they desiccate rapidly in air. Mountain ranges, too, form effective barriers to dispersal because they present extremes of cold too great for many organisms. The amount of rainfall, the rate of evaporation of water from the soil sur-face, and light intensity are all critical factors limit-ing the distribution of most plants. But in all these cases, the ultimate barriers are not the hostile fac-tors of the environment but the species' own physi-ology, which has become adapted to a limited range of environmental conditions. In its distribution a species is therefore the prisoner of its own evolu-tionary history.

　　Take the palms, for example. Figure 3.1 shows

Fig. 3.2 The dwarf palm *Chamaerops humilis*, one of the two native palms found in Europe.

the global distribution of this plant family, and it can be seen that members are found in all areas of the tropics and in many subtropical regions too. When one comes to the temperate areas, however, such as Europe, there are very few species of palm that can be regarded as native. Indeed, there are only two truly native palms in Europe. One of these, *Chamaerops humilis*, is a very small species which grows in sandy soils in southern Spain and Portugal, eastwards to Malta (Fig. 3.2). The second species is found on certain Mediterranean islands, mainly on Crete. So a family which is extremely successful and widespread in the tropics has failed

Fig. 3.3 The date palm, *Phoenix dactylifera*, at its most northerly site in the Great Kavir Desert of Iran.

to achieve similar success in the temperate regions. The real problem with the palms is the way they grow: they have only a single growing point at the apex of their upright stems, and if this is damaged by frost then the whole stem perishes. This weakness has even limited the use of palms as domesticated crop plants, for species such as the date palm cannot be grown in areas with frequent frosts. Even in the deserts of northern Iran where the summers are hot and dry, the date palm (*Phoenix dactylifera*) is a rare sight because of the intense cold in these high-altitude drylands during winter (Fig. 3.3). Perhaps the most successful palms in the temperate regions are the *Serenoa* species, which reach 30°N in the United States, and *Trachycarpus*, which attains an altitude of 2400 m in the Himalayas. But the family as a whole is limited geographically by its frost sensitivity.

Some plants and animals are confined in their distribution to the areas in which they evolved; these are said to be 'endemic' to that region. Their confinement may be due to physical barriers to dispersal as in the case of many island faunas and floras (palaeoendemics), or to the fact that they have only recently evolved and have not yet had time to spread from their centres of origin (neoendemics).

At the habitat level, the microhabitats of organisms are surrounded by areas of small-scale variation of physical conditions, or microclimates—similar

to, but on a much smaller scale than, geographical variations in climate—and of food distribution. These form barriers restricting species to their microhabitats. The insects that live in rotting logs, for instance, are adapted by their evolution to a microhabitat with a high water content and relatively constant temperatures. The logs provide the soft woody materials and microorganisms the insects may need for food, and also provide good protection from predators. Around the logs are areas with fewer or none of these desirable qualities and, for many of the animals, attempts to leave their microhabitat would result in death by desiccation, starvation or predation.

Overcoming the barriers

Nevertheless, a few inhabitants of rotting logs do occasionally make the dangerous journey from one log to another, for few environmental factors are absolute barriers to the dispersal of organisms and these factors vary greatly in their effectiveness. Most habitats and microhabitats have only limited resources, and the organisms living in them must have mechanisms enabling them to find new habitats and resources when the old ones become exhausted. These mechanisms often take the form of seeds, resistant stages, or (as in the case of the insects of the rotting-log microhabitat) flying adults with a fairly high resistance to desiccation. There is good evidence that geographical barriers are not completely effective either. When organisms extend their distribution on a geographical scale, it is likely that they are taking advantage of temporary, seasonal or permanent changes of climate or distribution of habitats that allow them to cross barriers normally closed to them. The British Isles, for instance, lie within the geographical range of about 220 species of birds, but a further 50 or 60 species visit the region as so-called 'accidentals'—these birds do not breed in Britain, but one or two individuals are seen by ornithologists every few years [4]. They come for a variety of reasons: some are blown off course by winds during migration, others are forced in certain years to leave their normal ranges when numbers are especially high and food is scarce. Many of these accidentals have their true home in North America, such as the pectoral sandpiper (*Calidris melanotos*), a few of which are seen every year, but some come from eastern Asia, such as the olive-backed pipit (*Anthus hodgsoni*) or even from the South Atlantic, such as the black-browed albatross (*Diomedea melanophris*). It is possible, though not very likely, that a few of these chance travellers may in time establish themselves permanently in Europe, as did the collared dove (*Streptopelia decaocto*) which since about 1930 has spread from Asia Minor and southern Asia across central Europe and into the British Isles and Scandinavia—perhaps the most dramatic natural change in distribution recorded for any vertebrate in recent times. This species is now common around the edges of towns and settlements in western Europe, and seems to depend for food largely on the seeds of weed species common in farms and gardens, together with the bread that humans often put out for garden birds. Several factors may have interacted to permit this extension of range of the collared dove. Increased human activity during the last century, involving extensive changes in the environment, has produced new habitats and food resources, and it is possible, too, that small changes in climate may have significantly favoured this species. It is, however, considered unlikely that the collared dove would have been able to take advantage of these changes without a change in its own genetic make-up, perhaps a physiological one permitting the species to tolerate a wider range of climatic conditions or to utilize a wider range of food substances. Its behaviour patterns have also changed, from nesting largely on buildings to nesting in trees, which may have favoured it in temperate Europe [5].

Biogeographers commonly recognize three different types of pathway by which organisms may spread between one area and another. The first, easiest pathway is called a *corridor*; such a pathway may include a wide variety of habitats, so that the majority of organisms found at either end of the corridor would find little difficulty in traversing it. The two ends would therefore come to be almost identical in their biota (i.e. the fauna plus the flora); for example, the great continent of Eurasia that links western Europe to China has acted as a corridor for the dispersal of animals and plants, at least until the recent climatic changes of the Ice Ages. In the second type of dispersal pathway the interconnecting region may contain a more limited variety of habitats, so that only those organisms that can exist in these habitats will be able to disperse through it. Such a dispersal route is known as a *filter*; the exclusively tropical lowlands of Central America provide a good example. Finally, some areas are completely surrounded by totally different environments, so that it is extremely difficult for any organism to reach them. The most obvious example is the isolation of islands by wide stretches of sea, but the specially adapted biota of a high mountain peak, of a cave or of a large, deep lake is also extremely isolated from the nearest similar habitat from which colonists might originate. The chances of such a dispersal are therefore extremely low, and largely due to chance combinations of favourable circumstances, such as high winds or floating rafts of vegetation. Such a dispersal route is therefore known as a sweepstakes route. It differs from a filter in kind, not merely in degree, for the organisms that traverse a *sweepstakes route* are not normally able to spend their whole life histories en route. Such organisms are alike only in their adaptations to traversing the route, such as those aerial adaptations of spores, light seeds or flight in the case of insects and birds, that enable them to disperse from island to island. Such a biota is therefore not a representative sample of the ecologically integrated, balanced biotas of a normal mainland area, and is said to be disharmonic.

A discussion of some patterns of distribution shown by particular species of animals and plants will show how varied and complex these may be, and will help to emphasize the various scales or levels on which such patterns may be considered. In fact, the number of examples that we can choose is quite limited, because the distribution of only a very small number of species has been investigated in detail. Even amongst well-known species, chance finds in unusual places are constantly modifying known distribution patterns, thereby demanding changes in the explanations that biologists give of these patterns.

Some existing patterns are continuous, the area occupied by the group consisting of a single region or of a number of regions which are closely adjacent to one another. These patterns can usually be explained by the distribution of present-day climatic and biological factors; the detailed distributions of several species of dragonfly provide good examples (below). Other existing patterns are discontinuous or disjunct, the areas occupied being widely separated and scattered over a particular continent, or over the whole world. The organisms which show such a pattern may, like the magnolias, be evolutionary relics, the scattered survivors of a once-dominant and widespread group, now unable to compete with newer forms. Others, the climatic relics or habitat relics, appear to have been greatly affected by past changes in climate or sea level. Finally, as will be shown in Chapters 7 and 8, the disjunct patterns of some living groups, and of many extinct groups, have resulted from the physical splitting of a once-continuous area of distribution by the process of continental drift.

A successful family: the daisies (Asteraceae)

The daisy family provides a useful example of the way in which we need to invoke different explanations for distribution patterns at different geographical and taxonomic scales. The daisy family is extremely large (having over 22 000 species) and extremely successful, if you measure biogeographical success by the areal extent of distribution. It is a cosmopolitan family, i.e. it is found throughout the world when we look at it on a global scale. In fact, the term cosmopolitan when used of the flowering plants is usually a slight exaggeration, since very few species of flowering plant have managed to establish themselves in Antarctica; even the Asteraceae have not achieved that, but they are present on all other continents. There has clearly been no insuperable barrier to the

geographical spread of the family during its evolutionary history.

When we look at those areas of the world where the Asteraceae are most abundant and diverse, we find that the mountainous regions of the tropics and subtropics, together with some of the semi-arid regions of the world and those with Mediterranean climates (hot, dry summers and mild, wet winters) are the richest in members of this family. The equatorial rainforests are actually rather poor in daisy family species. Often, biogeographers use such information in trying to reconstruct the evolutionary origins of a group. A great deal of generalization is involved, but it does seem as though this family has been most successful away from the competition of tall trees, in the more drought-prone habitats where their general adaptability and very diverse fruit-dispersal systems have given them many advantages.

Taking just one genus from within the family, the groundsel genus *Senecio*, we find that it reflects the whole family in many ways, being large (about 1250 species) and widely dispersed (cosmopolitan apart from Antarctica). Many members are efficient weeds, being short-lived, having efficiently dispersed air-borne fruits and having wide ecological tolerances of climate and soils. Some taxonomists prefer to split this very large genus up into subgenera, and one of these, the subgenus *Dendrosenecio*, is remarkable both for its form (Fig. 3.4) and its restricted distribution pattern. This genus consists of just 11 species, the giant tree-groundsels—stocky, woody plants up to 6 m in height often with branched upper sections bearing terminal clusters of tough, leathery leaves. In distribution, this subgenus is restricted to East Africa and, moving in on the geographical scale, only on the high mountains of East Africa (Fig. 3.5) above the forest limits of bamboos and tree-heathers [6]. If we focus in from the taxonomic level of subgenus to species, we find that each of the major mountains of East Africa has its own group of endemic species of *Dendrosenecio*, with never more than three species on any particular mountain.

Detailed analysis of the genetic material, the

Fig. 3.4 Giant tree-groundsels from East African mountains. Family Asteraceae, genus *Senecio*, subgenus *Dendrosenecio*. (a) Branched form; (b) unbranched form.

(a)   (b)

DNA, of the tree-groundsel species by Eric Knox at Kew in London, and Jeffrey Palmer at Indiana University [7], has shown that each species is more closely related to its neighbouring species on its own mountain than to the species of other mountains, despite the fact that in form, such as branching pattern (Fig. 3.4), it may more closely resemble the giant groundsels from elsewhere. It seems that chance has led to the colonization of each mountain peak (with the exception of one, Mount Meru, which is devoid of giant groundsels), and that the chance invader has in the course of time evolved into two or three species. (The time involved, incidentally, cannot be very long since Mount Kilimanjaro is only a million years old—quite young by geological standards.)

If we take the spatial scale of analysis one step lower and look at separate species on just one of the mountains, then further factors come into play in the interpretation of distribution patterns. On Mount Elgon (4300 m), situated on the border between Uganda and Kenya, north of Lake Victoria (Fig. 3.5), two species of the tree-groundsels are found, *Senecio elgonensis* and *S. barbatipes*. In the open, alpine zone where these trees are found,

S. elgonensis predominates below 3900 m and *S. barbatipes* above this level, so there is an altitudinal differentiation in their ranges on the mountain. Precisely what features of the morphology or the physiology of the two species leads to these climatic preferences is not known, and there are no detailed meteorological measurements for the mountain, but temperature differences with altitude and, in particular, the incidence of frost, is likely to be the most important factor affecting the distribution of the two species. The giant senecios are more frost-tolerant than most tropical plants, being insulated by thick layers of leaves and leaf bases (Fig. 3.6). When the night air temperature drops to –4°C, the temperature within the leaves only falls to 2°C. But the different sensitivities of the individual species may affect altitudinal limits, perhaps via seed production or germination. The two species may also be in competition with one another for space or some other resource (see examples later in this chapter).

Taking a final and even more detailed look at the distribution of *S. elgonensis* in a small valley within the lower part of the alpine zone of Mount Elgon, we find that the population is most dense

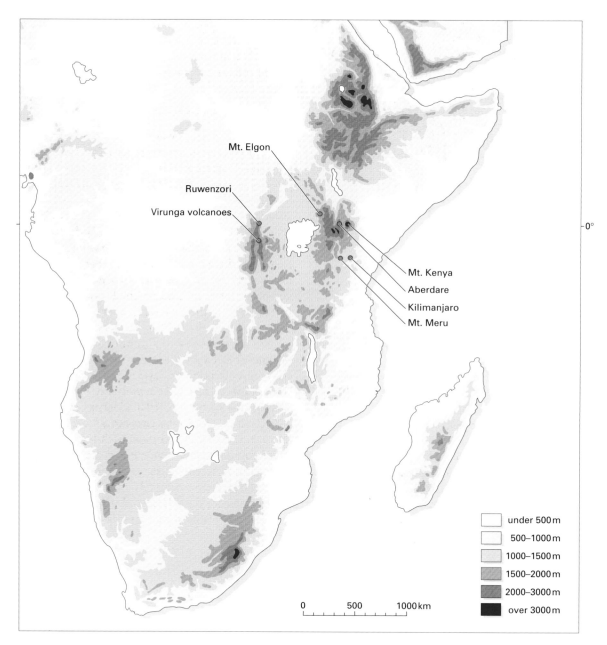

Fig. 3.5 Map of eastern and southern Africa showing the high mountain peaks on which tree-groundsels are found. Mt. Meru is an exception, having no tree-groundsels.

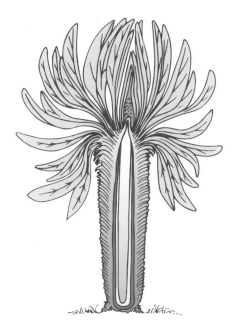

Fig. 3.6 Cross-section of a tree-groundsel showing its thick central pith, surrounded by wood and cortex, together with the outer layer of dead leaves and leaf bases, forming an insulating sheath that protects the living tissues from frost.

demonstrates the way in which we must invoke different factors to account for the distribution patterns of organisms depending upon both the taxonomic and the geographical scale we are using.

Patterns of dragonflies

One group of species whose distributions are quite well known, at least in western Europe, are the Odonata, dragonflies and damselflies [8]. The common blue damselfly, *Enallagma cyathigerum*, is possibly one of the most abundant and widely distributed dragonfly species (Fig. 3.8a). The adults are on the wing in midsummer, around bodies of fresh water. The female lays eggs in vegetation below the surface of the water and the larvae hatch in a week or two. These live on the bottom of the pond, stream or lake and feed on small crustaceans and insect larvae until they reach a size of 17–18 mm, which may take from 2 to 4 years, depending on the quality and quantity of food available. In the May or June after reaching full size, the larvae climb up the stems of emergent vegetation (plants rooted in the mud with stems and leaves emerging from the water surface), cast their larval skins, and emerge as winged adults. *E. cyathigerum* is found in a wide range of freshwater habitats, including both still and moving water, although it is perhaps least common in fast-moving water, or in places where silt is being deposited. Probably the ideal habitat for this species is a fairly large body of still water with plenty of floating vegetation. *E. cyathigerum* is

around the valley floor (Fig. 3.7) where a damp area of water seepage exists. The species is evidently affected at this habitat scale by the availability of deeper, moist soils, preferring these to the shallow, free-draining soils of the alpine slopes and ridges.

This analysis of distribution patterns at gradually reducing taxonomic levels within the Asteraceae

Fig. 3.7 Diagram of a cross-section of a small valley on Mt. Elgon, Uganda, showing the higher density of the tree-groundsels in the valley bottom.

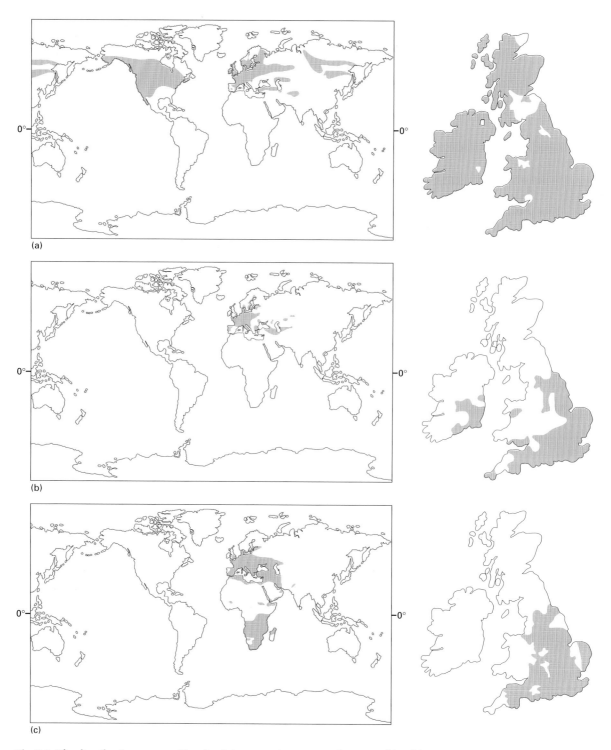

Fig. 3.8 The distribution on a world scale of three species of dragonfly. (a) Common blue damselfly, *Enallagma cyathigerum*; (b) ruddy sympetrum, *Sympetrum sanguineum*; and (c) emperor dragonfly, *Anax imperator*.

found in both acid and alkaline waters, and often occurs in brackish pools on salt marshes. As a result, the species can be regarded as having very wide ecological tolerances; it is said to be eurytopic.

The geographical distribution of this damselfly is also very wide. In Europe the distribution lies mostly between 45°N and the Arctic Circle, although it includes some of the wetter parts of Spain, and is rather scattered in northern Scandinavia; it is not found in Greenland or Iceland. The species is found in a few places in North Africa, in Asia Minor and around the Caspian Sea. Large areas of Asia south of the Arctic Circle also fall within its range. In North America it is found everywhere north of about 35–40°N to the Arctic Circle, except Labrador and Baffin Island. Populations also occur in the ideal habitats provided by the Everglades of Florida, which mark the species' furthest southward expansion. The broad geographical distribution of this species is almost certainly due to its ecological tolerance and ability to make use of a wide range of habitats in very different climates. This type of distribution pattern, a belt around the Northern Hemisphere, is shown by many species of animals and plants, and is termed circumboreal—'around the northern regions'. The frequency of this pattern in very different organisms suggests that the two northern land masses may once have been joined, enabling certain species to spread right around the hemisphere.

As *E. cyathigerum* is so successful, one might ask why it has not spread further southward. One reason may be the relative scarcity of watery habitats in the subtropical regions immediately to the south of its present range—the arid areas of Central America, North Africa and central Asia. The species is perhaps not robust enough for the long migrations that would be needed to reach suitable habitats in the Southern Hemisphere (there are very few wind belts that might assist such a migration). Another possibility is that there are other species already occupying all the habitats that *E. cyathigerum* could colonize further south. These species may be better adapted to the physical conditions of their habitats than *E. cyathigerum*, and could therefore compete successfully with it for the available food

resources; this might exclude the species from these areas. In fact in the Southern Hemisphere there are many species of the genus *Enallagma* and the closely related genus *Ischnura* that might be expected to have similar habitat and food requirements *to E. cyathigerum*; there are at least eight species of *Enallagma* in South Africa alone. Any of these explanations, or a combination of them, would explain why the common blue damselfly is confined to the high latitudes of the Northern Hemisphere.

The distribution of *E. cyathigerum* may be contrasted with that of the beautiful dragonfly *Sympetrum sanguineum*, sometimes called the ruddy sympetrum (Fig. 3.8b). This species has a limited distribution in western Europe, parts of Spain, a few places in Asia Minor and North Africa, and around the Caspian Sea; it is not found in eastern Asia or North America. The reason for the limited distribution of this dragonfly is almost certainly that the larva has very precise habitat requirements.

The larva is found in ditches and ponds with still waters, but only where certain emergent plants—the great reedmace (*Typha latifolia*) and horsetails (*Equisetum* species)—are growing. Why the larva should have these precise requirements is not clear, because it is certainly not a herbivore, feeding on insect larvae and crustaceans, but so far the larvae have never been found away from the roots of these plants. *S. sanguineum* could therefore be described as a stenotopic species—one with very limited ecological tolerance. The fact that it can colonize only a very few habitats must certainly have limited its distribution, but other, unknown, factors are also at work, for the species is often absent from waters in which reedmace or horsetails are present.

The northern distribution of these two species may be contrasted with that of the emperor dragonfly, *Anax imperator* [8,9] (Fig. 3.8c). The adults of this species are 8–10 cm long, and the larva is found typically in large ponds and lakes and in slow-moving canals and streams; it is a voracious predator and can eat animals as large as fish larvae. The distribution covers a band of Europe between about 50°N and 40°N but, unlike the other two species,

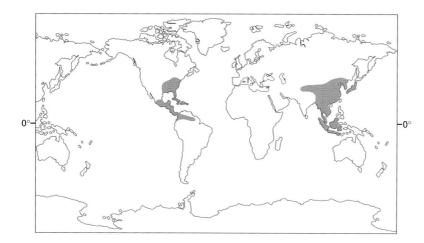

Fig. 3.9 World distribution map of the magnolias, illustrating a disjunct distribution.

it is well distributed on the North African coast and the Nile Valley and stretches across Asia Minor to north-west India. It even spreads across the Sahara Desert down into Central Africa, where there are suitable habitats such as Lake Chad and the lakes of East Africa, and it is found in most parts of South Africa except the Kalahari Desert. (It is possible that the South African population may belong to a separate subspecies from the European forms.)

The distribution of *A. imperator* is therefore confined to the Old World and does not extend far into Asia. It appears to be basically a Mediterranean and subtropical species whose good powers of flight and fairly broad ecological tolerance have enabled it to cross the unfavourable areas of North Africa to new habitats in southern Africa. No doubt favourable habitats for *A. imperator* do occur in the other land masses (although there may be potential competitors there, of course), but the dragonfly cannot now reach them because the land connections lie to the north, where its distribution seems limited.

In most tropical dragonfly species the larvae emerge from the water and metamorphose to the adult at night; they are very vulnerable to predators at the time of emergence, and darkness probably affords some protection from birds. But the process of metamorphosis is inhibited by cold temperatures, and in northern Europe many species are compelled by low night temperatures to undergo at least part of their emergence in daylight, when birds eat large numbers of them. This probably imposes a northern limit to the distribution of many species of dragonfly, including *A. imperator*, which would explain why this species has not been able to invade the Americas or eastern Asia.

Magnolias: evolutionary relicts

The magnolias (family Magnoliaceae, genus *Magnolia*) have a very interesting modern distribution, as shown in Fig. 3.9. Of the 80 or so species of *Magnolia*, the majority are found in South-East Asia and the remainder, about 26 species, in the Americas—ranging from Ontario in the north, through Mexico, down into the northern regions of South America [10]. Their distribution is clearly disjunct, being separated into two main centres in this case. Unlike the palms, we cannot explain their distribution pattern simply in terms of the climatic sensitivities of the plants concerned, for the magnolias are reasonably hardy in comparison; they can be cultivated well into the north of the temperate area. Nor would climatic constraints explain why they are not found in intermediate tropical and subtropical regions, as are the palms.

To understand the distribution of the magnolias, we need to look at their evolutionary history. Fossils

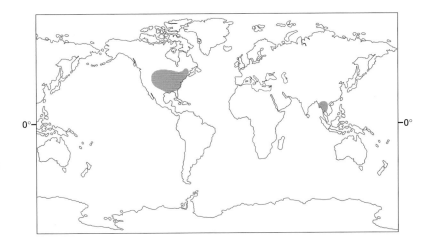

Fig. 3.10 World distribution of the tulip trees (*Liriodendron* species). Only two species now survive, in widely separated localities, although it was once a widespread genus.

of magnolia-like leaves, flowers and pollen grains are known from Mesozoic times—the age of the dinosaurs. Indeed, the magnolia family is regarded by botanists as one of the most primitive families of flowering plant groups. Its showy flowers were attractive to the rapidly evolving insects and together they coevolved into a most successful team in which the insect visited the flowers for food and, in doing so, ensured the passage of pollen from one plant to another, thus taking the chance and the waste out of the highly risky wind-pollination process. The magnolias spread and must have formed a fairly continuous belt around the tropical, subtropical and temperate parts of the world, for their fossil remains have been found through Europe and even in Greenland. For perhaps as long as 70 million years, the magnolias remained widespread, right up to the last two million years, during which they have been lost from such intermediate geographical areas as Europe.

Being small, slow-growing shrubs and trees, they were not strong competitors for the more robust and fast-growing tree species and, when the climatic fluctuations of the last two million years began to disturb their stable woodland environment, they succumbed to the pressures and soon became extinct across much of their former range. Only in two parts of the world have they managed to escape and survive, as evolutionary relicts.

It is interesting that another genus of the magnolia family, the tulip tree, *Liriodendron*, has a very similar distribution to that of the *Magnolia* genus and in all probability shares a similar fossil history. But, in the case of the tulip trees, only two species have survived, *L. tulipifera* being a successful component of the deciduous, temperate forests of North America, and *L. chinense* surviving only in a restricted area of South-East Asia (Fig. 3.10).

Climatic relicts

Many other species, which in the past were widely distributed, were affected by climatic changes and survive now only in a few 'islands' of favourable climate. Such species are called climatic relicts—they are not necessarily species with long evolutionary histories, since many major climatic changes have occurred quite recently. The Northern Hemisphere has an interesting group of glacial relict species whose distributions have been modified by the northward retreat of the great ice sheets that extended as far south as the Great Lakes in North America, and to Germany in Europe, during the Pleistocene Ice Ages (the last glaciers retreated from these temperate areas about 10 000 years ago). Many species that were adapted to cold conditions at that time had distributions to the south of the ice sheets almost as far as the Mediterranean in Europe. Now that these areas are much warmer, such species survive there only in the coldest places, usually at high altitudes in mountain ranges, and the greater part of their distribution lies far to the north in Scandinavia, Scotland or Iceland. In some cases, species even appear to have become extinct in northern regions and are represented now only by glacial relict populations in the south, such as in the Alpine ranges.

An example of a glacial relict is the springtail *Tetracanthella arctica* (Insecta, Collembola). This dark-blue insect, only about 1.5 mm long, lives in the surface layers of the soil and in clumps of moss and lichens, where it feeds on dead plant tissues and fungi. It is quite common in the soils of Iceland and Spitzbergen, and has also been found further west in Greenland and in a few places in Arctic Canada. Outside these truly Arctic regions it is known to occur in only two regions; in the Pyrenean Mountains between France and Spain, and in the Tatra Mountains on the borders of Poland and Czechoslovakia (with isolated finds in the nearby Carpathian Mountains) (Fig. 3.11). In these mountain ranges the species is found at altitudes of around 2000 m in arctic and subarctic conditions. It is hard to imagine that the species can have colonized these two areas from its main centre further north, because it has very poor powers of distribution (it is quickly killed by low humidity or high temperatures) and is not likely to have been transported there accidentally by humans. The likely explanation for the existence of the two southern populations is that they are remnants of a much wider distribution in Europe in the Ice Ages. But it is surprising that *T. arctica* has not been found at high altitudes in the Alps, despite careful searching by entomologists. Perhaps it has simply not yet been noticed, or perhaps it used to occur there but has since died out. One interesting feature of this species is that whereas animals from the Arctic and the Tatras have eight small eyelets (ocelli) on either side of the head, specimens from the Pyrenees have only six. This suggests that the Pyrenean forms have undergone some evolutionary changes since the end of the Ice Ages while they have been isolated from

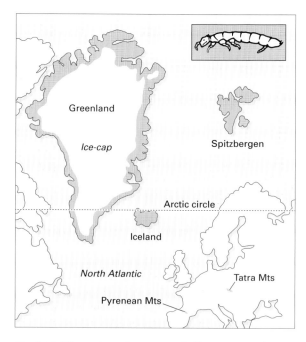

Fig. 3.11 The springtail *Tetracanthella arctica*, and a map of its distribution. It is found mostly in northern regions, but populations exist in the Pyrenees and in mountains in central Europe. These populations were isolated at these cold, high altitudes when the ice sheet retreated northwards at the end of the Ice Ages.

the rest of the species, and perhaps they should be classified as a separate subspecies.

A plant example of a glacial relict (Fig. 3.12) is the Norwegian mugwort (*Artemisia norvegica*), a small alpine plant now restricted to Norway, the Ural Mountains and two isolated localities in Scotland. During the last glaciation and immediately following it, the plant was widespread, but it became restricted in distribution as forest spread.

There are probably several hundred species of both animals and plants in Eurasia that are glacial relicts of this sort, and they include many species that, in contrast to the springtail, have quite good powers of dispersal. One such species is the mountain or varying hare, *Lepus timidus*, a seasonally variable species (its fur is white in the winter and bluish for the rest of the year), which is closely related to the

Fig. 3.12 The Norwegian mugwort, *Artemisia norvegica*: (a) the plant; (b) distribution map showing its restricted range in only two mountainous areas of Europe.

(a)

(b)

more common brown hare, *L. capensis*. The varying hare has a circumboreal distribution, including Scandinavia, Siberia, northern Japan and Canada (although the North American form, the snowshoe hare, is thought by many zoologists to belong to a separate species, *L. americanus*). The southernmost part of the main distribution is in Ireland and the southern Pennines of England, but there is a glacial relict population living in the Alps that differs in no important features from those in the more northerly regions. There is, however, an interesting complication—*L. timidus* is found all over Ireland, thriving in a climate that is no colder than that of many parts of continental western Europe. There seems to be no climatic reason why this hare should not have a wider distribution in many parts of the world, but it is probably excluded from many areas by its inability to compete with its close relatives, the brown hare (*L. capensis*) in Europe and the jack rabbit (*L. americanus*) in North America, for food resources and breeding sites. Relict populations of the varying hare survive in the Alps because, of the two species, it is the better adapted to cold and snowy conditions [11].

One very remarkable example of a glacial relict is the dung beetle species *Aphodius holdereri* (Fig. 3.13). This beetle is now restricted to the high Tibetan plateau (3000–5000 m), having its southern limit at the northern slopes of the Himalayas. In 1973

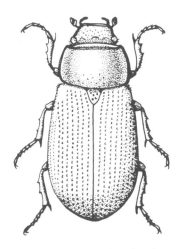

Fig. 3.13 *Aphodius holdereri*, a dung beetle now found only in the high plateau of Tibet.

G. Russell Coope, of London University, found the remains of at least 150 individuals of this species in a peaty deposit from a gravel pit at Dorchester-on-Thames in southern England [12]. The deposit dated from the middle of the last glaciation, and subsequently 14 sites have yielded remains of this species in Britain, all dated between 25 000 and 40 000 years ago. Evidently *A. holdereri* was then a widespread species, but climatic changes have severely restricted the availability of suitable habitats for its survival. Only the remote Tibetan

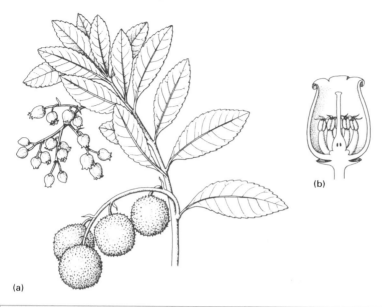

(a)

(b)

(c)

Fig. 3.14 The strawberry tree, *Arbutus unedo*: (a) plant showing leathery leaves, and swollen fruit which are red in colour; (b) cross-section of a flower; (c) map of European distribution, showing relict population in Ireland.

mountains now provide *Aphodius* with the extreme climatic conditions necessary for its survival.

The strawberry tree in Europe is a good example of what may be termed a postglacial relict (Fig. 3.14). The strawberry trees of North America have a fairly continuous distribution, but the western European species *Arbutus unedo* is disjunct, having its main centre of distribution in the Mediterranean region but with outliers in western France and western Ireland.

The Ice Age closed with a sudden warming of the climate, and the glaciers retreated northwards; behind them came the plant and animal species that had been driven south during glacial times. Warmth-loving animals, particularly insects, were able to move northward rapidly but plants were slower in their response, because their rate of spread is slower. Seeds were carried northward, germinated, grew, and finally flowered and sent out more seeds to populate the bare northlands. As this migration continued, melting glaciers produced vast quantities of water that poured into the seas, and the ocean levels rose. Some of the early colonizers reached areas by land connections that were later severed by rising sea levels.

The maritime fringe of western Europe must have provided a particularly favourable migration route for southern species during the period following the retreat of the glaciers. Many warmth-loving plants and animals from the Mediterranean region, such as the strawberry tree, moved northward along this coast and penetrated at least as far as the south-west of Ireland, before the English Channel and the Irish Sea had risen to form physical barriers to such movement. The nearness of the sea, together with the influence of the warm Gulf Stream, gives western Ireland a climate that is wet, mild and frost-free, and this has allowed the survival of certain Mediterranean plants that are scarce or absent in the rest of the British Isles.

Like many Mediterranean trees and shrubs, the strawberry tree is sclerophyllous, which means it has hard, leathery leaves (Fig. 3.14). This is a plant adaptation often associated with arid climates and

seems out of place in the west of Ireland. Flowering in many plant species is triggered by a response to a particular daylength—this is called photoperiodism. *Arbutus* flowers in late autumn, as the length of night is increasing, and this is an adaptation which is again associated with Mediterranean conditions, since at this season the summer drought gives way to a warm, damp period. The flowers, which are cream-coloured, conspicuous and bell-shaped, have nectaries that attract insects, and in Mediterranean areas they are pollinated by long-tongued insects such as bees, which are plentiful in late autumn. In Ireland, however, insects are scarce in the autumn and pollination is therefore much less certain. Thus, the strawberry tree reached Ireland soon after the retreat of the glaciers and has since been isolated there as a result of rising oceans. Although the climate has steadily grown colder since its first colonization, *Arbutus* has so far managed to hold its own and survive in this outpost of its range, despite having features in its structure and life-history that seem ill-adapted to western Ireland.

Another example of a disjunction that has taken place in relatively recent times is the gorilla (*Gorilla gorilla*). In West Africa it is found in an area of lowland tropical rain forest, but in the east it is not limited to lowland forest, being also found in mountains (Fig. 3.15). Some zoologists regard the two populations as separate subspecies. It is thought that the populations have diverged as a result of the changing patterns of vegetation in central Africa over the last two million years or so [13]. This pattern of disjunction is reflected in the distributions of many African plants and animals [14].

Endemic organisms

Because each new species of organism evolves in one particular, restricted area, its distribution will be limited by the barriers that surround its area of origin. Each such area will therefore contain organisms that are found there and nowhere else; these organisms are said to be endemic to that area. As time goes by, increasing numbers of organisms

Fig. 3.15 Distribution map of the gorilla, a mammal with a disjunct distribution.

will evolve within the area, and the percentage of its biota that is endemic is therefore a good guide to the length of time for which an area has been isolated.

As these organisms continue to evolve they will also become progressively different from their relatives in other areas. Taxonomists are likely to recognize this by giving higher taxonomic rank to the organisms concerned. So, for example, after two million years the biota of an isolated area might contain only a few endemic species. After 10 million years the descendants of these species might be so unlike their nearest relatives in other areas that they might be placed in one or more endemic genera. After 35 million years these genera might appear to be sufficiently different from their nearest relatives as to be placed in a different family, and so on. (The absolute times involved would, of course, vary and depend upon the rate of evolution of the group in question.) Therefore, the longer an area has been isolated, the higher the taxonomic rank of its endemic organisms is likely to be, and vice versa.

Figure 3.16 shows the proportion of the mountain flora in various European mountain ranges that are

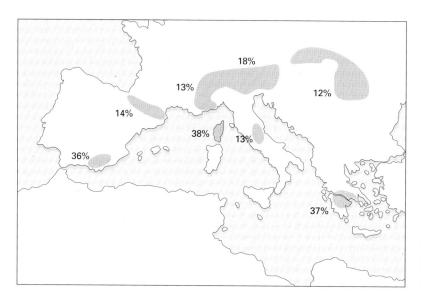

Fig. 3.16 The percentage of endemic plants in the floras of the mountain ranges of southern Europe. From Favarger [15].

endemic to their particular area. It is evident that the more northerly of the mountain ranges shown have a lower proportion of their flora that is endemic whereas the southern, Mediterranean mountains have higher proportions [15].

The montane plants, like the glacial relics described above, are now limited in range because of the increasing warmth of the last 10 000 years. The northern mountains may be poorer in endemics simply because local glaciation there was more severe and some of the species surviving further south became extinct. On the other hand, the richness of the southern mountains could be explained by the fact that the geographical barriers between the northern montane blocks are less severe (less distance, no sea barriers) and hence migration and sharing of mountain floras is more likely than in the south, where barriers are considerable.

In general there are two major factors influencing the degree of endemism in an area; these are isolation and stability. Thus, isolated islands and mountains are often rich in endemics. Long-term climatic stability is rather rare, but there is evidence that some parts of the earth have been more stable than others. For example, the Cuatro Ciénegas basin in Mexico appears to have retained a very stable vegetation over the last 40 000 years, judging from the evidence of pollen from lake cores collected in the area (see Chapter 10). This basin is also rich in endemic organisms [16]. This kind of 'fossil endemism' is called palaeoendemism, in contrast to neoendemism resulting from recent surges in the evolutionary process and the generation of new species that have not yet had an opportunity to spread beyond their current limits.

California is rich in neoendemics, including such plant genera as *Aquilegia* and *Clarkia*, which are undergoing rapid evolution. The richness of the flora of California, however, is typical of many regions of the earth with a 'Mediterranean' type of climate, including the Mediterranean basin itself, Chile, the southern tip of South Africa and the south-western extremity of Australia. Much debate has surrounded

the floral richness of these regions, and it may well be that the long history of recurrent fires has created conditions under which small, isolated populations of plants have diversified, leading to a high density of species, many with restricted distributions [17].

Physical limitations

The geographical range of a species is not always determined by the presence of topographic barriers preventing its further spread. Often a species' distribution is limited by a particular factor in the environment that influences its ability to survive or reproduce adequately. These limiting factors in the environment include physical factors such as temperature, light, wetness, and dryness, as well as biotic factors such as competition, predation, or the presence or absence of suitable food. In the remainder of this chapter the ways in which such factors influence organisms will be described in more detail.

First, though, the meaning of the term 'limiting factor' must be understood. Anything that tends to make it more difficult for a species to live, grow or reproduce in its environment is a limiting factor for the species in that environment. To be limiting, such a factor need not necessarily be lethal for a species; it may simply make the working of its physiology or behaviour less efficient, so that it is less able to reproduce or to compete with other species for food or living space. For instance, we suggested earlier that a northern limit may be set to the distribution of certain dragonflies by low night-time temperatures. In the more southerly parts of northern regions, at least, temperatures are not so low that they kill dragonflies directly, but they are low enough at night to force the insects to metamorphose during the day, when they are more vulnerable to predatory birds. In this case, then, the limiting factor of temperature does not operate directly but is connected with a biotic environmental factor, that of predation. Many other limiting factors act in a similar way.

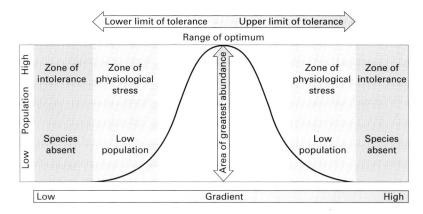

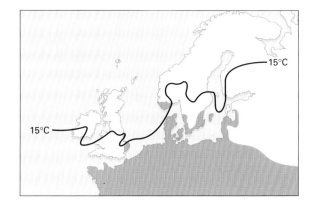

Fig. 3.17 Graphical model of the population abundance maintained by a species of animal or plant along a gradient of a physical factor in its environment.

Environmental gradients

Many physical and biotic factors affect any species of organism, but each can be considered as forming a gradient. For example, the physical factor of temperature affects species over a range from low temperatures at one extreme to high temperatures at the other, and this constitutes a temperature gradient. These gradients exist in all environments and affect all the species in each environment. As seen earlier, different species vary in their tolerance of environmental factors, being either eurytopic (ecologically tolerant) or stenotopic (ecologically intolerant), but each species can function efficiently over only a more or less limited part of each gradient. Within this range of optimum the species can survive and maintain a large population; beyond it, toward both the low and the high ends of the gradient, the species suffers increasing physiological stress—it may stay alive, but because it cannot function efficiently it can maintain only low populations. These areas of the gradient are bordered by the upper and lower limits of tolerance of the species to the environmental factor. Beyond these limits the species cannot survive because conditions are too extreme; individuals may live there for short periods but will either die or pass quickly through to a more favourable area (see Fig. 3.17). A species may not achieve its full potential distribution in the field because of competitive interactions with other organisms. When under conditions of

Fig. 3.18 Distribution of the grey hair grass (*Corynephorus canescens*) in northern Europe (shaded) and its relationship to the 15°C July mean isotherm. From Marshall [18].

physiological stress, a species easily succumbs to such competition.

The grey hair grass (*Corynephorus canescens*) is widespread in central and southern Europe and reaches its northern limit in the British Isles and southern Scandinavia (Fig. 3.18). J.K. Marshall has examined the factors that may be responsible for maintaining its northern limit, and he found that both flowering and germination were affected by low temperature [18]. The grass has a short life span (about 2–6 years), so it relies upon seed production to maintain its population. Any factor interfering with flowering or with germination could therefore limit its success in competitive situations. At its northern limit, low summer temperature delays its flowering with the result that the season is already well advanced when the seeds are shed. Seed germination is slowed down at temperatures below 15°C and seeds sown experimentally after October had a very poor survival rate. This may explain why its northern limit in Europe so closely matches the 15°C July mean isotherm. Other factors, however, must be in operation to prevent its spread in southern and central Britain.

Even mobile animals, like birds, may have their distributions closely linked to temperature, as in the case of the eastern phoebe (*Sayornis phoebe*), a migratory bird of eastern and central North America. Analysing data collected by ornithologists of the National Audubon Society during the Christmas period, ecologist Terry Root has been able to check the winter distribution of this bird against the climatic conditions [19]. She found that the wintering population of the eastern phoebe was confined to that part of the United States in which the mean minimum January temperature exceeded −4°C. The very close correspondence of the bird's winter range to this isotherm (Fig. 3.19) probably relates to the energy balance of the birds. Warm-blooded animals, such as birds, use up large quantities of energy in maintaining their high blood temperature and in cold conditions they can lose a great deal of energy in this way and therefore have to eat more. Terry Root found that birds in general do not occupy regions where low temperature forces them to raise their resting metabolic rate (i.e. their energy consumption) by a factor of more than 2.5. In the case of the eastern phoebe this critical point is reached when the temperature falls below −4°C, so the bird fails to occupy colder regions. Other birds have different temperature limits because they have different efficiencies in their heat generation and conservation, but they still seem to draw the line at raising their resting metabolism by a factor of more than 2.5.

Many plants have their seeds adapted to a specific temperature for germination, and this often relates to conditions prevailing when germination is most

Fig. 3.19 The northern boundary of the distribution (solid line) of the eastern phoebe in North America in December/January, compared with the −4°C January minimum isotherm. From Root [19].

appropriate for the species. P.A. Thompson of Kew Gardens, England, has devised a piece of apparatus for examining the effect of temperature on germination [20]. It consists of a metal bar, one end of which is maintained at −40°C and the other at 3°C; between is a gradient of temperatures. Groups of seeds of the species to be examined are placed along the bar and kept moist, and a record is kept of the number of days required for 50% of the seeds within each group to germinate. The results are expressed on graphs, and the lowest point on the U-shaped curves shows the optimum temperature for germination.

In Fig. 3.20 the germination responses of three members of the catchfly family are shown, together with their geographical ranges. The catchfly (*Silene secundiflora*) is a Mediterranean species, so the optimum time for germination is the autumn, when the hot, dry summer is over and the cool, moist winter is about to begin. Its optimum germination occurs at about 17°C. The ragged robin (*Lychnis*

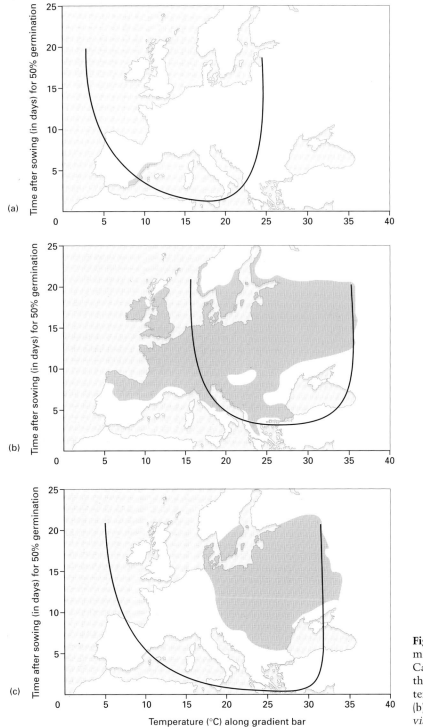

Fig. 3.20 Distribution maps of three members of the plant family Caryophyllaceae, together with their germination responses to temperature: (a) *Silene secundiflora*; (b) *Lychnis flos-caculi*; and (c) *Silene viscosa*. From Thompson [20].

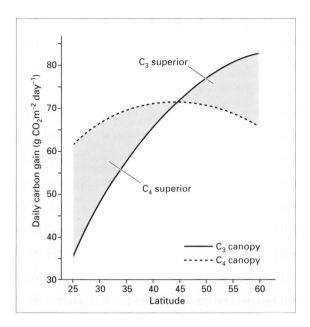

Fig. 3.21 Predicted levels of photosynthesis for C_3 and C_4 species over a range of latitudes in the Great Plains during July. The C_4 advantage is lost in latitudes greater than 45°N. From Ehleringer [22].

Fig. 3.22 Proportion of C_4 species in the grass flora of various parts of North America. From Teeri & Stowe [23].

flos-cuculi) occurs throughout temperate Europe and here the cold winter is the least favourable period for growth, hence there are advantages to be gained by germinating in the spring. Optimum germination occurs at about 27°C. The third species, the sticky catchfly (*Silene viscosa*), is an eastern European steppe species. The invasion of open grassland is an opportunistic business; each chance that offers itself must be taken, so any temperature limitation is likely to be an unacceptable restriction on a plant in its struggle for space. Wide tolerance of temperature is thus an advantage and *S. viscosa* seeds germinate well over the range 11–31°C.

In most plants the first product of photosynthesis is a sugar containing three carbon atoms; these plants are known as C_3 plants. In some plant species, however, there is a supplementary mechanism at work in which carbon dioxide is temporarily fixed into a four-carbon compound, later to be fed into the conventional fixation process in specialized cells around the bundles of conducting tissue in the leaf [21].

These so-called C_4 plants occur in a number of different flowering plant families, mostly from tropical and subtropical regions. Some important tropical crop species, such as sugar cane, are C_4 plants.

For a variety of biochemical reasons the C_4 mechanism is most advantageous under conditions of high light intensity and high temperature, whereas it may be disadvantageous at low light and low temperature. J.R. Ehleringer has calculated that, at a latitude of about 45° in an area like the great plains of North America, the relative advantages and disadvantages of each system are roughly in balance (Fig. 3.21) [22]. If one examines the proportion of C_3 and C_4 species among the grasses in different. sites in North America, then one finds that the C_4 system occurs in more than 50% of the grasses in most areas south of 40°N and in fewer than 50% north of that latitude (Fig. 3.22) [23]. Competitive interaction between species of grass has therefore led to the selection of the photosynthetic mechanism most appropriate to the needs of any given locality. Those C_4 species found north of the critical line are often associated with particular circumstances that favour them; for example, they may have their maximum growth rate in late summer when temperatures are highest, whilst the C_3 species grow best in the cooler conditions of spring and early summer [24]. As the climate is now becoming warmer, it will be interesting to observe whether there are shifts in the balance of biogeographic advantages in C_3 and C_4 photosynthesis and consequent changes in the distribution patterns of some species.

Although temperature is one of the most important environmental factors, because of its effect on the metabolic rate of organisms, many other physical factors in the environment are limiting ones. A whole family of factors is related to the amount of water present in the environment. Aquatic organisms obviously require water as the basic medium of their existence, but most terrestrial animals and plants, too, are limited by the wetness or dryness of the habitat, and often also by the humidity of the atmosphere, which in turn affects its 'drying power' (or, more precisely, the rate of evaporation of water from the ground and from animals and plants). Light is of fundamental importance because it provides the energy that green plants fix into carbohydrates during photosynthesis, thus obtaining energy for themselves (and ultimately for all other organisms).

Light in its daily and seasonal fluctuation also regulates the activities of many animals. The concentrations of oxygen and carbon dioxide in the water or air surrounding organisms are also important. Oxygen is essential to most animals and plants for the release of energy from food by respiration, and carbon dioxide is vital because it is used as the raw material in the photosynthesis of carbohydrates by plants. Many other chemical factors of the environment are of importance, particularly soil chemistry where plants are concerned. Pressure is important to aquatic organisms; deep-sea animals are specially adapted to live at high pressures, but the tissues of species living in more shallow waters would be easily damaged by such pressures.

In marine environments, variation in the salinity of the water affects many organisms, because many marine organisms have body fluids with much the same salt concentration as sea water (about 35 parts per thousand), in which their body tissues are adapted to function efficiently. If they become immersed in a less saline medium (in estuaries, for instance), water moves into their tissues due to the physical process called osmosis, by which water passes across a membrane from a dilute solution of a salt to a concentrated one. If the organisms cannot control the passage of water into their bodies, the body fluids are flooded and their tissues can no longer function. This problem of salinity is an important factor in preventing marine organisms from invading rivers, or freshwater ones from invading the sea and spreading across oceans to other continents.

In an estuary, the salinity varies both in space and in time. The distance from the sea influences salinity as the input of sea water becomes less, but salinity at any given location will vary with time because of the impact of tidal flows. The crustacean genus *Gammarus* is found in estuaries, but is represented by different species according to the nature of the salinity conditions (Fig. 3.23). Each species has its optimum set of conditions for salinity, but also has its distribution limits which result from a combination of its reduced tolerance and also competition from other species which may perform more efficiently under the new conditions.

Interaction of factors

The environment of any species consists of an extremely complicated series of interacting gradients of all the factors, biotic as well as physical, and these influence its distribution and abundance. Populations of the species can live only in those areas where favourable parts of the environmental gradients that affect it overlap. Factors that fall outside this favourable region are limiting ones for the species in that environment.

Some of the interactions between the various factors in an organism's environment may be very

Fig. 3.23 The distribution along a river of three closely related species of amphipod (Crustacea), relative to the concentration of salt in the water. *Gammarus locusta* is an estuarine species and is found in regions where salt concentration does not fall below about 25 parts per thousand. *G. zaddachi* is a species with a moderate tolerance of salt water and is found along a stretch of water between 11 and 19 km (8–12 miles) from the river mouth where salt concentrations average 10–20 parts per thousand. *G. pulex* is a true freshwater species and does not occur at all in parts of the river showing any influence of the tide or salt water [25].

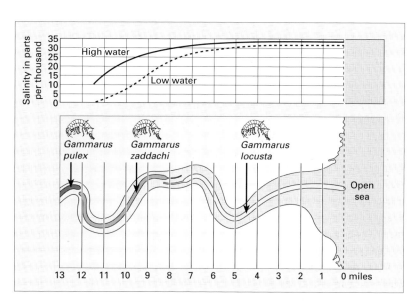

complex and difficult for the ecologist to interpret, or for the experimentalist to investigate. This is because a series of interacting factors may have more extreme effects on the behaviour and physiology of a species than any factor alone. To take a simple example, temperature and water interact strongly on organisms, because both high and low temperatures reduce the amount of water in an environment, high temperatures causing evaporation and low ones causing freezing, but it may be very hard to discover if an organism is being affected by heat or cold or by lack of water. Similarly, light energy in the form of sunlight exerts a great influence on organisms because of its importance in photosynthesis and in vision, but it also has a heating effect on the atmosphere and on surfaces, and therefore raises temperatures. In natural situations it is often almost impossible to tell which of many possible limiting factors is mainly responsible for the distribution of a particular species.

An interesting example of the complexity of interaction between environmental factors was studied by the American ecologist M.R. Warburg, in his work on two species of woodlice (sowbugs or slaters, Crustacea, Isopoda) living in rather dry

habitats in southern Arizona [26]. One species, *Armadillidium vulgare*, is found mostly in grasslands and scrubby woodland and is also widely distributed in similar habitats elsewhere in North America and in Europe. The other, *Venezilla arizonicus*, is a rather rare species, confined to the south-western United States and found in very arid country, with stony soil and a sparse vegetation of cactus and acacia. Warburg investigated the reactions of these two species to three environmental factors: temperature, atmospheric humidity and light. His experimental techniques involved the use of a simple apparatus, the choice chamber or preferendum apparatus, in which animals may be placed in a controlled gradient of an environmental factor. The behaviour of the animals, particularly the direction in which they move and their speed, can then be used to suggest which part of the gradient they find most satisfactory; this is termed the preferendum of these particular animals in this particular gradient. Warburg's method for testing the interactions of light, temperature and humidity on the woodlice was the classic scientific approach of isolating the effects of each factor separately, and then testing them two or three at a time in all possible combinations. For instance, he might set up a gradient of temperature between hot and cold and test the reactions of the animals to this, either with the whole gradient at a low humidity (dry) or the whole gradient at a high humidity (wet), or with the hot end of the gradient dry and the cold wet, or with the cold end dry and the hot end wet. He might then test the effect of light on these four situations by exposing each in turn to constant illumination, constant darkness, or one end of the gradient in darkness and the other in light. Such work is extremely time-consuming and requires great patience. It is similar to Thompson's work on seed germination, but permits the examination of several factors in combination.

Warburg found that, in general, *A. vulgare* prefers low temperatures (around 10–15°C), high humidities (above 70% relative humidity, i.e. the air is 70% saturated with water vapour) and is rather weakly attracted to light. This accords well with what is known of the species' habitat and habits—it lives in fairly humid, cool places and is active during the day. *V. arizonicus*, on the other hand, prefers lower humidities (around 45%), higher temperatures (20–25°C) and will generally move away from the light. Again, this accords well with the species' habits, since it lives in rather dry, warm places and is active at night. The reactions of the species change, however, and become harder to interpret when they are exposed to more extreme conditions. For instance, at high temperatures (35–40°C), *A. vulgare* tends to choose lower humidities, irrespective of whether these are in light or dark. One of several possible explanations for this behaviour is that, at these high temperatures, the species' physiological processes can be maintained only if body temperatures are lowered by permitting loss of water vapour from the body surface, which is more rapid at lower humidities. The normal reaction of *V. arizonicus* changes if the species is exposed to very high humidities; it then tends to move to drier conditions even if these are in the light. Warburg concludes that, for these two species, light is not really an important physiological factor and acts mostly as a 'token stimulus', a clue to where optimum conditions of humidity and temperature may

be found. For *V. arizonicus*, which lives in a dry or xeric habitat, darkness indicates the likely presence of the high temperatures and low humidities it prefers. For *A. vulgare*, in its cooler, more humid or 'mesic' habitat, there is little risk of desiccation except in the most exposed situations, and the species can afford to be relatively indifferent to light.

Warburg's study indicates the great complexity of the reactions of even relatively simple invertebrate animals such as Crustacea to the physical factors of their environment. If we analyse also the biotic factors of the animal's environment, food and enemies, the picture becomes even more complex. Other studies of *A. vulgare*, for instance, in California, indicate that the species shows quite strong preferences for different types of food (mostly various types of dead vegetation) and these also influence its distribution.

Knowing the physical requirements of a species can be of great economic significance, especially when an animal or plant is to be taken from one part of the world to another. The acacia trees of Australia, for example, are useful because many of them can grow in hot, dry conditions and they can provide a valuable resource of fuel wood for human populations in the dry regions of the world. But there are many acacia species and each has its own peculiar climatic requirements. Trevor Booth and his colleagues at Canberra, Australia, have been documenting as many as possible of these requirements, based on the analysis of species distribution in Australia [27]. The climate of the region in which it is intended to introduce the acacia tree is then analysed and a computer match can be generated in which the ideal acacia for a particular region is selected. This has saved much wasted money and effort in avoiding the old trial-and-error style of forestry. Figure 3.24 shows those parts of Australia with a climate most similar to a proposed site for acacia introduction in Zimbabwe. The best acacia match for such requirements was found to be *Acacia holosericea*, the distribution map of which clearly corresponds very closely to this climatic requirement.

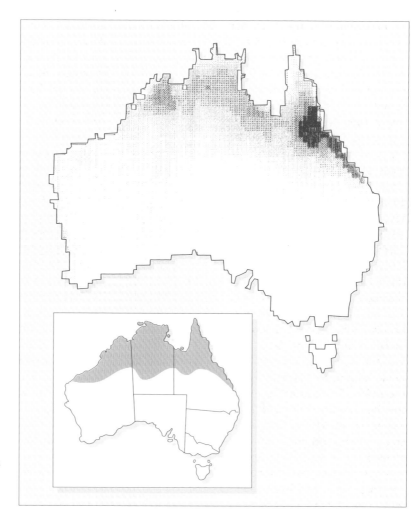

Fig. 3.24 Computer analysis of the climate of Australia to detect greatest similarity with that of Kadoma, Zimbabwe, which is to be the site of introduction of an acacia species for fuelwood production. Darkest areas have a climate most similar to that of the proposed site of introduction. Inset shows the distribution pattern in Australia of *Acacia holosericea*, a species which seems to match the climatic requirements for introduction. From Booth [27].

Species interaction

Physical factors evidently play an important part in determining the distribution limits of many plants and animals, but organisms also interact with one another, and this can also place constraints on geographical ranges. One species may depend strictly on another for food, as in the case of some butterflies which may be limited to a single food plant, or a parasite may be limited to a specific host. Some species may be unable to colonize an area because of the existence of certain efficient predators or parasites in that area, or because some other species that is already established there can compete more efficiently for a particular resource that is in demand. These are biotic factors, and they are often responsible for limiting the geographical extent of a species within its potential physical range.

When a species is prevented from occupying an area by the presence of another species, this is termed competitive exclusion. It is not always easy to observe this ousting of one species by another in nature, but an example of its occurrence is in the barnacle species that occupy the rocky sea shores of western Europe and north-eastern North America. Adult barnacles are firmly attached to rocks and feed by filtering plankton from the water when the tide is in. Two species that are common are *Chthamalus stellatus*, which is found within an upper zone of the shore just below high-tide mark, and *Balanus balanoides*, which occupies a much wider zone below that of *C. stellatus*, down to low water mark. The distribution of the two species does not overlap by more than a few centimetres. This situation was analysed by the ecologist J.H. Connell [28], who found that when the larvae of *C. stellatus* ended their free-swimming existence in the sea and settled down for life, they did so over the upper part of the shore above mean tide level. The larvae *of B. balanoides* settled over the whole zone between high and low water, including the area occupied by the adults of *C. stellatus*. Despite overlapping patterns of distribution of the larvae, different distributions of the adults of the two species result from two separate processes. One process acts on the zone at the top of the shore. The young *B. balanoides* are eliminated from this region because they cannot survive the long period of desiccation and the extremes of temperature to which they are exposed at low tide. *C. stellatus* are more resistant to desiccation and survive. Lower down the rocks the *B. balanoides* persist because they are not exposed for so long, and here the larvae of *C. stellatus* are eliminated by direct competition from the young *B. balanoides*. These grow much faster and simply smother the *Chthamalus* larvae or even prise them off the rocks. Connell also performed experiments on these species and found that, if adult *B. balanoides* were removed from a strip of rock and young ones prevented from settling, the *C. stellatus* were able to colonize the full length of the strip right down to low tide level. This showed that the competition with *B. balanoides* was the main factor limiting the distribution of *C. stellatus* to the upper part of the shore.

This example is a relatively simple one because only two species are involved. In most communities of animals and plants many species interact, and this makes it extremely difficult to sort out the full picture of the relationships between species. One approach to the problem is to remove one species from the community and to observe the reaction on the part of the others. This method has been tried in salt-marsh communities in North Carolina by J.A. Silander and J. Antonovics [29], who removed selected plant species and recorded which of the other plants present in the community expanded into the spaces left behind (Fig. 3.25). They found a great range of responses. The removal of one grass species, *Muhlenbergia capillaris*, resulted in an equal expansion on the part of five other plants, suggesting that the grass was in competition with many other species. In the case of the sedge *Fimbristylis spadiceae*, however, removal led to the expansion of only one other plant, the chord grass species *Spartina patens*. The reciprocal experiment in which *Spartina* was removed similarly led to *Fimbristylis* taking full advantage of the new opportunity. In this case we seem to have only two species that are competing for this particular niche.

Selective removal of species in this way, however, is somewhat artificial and can result in disturbance to the physical environment that alters the very nature of the habitat, so it can only provide a preliminary guide to the relationships of species in the community.

There are many examples of the displacement of native species by an invader. The European starling, *Sturnus vulgaris*, for instance, was introduced into Central Park, New York, in 1891. Since then it has spread widely and is now found in all of the States (Fig. 3.26). It is mostly found in urban areas, and in the east has largely displaced the bluebird (*Sialia sialis*) and the yellow-shafted flicker (*Colaptes auratus*). These species nest in tree holes or in man-made holes, and starlings can occupy and hold most of the limited supply of these nest sites. In the towns, then, the starling successfully competes with the native species for living space. But when flocks of starlings invade the countryside, they compete for food, insects and seeds with the meadow larks (*Sturnella*) and these birds also have declined in some areas. On the other hand, some North American species have been successful in new

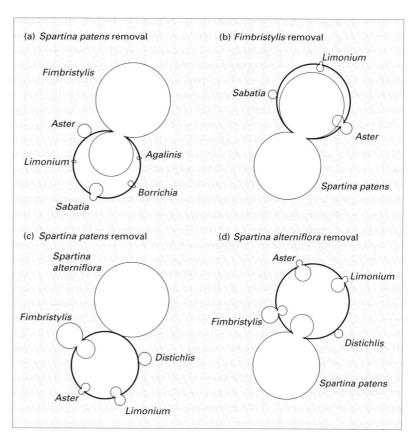

Fig. 3.25 A graphic illustration of the effect of removing a single plant species from a salt-marsh community. The sizes of the circles represent the abundance of the plant species, and the heavy circle refers to the species that has been removed. Circles intruding into the heavy circle denote the response of different species to the perturbation. (a) High marsh site, where removal of *Spartina patens* results mainly in expansion of *Fimbristylis*, and removal of *Fimbristylis* results in expansion of *Spartina*. The two species seem to be in competition and the effects are roughly reciprocated. (b) Low marsh site where two species of *Spartina* predominate. Removal of *S. patens* results in no response by the other *Spartina* species, whereas removal of *S. alterniflora* does permit some expansion of *S. patens*. Competition here is not reciprocal. From Silander & Antonovics [29].

habitats in Europe and Australia. An example is the American grey squirrel (*Sciurus carolinensis*), which was introduced into the British Isles in the nineteenth century. Between 1920 and 1925 the native red squirrel (*S. vulgaris*) suffered a dramatic decline in numbers in Britain, largely due to disease. The spread of the grey squirrel has been accompanied by the disappearance of the red squirrel from many areas, particularly those in which the red squirrel's numbers were reduced by disease and those into which the grey squirrel subsequently first spread and established itself. Where the grey squirrel has replaced the native red, it probably has done so by virtue of its superior adaptability to the niche of herbivore at canopy level in deciduous woodland. In the few locations where the grey squirrel has not succeeded in invading, such as the

Isle of Wight, an island off the south coast of Britain, the red squirrel still thrives, so disease is not the sole cause of red squirrel decline; the presence of the new invader is evidently involved also, possibly by means of competition for food or habitat.

Invasion and displacement of species in recent times is usually a consequence of human carriage of organisms from one part of the world to another. Colonists of new lands have often carried with them the familiar plants and animals from the old country and, where the climate of the new lands have proved appropriate for their survival, these species have often gained a permanent foothold and thus extended their geographical range. Table 3.1 shows the numbers of alien plant species that have established themselves in different parts of the world, compared with the number of species that

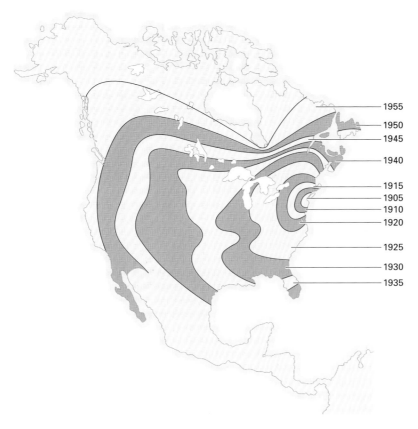

1955
1950
1945
1940
1915
1905
1910
1920
1925
1930
1935

Fig. 3.26 Map showing the range extension of the European starling in North America following its introduction late in the nineteenth century.

Region	Native species	Alien species	% Aliens
Hawaii	1 143	891	44
British Isles	1 255	945	43
New Zealand	2 449	1 623	40
Australia	15 638	1 952	11
United States	17 300	2 100	1
Europe (continental)	11 820	721	6
Southern Africa	20 573	824	4

Table 3.1 The proportion of introduced plant species in the floras of various regions of the world. Data based on Lovei [30].

were already present. It can be seen that the localities with the highest proportion of newcomers are the islands; these seem to be particularly sensitive to invasion by aliens [30].

There is good evidence that, in the past, whole faunas have invaded new areas and eliminated the native species by successful competition. For example, North and South America were separated by sea until the end of the Pliocene Period (about two million years ago). The Isthmus of Panama then came into being and enabled South American species to invade northwards and North American species southwards. In general, the northern species proved to be the more successful in competition, and most of the characteristic South American fauna of this time became extinct; but very few

North American species were wiped out by the South American invaders—probably because the North American herbivores were more efficient and the predators more successful [31].

Despite these dramatic examples of invasion and competitive displacement, it is most likely that, in natural situations, species that compete for food or other resources have evolved means of reducing the pressures of competition and of dividing up the resources between them. This is mutually advantageous since it reduces the risk of either species being eliminated and made extinct by competition with the others. This is an advantage not only to the species directly involved but to the whole community of species in the habitat, since it results in more species, depending on as many different sources of food as possible. Over evolutionary time scales, one might expect the species richness of an area to increase as this process of division of the resources gradually takes place. In such communities, competition occurs between so many different species, each with its own specialized adaptations, that no single species can become so numerous as to displace others. This may result in a greater degree of stability for the community, and stable communities are strongly resistant to the invasion of new species that might disrupt the highly evolved pattern of competition within them. But the functional consequences of high species diversity is hotly debated and will be considered in the next chapter.

Reducing competition

Many different ways of reducing competition between species have evolved. Sometimes species with similar food or space requirements exploit the same resources at different seasons of the year, or even at different times of day. A common system amongst predatory mammals and birds is for one species (or a group of them) to have evolved specialized night-time activity whilst another species or group of species are daytime predators in the same habitats. An example amongst birds is the owls on the one hand, many species of which hunt at night, judging distance mostly by ear, and the hawks and falcons on the other, which are daytime hunters, with extremely keen eyesight, especially adapted for judging distances accurately. Thus, both groups of predators can coexist in the same stretch of country, and prey on the same limited range of small mammals. Cases of this sort are described as temporal separation of species, and this is an effective method of eking out limited food resources amongst several species. Among plants this process can be seen operating in deciduous forest habitats, where many woodland floor herbs flower and complete the bulk of their annual growth before the leaf canopy emerges on the trees. In this way the light resources of the environment are used most efficiently.

A different type of temporal separation is shown within the complex grazing community of the East African savanna [32]. During the wet season all the five most numerous grazing ungulates (buffalo, zebra, wildebeest, topi and Thomson's gazelle) are able to feed together on the rich forage provided by the short grasses on the higher ground. At the beginning of the dry season, plant growth ceases there. The herbivores then descend to the lower, wetter ground in a highly organized sequence. First are the buffalo, which feed on the leaves of very large riverine grasses which are little used by the other species. The zebra, which are highly efficient at digesting the low-protein grass stems, move down next. By trampling the plants and eating the grass stems, they make the herb layer suitable for the next arrivals, topi and wildebeest. These two are found in slightly different areas. The jaws and teeth of the topi are adapted to the short, mat-forming grasses common in the north-western part of the Serengeti. Those of wildebeest are instead adapted to eating the leaves of the upright grasses commoner in the south-eastern Serengeti. These two species reduce the amount of grass, facilitating the grazing of the last species, Thomson's gazelle, which prefers the broader-leaved dicotyledonous plants to the narrow-leaved monocotyledonous grasses. The whole community therefore interacts in a complex manner, utilizing the pasture in a highly organized and efficient fashion.

Probably much more common, however, are cases where the resources of a habitat are divided up between species by the restriction of each of them to only part of it, to specialized microhabitats. This is called spatial separation of species; it means that each species must be adapted to live in the fixed set of physical conditions of its particular microhabitat. It also means that such a species is not adapted to live in other microhabitats, and may find it difficult to invade them even if they were for some reason vacant and their food resources untapped.

An example of spatial separation has been described in the extensive marshlands of the Camargue in southern France, where different wading birds have different preferences for the various available feeding areas. The flamingo (*Phoenicopterus ruber*) has very long legs and is thus able to wade into deep water where it can sieve planktonic organisms with its highly specialized bill. In shallower water the avocet (*Recurvirostra avosetta*) and the shelduck (*Tadorna tadorna*) feed in a similar way, by sweeping, side-to-side actions of their necks. On the water's edge it is the Kentish plover (*Charadrius alexandrinus*) that feeds predominantly, being restricted to these regions by its shorter legs.

The spatial patterns of distribution and feeding of predatory birds, such as these waders, sometimes reflect the patterns of their preferred food species. For example, the oystercatcher (*Haematopus ostralegus*) has a strong predilection for the bivalve mollusc *Cardium edule*, the cockle, and this is found mainly on sandy and muddy shores just below the mean high water mark of neap tides [33]. Therefore, this is the favourite feeding zone of the oystercatcher. Similarly, the mud-dwelling crustacean *Corophium volutator* is a favoured food species for the redshank (*Tringa totanus*) and, since it thrives best in the upper regions of mudflats, usually above the mean high water mark of neap tides, this is often where large numbers of feeding redshanks can be found.

One can find separation of feeding behaviour even within a single species. In the oystercatcher, for example, even though the cockle is the main prey animal, individual birds within a flock may

Fig. 3.27 Bill size variation within one species, the oystercatcher (*Haematopus ostralegus*). The upper bird is a 'stabber', and extracts its shellfish prey by pushing the bill between the valves of the shell and prising it open. The lower bird is a 'hammerer' and breaks the shell by violent blows of the bill on one of the valves [34].

use different techniques for extracting the food from the protective cover of its shell [34]. Some adopt a hammering technique, beating their bills against one of the valves until it breaks, while others use a more subtle approach, forcing their bills between the valves and prising them apart. Even among the hammerers there is further specialization, some attacking only the upper valve and others only the lower valve. These behavioural techniques are reflected in the bill structures of the individuals demonstrating such specializations. Careful observation of tagged birds has demonstrated that the hammerers have blunter bills than the stabbers (Fig. 3.27), but this is quite possibly an effect of bill wear and tear in the case of the hammerers, so there is not necessarily a genetic basis for the beak-shape separation of specialists. But the advantage of this behaviour is that different oystercatchers are seeking prey with different weaknesses (either thin shells, or weak muscles holding the valves together), so at least some of the competition for food has been removed.

This type of feeding specialization within a species can under some circumstances lead to the development of distinctive races. In the case of

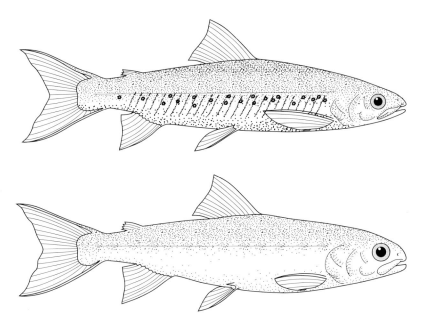

Fig. 3.28 Two different forms of the arctic charr from Loch Rannoch in Scotland. The upper fish (a) has narrow jaws and small teeth, and feeds mainly on plankton. The lower fish (b) has heavier jaws and larger teeth, and feeds on a wider range of prey, including invertebrates and other fish. From Walker *et al.* [36].

oystercatchers, for example, it is possible that certain differences in shape of bill could have a genetic basis which would equip some individuals for a slightly different diet or hunting technique. In such a situation, evolution could lead to the splitting of a population, especially if the structural differences were accompanied by behavioural differences in, for example, mate choice. A species could divide in two even in the absence of any geographical barrier to breeding. This happened with the hawthorn maggot, a native American fly which, on first encountering apple trees introduced from Europe, developed a taste for them. The new food option led to the development of a new race of the fly which eventually showed different mating preferences as well as food tastes, hence permitting the evolution of a new species, the apple maggot. This type of evolution, where no spatial barriers to interbreeding occur, is termed sympatric evolution, as opposed to the allopatric evolution taking place among separated populations [35].

In the arctic charr, a freshwater fish found in the lakes of mountainous regions of Europe, one may find two or more distinct forms in a single lake [36]. Figure 3.28 shows two forms of the charr discovered in Lake Rannoch in Scotland. The main difference is in the mouth structure, the upper fish the illustration having a relatively small mouth with small teeth. This animal is found to consume largely the zooplanktonic life found in the upper layers of the lake (the pelagic region). The lower fish has a larger, rounder mouth and bigger teeth. It eats a range of invertebrates and even fish, feeding mainly near the bottom (the benthic region). Perhaps this is the first stage in the sympatric evolution of two species from a single stock, the driving force being the advantage gained by avoiding competition for food.

This process has reached bizarre levels in Lake Victoria [37], in East Africa, where the number of forms of the 14 species of cichlid fish runs to 800. The sizes and shapes of these different forms are so varied that until very recently they were considered to be different species. Detailed study of their genetics has shown that they all, in fact, belong to a very restricted number of species. The avoidance of competition, combined with mating behaviour, must be one of the factors leading to the evolution of new species and thus helping to increase the biodiversity of the planet. An unfortunate consequence of human activity around Lake Victoria, however, is that nutrients and silt are draining into the lake from surrounding farmland and the sewage

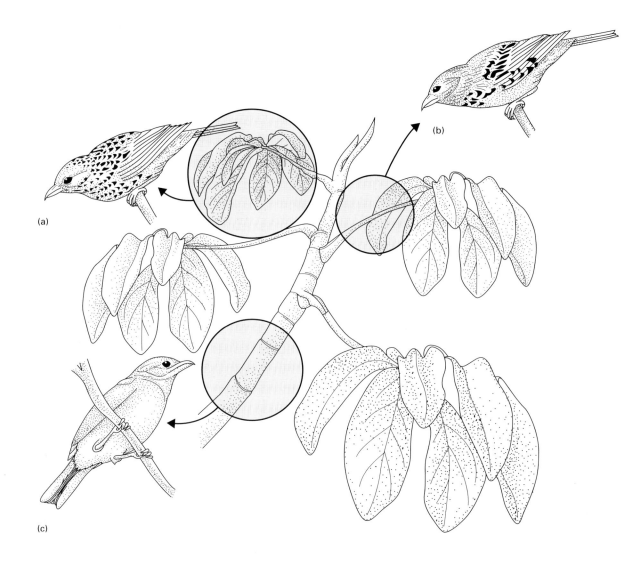

Fig. 3.29 Three species of tanager that coexist in the same forest on Trinidad. All feed on insects, but they exploit different microhabitats within the canopy and thus avoid direct competition. The speckled tanager (a) takes insects from the underside of leaves; the turquoise tanager (b) obtains its insects from fine twigs and leaf petioles; and the bay-headed tanager (c) preys upon insects on the main branches.

from urban settlements, causing turbidity and poor light penetration in the waters. The cichlid fish are mainly brightly coloured and depend on vision for their selection of mates, and without the aid of good visibility, they are unable to select the most appropriate mate for the continuance of their individual lines of evolutionary development. Human activity could thus bring the sympatric diversification of fish in this lake to a halt [38].

It is this kind of evolution which has led to the high level of specialization found within some groups of species, such as the tanagers of Central America (Fig. 3.29). Three closely related species of tanager, the speckled tanager (*Tangara guttata*), the

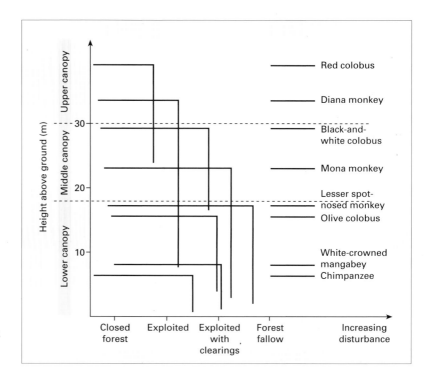

Fig. 3.30 Diagram to illustrate the different niche requirements of a range of primate species in the tropical forest of West Africa. Although the demands of the various species overlap, each has a particular height in the canopy or a type of site where it is most efficient and successful [39].

bay-headed tanager (*T. gyrola*) and the turquoise tanager (*T. mexicana*) may be found coexisting and feeding alongside one another without any apparent competitive interaction. The reason for this harmony is that each feeds in a slightly different location in the forest canopy. The speckled takes insects from the underside of leaves, the turquoise from fine twigs and the bay-headed from main branches. Each occupies its own niche and there is little overlap between them.

Even where overlap of niches does occur, animals may often coexist because they have their own distinctive location or way of life which they share with no other. This can be seen in Fig. 3.30, which displays diagrammatically the niches of various primate species in the tropical forest of Ghana [39]. Each species has its own preferred position in the forest canopy, some preferring undisturbed forest and others coping with exploited and cleared areas. There is considerable overlap in their tolerances, but each has its own specialist location where it can hold its own.

Predators and prey

Predators may be another biological factor influencing the distribution of species, but their effects have been much less studied than those of competition. The simplest influences that predators might have is to eliminate species by eating them or, alternatively, to prevent the entry of new ones into a habitat. There is very little evidence that either of these processes is common in nature. One or two experimental studies have shown that predators sometimes eat all the representatives of a species in their environment, particularly when the species is already rare. However, all such studies have been made in rather artificial situations in which a predator is introduced into a community of species that have reached some sort of balance with their environment in the absence of any predator; such communities are not at all like natural communities which already include predators. In general, it is not in the interests of predatory species to eliminate a prey species, because if they do this they destroy a potential source of food. Probably most natural communities have evolved so that there is a great number of potential prey species available to each predatory species. Thus, no species is preyed upon too heavily, and the predators can always turn to alternative food species if the numbers of their usual prey should be reduced by climatic or other influences.

Prey switching of this type has been described on the island of Newfoundland, where the wolf and the lynx were major predators last century, but where the wolf is now extinct. The lynx was a rare animal until a new potential prey animal was introduced to the island in 1864—the snowshoe hare. These multiplied rapidly, and so did the lynxes in response to the newly available food source. But the snowshoe hare population crashed to low levels in 1915 and the lynx, faced with starvation, switched its attentions to caribou calves, which had once been a major food source for the wolf. The snowshoe hare has now developed a 10-year cycle of high and low population levels and the lynx has continued to switch between hare and caribou depending upon whether the former is in a peak or a trough [40] This pattern of prey switching allows the lynx to maintain a fairly stable population level and, as a consequence, it also permits the recovery of snowshoe hare populations.

This type of behaviour on the part of a predator may thus serve to prevent the extinction of its prey. Sometimes the relationship is even more complex; the predator may prevent the invasion of more efficient or voracious predators which could reduce the prey population yet further. A good example of this is the Australian bell miner (*Marlorina melarlophrys*) [41]. This is a communal and highly territorial bird which occupies the canopy of eucalypt forests and which feeds largely on the nymphs, secretions and scaly covers of psyllid plant bugs. Yet the psyllid bugs survive well under this predation and, what is even more remarkable, they seem to require the attentions of the bell miners, for when these birds are removed, the populations of bugs crash and the eucalypt trees become more healthy. It seems that the aggressive behaviour

of the miners towards other birds prevents their entering bell miners' territories and eating all the psyllids. While the bell miners stay the psyllids are safe, but the trees suffer!

As mentioned earlier, competition may prevent two species from living together in a habitat, and may modify the distribution of species, because the resources of the habitat are inadequate to support both of them. Probably the most important effect of predators and of parasites and disease (which are effectively 'internal' predators) on the distribution of species is that, by feeding on the individuals of more than one species, they reduce the pressures of competition between them. Thus, by reducing pressures on the resources of the habitat, predators may allow more species to survive than would be the case if the predators were not there. It has been shown by laboratory experiments that if two species of seed-eating beetle (weevils) were kept together in jars of wheat, one species always eliminated the other within five generations. One of the two species always multiplied faster than the other, and this species won in the competition for food and places to lay eggs. But if a predator was introduced, such as a parasitic wasp, whose larvae feed inside the bodies of the beetle larvae of either species and eventually kill them, both species persisted. The numbers of both species were kept so low by the predator that competition for food, which would have caused one or other to be eliminated, never occurred.

Other studies of natural communities have largely confirmed the hypothesis that predators may actually increase the number of different species that can live in a habitat. The American ecologist Robert T. Paine made an especially fine study on the animal community of a rocky shore on the Pacific coast of North America [42]. The community included 15 species, comprising acorn barnacles, limpets, chitons, mussels, dog whelks and one major predator, the starfish *Pisaster ochraceus*, which fed on all the other species. Paine carried out an experiment on a small area of the shore in which he removed all the starfish and prevented any others from entering. Within a few months 60–80% of the available space in the experimental area was occupied by newly settled barnacles, which began to grow over other species and to eliminate them. After a year or so, however, the barnacles themselves began to be crowded out by large numbers of small, but rapidly growing mussels, and when the study ended these completely dominated the community, which now consisted of only eight species. The removal of predators thus resulted in the halving of the number of species and there was evidence, too, that the number of plant species of the community (rock-encrusting algae) was also reduced, because of competition with the barnacles and mussels for the available space.

A general conclusion, then, is that the presence of predators in a well-balanced community is likely to increase rather than reduce the numbers of species present, so that, overall, predators broaden the distribution of species. Only a few experiments similar to Paine's have been performed and so one must be cautious about applying this conclusion to all communities. There is some independent evidence, however, that herbivores, which act on plants as predators do on their prey, may similarly increase the number of plant species that can live in a habitat. In the last century, Charles Darwin noticed that in southern England, meadowland grazed by sheep often contained as many as 20 species of plants, while neglected, ungrazed land contained only about 11 species. He suggested that fast-growing, tall grasses were controlled by sheep grazing in the meadow, but that in ungrazed land these species grew tall so that they shaded the small slow-growing plants from the sun and eliminated them. A comparable process occurred in chalk grassland areas in Britain, when the disease myxomatosis caused the death of large numbers of rabbits; the resulting reduction in grazing allowed considerable invasion by coarse grasses and scrub. As a result, many of these areas are much less rich in species than they were under heavy 'predation'.

On the Washington coast, Paine performed another series of experiments in which he removed the sea urchin *Strongylocentrotus purpuratus*, which grazes on algae [43]. Initially, there was an increase in the number of species of algae present; the six or so new species were probably ones that were normally grazed too heavily by the sea urchin to survive in the habitat. But over 2 or 3 years the picture changed as the community of algae gradually became dominated by two species, *Hedophyllum sessile* on exposed parts of the shore, and *Laminaria groenlandica* in the more sheltered regions below low water mark. These two species were tall and probably 'shaded out' the smaller species, as did the tall grasses studied by Darwin. The total number of species present was in the end greatly reduced after the removal of the herbivores.

The activities of carnivorous predators in a community also have an effect on the plants since, by limiting to some extent the number of their herbivorous prey, they prevent overgrazing, and thus reduce the risk of rare species of plants being eliminated. Such interactions may be quite complex, however, as in the case of the Hawaiian damselfish, which is a predator of coral reef habitats. In an experimental study of the influence of this fish [44], plates were constructed which were suitable for algal colonization, and these were placed in three types of location: (i) within cages which excluded all herbivorous fishes; (ii) uncaged, but within the territories of the carnivorous damselfish; and (iii) uncaged and placed outside damselfish territories. The diversity of the colonizing algae was highest on the uncaged plates inside damselfish territories and least in the uncaged samples outside the territories. In other words, where there was no grazing at all the algal diversity was higher than when there was intense grazing, but the highest diversity was found in sites where grazing was controlled to an extent by the predation of the damselfish upon the grazers.

The complicated sets of interactions between predator, grazer and plant can lead to the development of a finely balanced and diverse ecosystem, as shown by this coral reef example. In all of these experiments in the manipulation of communities of organisms, it has been found that one species exerts a profound influence over many other components of the community, not just its prey organism. The removal of this one species can create effects far in excess of what may originally have been expected. Influential species of this kind are known as 'keystone species'. Identifying the keystone species in an ecosystem is clearly a very important task, especially if biodiversity is to be maintained. The loss of a keystone species can cause an avalanche of local extinctions.

Summary

1 Patterns of plant and animal distributions over the surface of the planet are very varied, and can be accounted for by a large number of different causes, sometimes in complicated combinations.
2 The causes of patterns vary according to the taxonomic level that we are dealing with.
3 The causes of patterns also vary with the spatial scale at which we are considering the organism, whether global, regional, or local—the habitat scale.
4 Factors that need to be considered in the explanation of patterns include geological history, climate and microclimate, availability of food, chemistry of the environment, and competition.
5 Human beings have created some global experiments in biogeographical patterns as a result of species introductions.
6 The spatial and temporal separation of organisms, as well as the specialization of groups within populations, can lead to the formation of new species and an increase in biodiversity.

Further reading

Begon M, Harper JL, Townsend CR. *Ecology: Individuals, Populations and Communities*, 3rd edn. Oxford: Blackwell Science, 1996.
Colinvaux P. *Ecology 2*. New York: John Wiley, 1993.
Ricklefs RE. *The Economy of Nature*, 4th edn. New York: W.H. Freeman, 1997.
Ricklefs RE, Schluter D. (eds) *Species Diversity in Ecological Communites: Historical and Geographical Perspectives*. Chicago: Chicago University Press, 1993.

References

1 Barton NH. Speciation. In: Myers AA, Giller PS, eds. *Analytical Biogeography*. London: Chapman & Hall, 1988: 185–218.
2 Gonder MK, Oates JF, Disotell TR, Forstner MRJ, Morales JC, Melnick DJ. A new west African chimpanzee subspecies? *Nature* 1997; 388: 337.
3 Rosen BR. Biogeographic patterns: a perceptual overview. In: Myers AA, Giller PS, eds. *Analytical Biogeography*. London: Chapman & Hall, 1988: 23–55.
4 Sharrock JTR, Sharrock EM. *Rare Birds in Britain and Ireland*. Berkhamsted: T. & A.D. Poyser, 1976.
5 Cramp S. *Handbook of the Birds of Europe, the Middle East and North Africa* IV. Oxford: Oxford University Press, 1985.
6 Hedberg O. Features of Afroalpine plant ecology. *Acta Phytogeogr Suecica* 1995; 49: 1–144.
7 Knox EB, Palmer JD. Chloroplast DNA variation and the recent radiation of the giant senecios (Asteraceae) on the tall mountains of eastern Africa. *Proc Natl Acad Sci USA* 1995; 92: 10349–53.
8 Corbet PS, Longfield C, Moore NW. *Dragonflies*. London: Collins, 1960.
9 Corbet PS. The life history of the emperor dragonfly, *Anax imperator* Leach (Odonata: Aeshnidae). *J Anim Ecol* 1957; 26: 1–69.
10 Dandy JE. Magnolias. In: Horai B, ed. *The Oxford Encyclopedia of Trees of the World*. Oxford: Oxford University Press, 1981: 112–14.
11 Corbet GB, Southern HN, eds. *The Handbook of British Mammals*, 2nd edn. Oxford: Blackwell Scientific Publications, 1977.
12 Coope GR. Tibetan species of dung beetle from Late Pleistocene deposits in England. *Nature* 1973; 245: 335–6.
13 Jolly D, Taylor D, Marchant R, Hamilton A, Bonnefille R, Buchet G, Riollet G. Vegetation dynamics in central Africa since 18,000 yr BP. Pollen records from the interlacustrine highlands of Burundi, Rwanda and western Uganda. *J Biogeogr* 1997; 24: 495–512.
14 Hamilton AC. *Environmental History of East Africa*. London: Academic Press, 1982.
15 Favarger C. Endemism in the montane floras of Europe. In: Valentine DH, ed. *Taxonomy, Phytogeography and Evolution*. London: Academic Press, 1972: 191–204.
16 Meyer EF. Late Quaternary paleoecology of the Cuatro Cienegas Basin, Coahuila. *Mexico Ecol* 1973; 54: 982–95.
17 Cowling RM, Rundel PW, Lamont BB, Arroyo MK, Arianoutsou M. Plant diversity in Mediterranean-climate regions. *Trends Ecol Evol* 1996; 11: 362–6.
18 Marshall JK. Factors limiting the survival of *Corynephorus canescens* (L.) Beauv. in Great Britain

at the northern edge of its distribution. *Oikos* 1978;
19: 206–16.

19 Root T. Energy constraints on avian distributions.
Ecology 1988; 69: 330–9.

20 Thompson PA. Germination of species of
Caryophyllaceae in relation to their geographical
distribution in Europe. *Ann Bot* 1978; 34: 427–49.

21 Moore PD. The varied ways plants tap the sun. *New
Scientist* 1981; 81: 394–7.

22 Ehleringer JR. Implications of quantum yield
differences on the distribution of C_3 and C_4 grasses.
Oecologia 1978; 31: 255–67.

23 Teeri JA, Stowe LG. Climatic patterns and the
distribution of C4 grasses in North America.
Oecologia 1976; 23: 1–12.

24 Williams GJ, Markley JL. The photosynthetic
pathway type of North American shortgrass prairie
species and some ecological implications.
Photosynthetica 1973; 7: 262–70.

25 Spooner GM. The distribution of *Gammarus* species
in estuaries. *J Mar Biol Assoc* 1974; 27: 1–52.

26 Warburg MR. Behavioural adaptations of terrestrial
isopods. *Am Zool* 1968; 8: 545–99.

27 Booth T. Which wattle where? Selecting Australian
acacias for fuelwood plantations. *Plants Today* 1988;
1: 86–90.

28 Connell J. The influence of interspecific competition
end other factors on the distribution of the barnacle
Chthamalus stellatus. *Ecology* 1961; 42: 710–23.

29 Silander JA, Antonovics J. Analysis of interspecific
interactions in a coastal plant community—a
perturbation approach. *Nature* 1982; 298: 557–60.

30 Lovei GL. Global change through invasion. *Nature*
1997; 388: 627–8.

31 Webb SD. A history of Savannah vertebrates in the
New World. Part II. South America and the Great
Interchange. *Annu Rev Ecol Syst* 1978; 9: 393–426.

32 Bell RHV. The use of the herb layer by grazing
ungulates in the Serengeti. In: Watson A, ed. *Animal
Populations in Relation to Their Food Resources.*
Oxford: Blackwell Scientific Publications, 1970:
111–27.

33 Hale WG. *Waders.* London: Collins, 1980.

34 Sutherland WJ. Why do animals specialize? *Nature*
1987; 325: 483–4.

35 Morell V. On the many origins of species. *Science*
1996; 273: 1496–502.

36 Walker AF, Greer RB, Gardner AS. Two ecologically
distinct forms of Arctic Charr *Salvelinus alpinus*
(L.) in Loch Rannoch. *Scot Biol Conserv* 1988; 43:
43–61.

37 Meyer A, Kocher TD, Basasibwaki P, Wilson AC.
Monophyletic origin of Lake Victoria cichlid fishes
suggested by mitochondrial DNA sequences. *Nature*
1990; 347: 550–3.

38 Seehausen O, van Alphen JJM, Witte F. Cichlid fish
diversity threatened by eutrophication that curbs
sexual selection. *Science* 1997; 277: 1808–11.

39 Martin C. *The Rainforests of West Africa.* Basel:
Birkhäuser Verlag, 1991.

40 Bergerud AT. Prey switching in a simple ecosystem.
Sci Am 1983; 249 (6): 116–24.

41 Lyon RH, Runnalls RG, Forward GY, Tyers J.
Territorial bell miners and other birds affecting
populations of insect prey. *Science* 1983; 221:
1411–13.

42 Paine RT. Food web complexity and species diversity.
Am Naturalist 1966; 100: 65–75.

43 Paine RT, Vadas RL. The effect of grazing in the sea
urchin *Strongylocentrotus* on benthic algal
populations. *Limnol Oceanogr* 1969; 14: 710–19.

44 Hixon MA, Brostoff WN. Damselfish as keystone
species in reverse: intermediate disturbance and
diversity of reef algae. *Science* 1983; 220: 511–13.

CHAPTER 4: *Communities and ecosystems*

No organism lives in total isolation from all others. We have seen in the previous chapter that different organisms interact with one another in both the long and the short term, competing for resources and sometimes excluding one another from certain areas. Over evolutionary time, this can lead to populations specializing in certain ways, perhaps in the way they obtain food, or the type of food they eat, or the type of microclimate in which they perform best. This can lead to the development of two or more species from a single original species, as in the case of the giant groundsels (see p. 38). This process will be examined in greater detail in the next chapter. However, there are other ways in which species interact. One animal may feed exclusively upon another, so that the consumer is associated with its food species in its distribution. This is particularly true of some butterfly species, whose caterpillars may have a very specific demand for a certain food plant, like the monarch butterfly and the milkweed in North America. The same is true of parasites and their hosts; the parasite is totally limited in its own distribution by its host's range. Some plants may be dependent on specific pollinators or seed-dispersal agents. Even human beings are constrained in their distribution to a certain extent by the climatic demands of our domestic plants and animals.

These are examples of dependencies, but there are other situations in which the organisms may not actually need one another, but have similar requirements from their environment. Many plant species, for example, demand moist conditions but cannot tolerate constant waterlogging, and they are frequently found together in specific habitats where these requirements are satisfied. Other species may require certain elements in their mineral supply, such as calcium, and these will be found together on calcareous soils. As long as competitive interactions do not lead to the elimination of one species or another, these may coexist in a distinct assemblage. A particular group of species may be so regular in its occurrence that one can predict from the presence of one or two of the species that certain others will also be found [1]. This is especially true of stenotopic (narrow ecological range) organisms that are confined to certain very distinct types of habitat. Plant ecologists and vegetation scientists take advantage of these associations between species to divide vegetation into convenient units, variously termed 'associations' or 'communities'.

The community

The use of the term 'plant community' by botanists is objectionable to some ecologists because it is based on just the plant component of the biota of an area. 'Community' should really refer to the total assemblage of living species found in a site, interacting in a whole range of different ways and forming a complex grouping of plant, animal and microbial components. Some animal species, such as those herbivores with very specific food requirements, may form close, dependent unions with certain plants, but others may have requirements for certain spatial architectural conditions that are best supplied by particular assemblages of plants. Again, the outcome of such associations is the existence of communities of plants and animals in nature which, within a geographical area, may be repeated in similar topographic and environmentally comparable sites, and which may be very predictable in their species composition.

The study of plant communities has developed as an independent area of ecology (sometimes termed

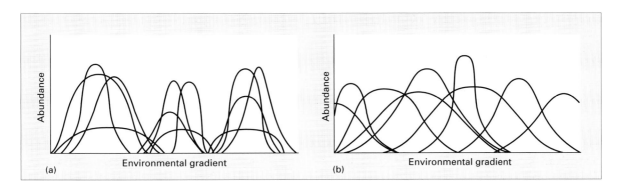

Fig. 4.1 Diagrammatic representation of two models of vegetation. (a) The model of Clements in which species' requirements coincide, leading to the separation of distinct 'communities'. (b) The 'individualistic' model of Gleason in which each species is distributed independently and no clear 'communities' are apparent.

'vegetation science') mainly because the plant components of communities are often the most evident features. In terrestrial sites, the plants generally occupy the most biomass, are static so can easily be sampled and counted, and are relatively easy to identify when compared with such groups as the invertebrate animals. The green plants are also all involved in the same basic exercise—trapping solar energy and fixing atmospheric carbon dioxide—and so can be treated as a distinctive set of organisms in the general community.

The idea of discrete plant communities that can be described, and even named and classified, is very attractive to the neatness of the human mind and is certainly valuable in the process of mapping areas and assessing their value for conservation and in determining plans for management. Whether plant communities have an objective reality as discrete entities has been energetically debated throughout the twentieth century, often turning upon the views of the two American ecologists, Frederic Clements and Henry Gleason, who began the discussion in the 1910s and 1920s. Essentially, Clements regarded the plant community as an organic entity in which the positive interactions and interdependencies between plant species led to their being found in distinct associations that were frequently repeated

in nature. The view proved both attractive and pragmatically useful, forming the basis for early attempts at describing and classifying vegetation by such ecologists as Braun Blanquet in France and Arthur Tansley in Britain.

Gleason's argument emphasized the individual ecological requirements of plant species, pointing out that no two species have quite the same needs. Very rarely do the distributional or ecological ranges of any two species coincide precisely, and the degree of association between ground flora and canopy is often weaker than one might assume from casual observation. The application of statistical techniques to the problem soon demonstrated that, although species often overlap in frequently occurring assemblages, the composition of these groups varies geographically as the physical limits of species are encountered. Studies of the past history of plant and animal species over the last 10 000 years or so have also demonstrated that they come into contact at certain times in their history, but have also been periodically separated as the climate changed. Often the associations we now observe are of relatively recent origin and should be regarded as transitory, a moment in history when certain species happen to have coincided. The concept of the community, according to this school of thought, must be looked upon as useful but somewhat artificial; vegetation is actually a continuum both in space and in time.

These two approaches to the description of plant communities can be expressed graphically in the form shown in Fig. 4.1, where the distributions of individual species are depicted along an environ-mental gradient. This may represent any environ-mental factor such as soil moisture, acidity, altitude, and so on. All species have their optimum for a given factor (as shown previously in Fig. 3.17) and have their ecological limits within which survival and growth is possible; some may have a narrow range of tolerance and others may have a wide range of tolerance and hence, potentially, a wider distribution. If the Clements model is correct then we would expect that a number of species might coincide in their optima and limits, as shown in Fig. 4.1(a). But if the individualistic concept of Gleason is preferred, the pattern should be as shown in Fig. 4.1(b), where there is little coincid-ence between species. Most studies have demon-strated that the Gleason model is closer to the truth, but the two concepts are not totally mutu-ally exclusive. It is possible that species may show a degree of clustering without attaining a complete separation of discrete units. Much depends upon whether certain sets of environmental variables are encountered more frequently than others; also, the existence of sharp boundaries between habitats in nature (as at the edge of a steep-sided lake) may lead to 'communities' being apparently more discrete. Many modern systems of vegetation classifica-tion, such as that devised by John Rodwell for the description of the vegetation types of Britain [2] are based on the idea of frequently repeated combina-tions being selected and described which can then be used as reference points. This type of scheme allows the possibility of a wide range of inter-mediate types, which is what one would expect if Gleason's ideas are valid.

The existence of discrete communities depends upon positive interactions between species, and this is not always easy to account for, especially in the case of plants that are basically all seeking the same resources of sun, space, water and soil nutri-ents. It is much easier to visualize competitive, negative interactions between species than mutu-ally beneficial ones. The Darwinian approach to ecology has also led us to be suspicious of any hint of altruism in a hypothesis involving the interaction of species. But positive relationships can occur [3]. Often we find, for example, that tree species sur-vive best as seedlings when under the cover of specific shrub plants, as in the case of the coast live oak (*Quercus agrifolia*) from California [4]. It has been found that 80% of the seedlings of this tree are located beneath the cover of just two species of shrub. This 'nurse' effect of one species upon another is not uncommon in nature and is an exam-ple of 'facilitation', which is an important element in the process of succession (see p. 27). In all vege-tation studies, the possibility of facilitation inter-actions that lead to positive associations between species must be balanced with the competitive interactions that lead to negative associations.

The importance of scale in this argument is very evident. As we increase the size of the area under study, increasing numbers of species are recorded together and thus appear positively associated, while reducing the sampling area must inevitably lead to negative associations. In addition, vegetation itself can often be visualized as a mosaic [5]. Patches of vegetation may be in different stages of recovery from disturbance, or regional catastrophe, or from the death of an old tree, or human clearance, or the vegetation may simply reflect underlying environ-mental patterns of geology, soils or hydrology. All of these can affect the pattern of species distribu-tions and associations, both in space and in time, and each type of disturbance leads to a different spatial scale of vegetation pattern (Fig. 4.2). If the boundaries between the different elements in the mosaic are sharp and well defined, then the dis-tinction between 'communities' will also be more clear-cut. Perhaps this is why the community concept, especially in vegetation studies, has proved more useful in countries like France and Switzerland which have very old cultural land-scapes that have evolved over millenia of intensive human fragmentation and cultivation. In coun-tries where such intensive agricultural activity is a relatively recent event, such as in America and Australia, landscapes still persist that are occupied by a continuum of vegetation rather than a clear patchwork mosaic.

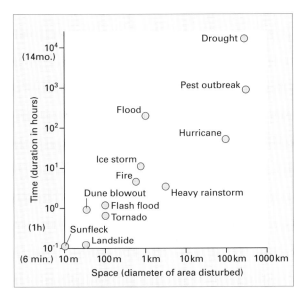

Fig. 4.2 Disruption of habitats by disturbance creates a mosaic of patches in different stages of recovery. This leads to an uneven landscape where separate 'communities' of organisms may be perceived. The spatial scale of patches varies with the type of disruption as shown here. Note the log/log scale on the axes. From Forman [5].

The ecosystem

The concept of community encompasses only living organisms and ignores the physical and chemical environment in which those organisms live. If we include these—the underlying rock and soil, the water moving through the habitat, and the atmosphere permeating the soil and surrounding the vegetation—we have an even more complex, interactive system that is called the ecosystem. Whereas the idea of community concentrates on the different species found in association with one another, the concept of the ecosystem is largely concerned with the processes that link different organisms to one another.

There are two fundamental ideas that underlie the ecosystem concept; these are energy flow and nutrient cycling. Energy, initially fixed from solar radiation into a chemical form by green plants, moves into herbivores as a result of their feeding

upon plants, and then moves on into carnivores as the herbivores are themselves consumed. Since herbivores rarely consume all the available plant material, and since carnivores do not eat every morsel of their prey organisms, some living tissues are allowed to die naturally before being eaten. These dead tissues constitute an energy resource which can be exploited by scavengers, detritus feeders and decomposers. Thus, we can classify all the organisms in any community in terms of their feeding relationships. In practice, a series of complex feeding webs are usually formed, relating each species to many others, whether as feeder or food.

Food webs can be very complicated, especially in ecosystems with a high biodiversity. Figure 4.3 shows the food web of a tropical rainforest at El Verde, Puerto Rico [6], and the complexity of interactions even within this simplified diagram are immediately apparent. Here the major groups of organisms are not split into their individual species, because to do so would make the system too complicated to be able to visualize. But the real ecosystem has an abundance of species within each of the boxes shown, each one with its own particular way of life, or niche, and each occupying one or more specific locations in the interactive net of feeding relationships.

The arrows shown in a diagram of this sort indicate the linkages between consumers and those they consume—grazer and plant, parasite and host, predator and prey. But they can also be regarded as routes along which energy flows. The base level in the diagram consists of the plants and their products that have captured the energy of sunlight in the process of photosynthesis, and have stored this energy in organic materials by taking carbon dioxide gas from the atmosphere and with it building large molecules. These compounds include the sugars of nectar, the lignins of wood, the cellulose from leaves, etc., and they are available to grazing animals or to detritus feeders and decomposers. The carbon derived from the atmosphere is also combined with elements obtained from the soil; nitrogen is combined to form amino acids and hence proteins, phosphorus is used in phospholipids that form an

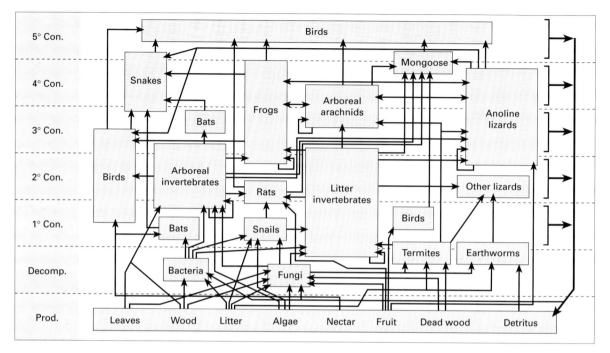

Fig. 4.3 Food web of a tropical rainforest derived from observations at El Verde, Puerto Rico [6]. Organisms are grouped first taxonomically (into boxes) and then arranged in a series of layers (trophic levels) according to their feeding positions in the system. Several taxonomic groups are represented in more than one trophic layer (e.g. birds) because of the varied feeding habits of their component species. Detritus from dead organisms in this diagram is brought back into the base layer with the primary producers (green plants).

essential component of living cell membranes. All of these form a food resource for consumers and decomposers. From the primary consumers, energy is passed through the feeding web, sometimes being stored for a short while in the bodies of the living and dead organisms, often being lost in the process of respiration as the animals conduct themselves in an energetic manner, and eventually all being dissipated in the respiratory heat output of the ecosystem. In Fig. 4.3 the dead residues of animals and plants are shown returning to the detritus that is available to decomposers and that leads back into the food web.

It is convenient to conceive the ecosystem as a series of layers, as represented here, since organisms obtain their energy in a kind of stratified sequence as it passes from one animal to another. But it is an oversimplification to place whole groups of animals within particular layers, hence birds are represented three times on the diagram according to whether they eat fruit (primary consumers), herbivorous and carnivorous invertebrates (secondary and tertiary consumers), or whether they feed as predators at the top of the food web. For these top predators, the energy they eventually receive has passed through many organisms and many feeding (trophic) levels. Since energy is lost at each transfer from one trophic level to the next, the energy reaching the top carnivores is more limited than that available at the base of the food web. This is one reason why top carnivores will generally be fewer in number than animals lower down the system.

Ecosystems have inputs and outputs of energy, but generally these are relatively simple. Solar energy is the principle source of energy for most ecosystems, and respiration and consequent energy dissipation as heat is the main loss. But some ecosystems are exceptional, such as the deep-sea vents where bacteria capture the energy released from reactions involving inorganic chemicals such as iron and sulphur and from the methane that belches out of the vents. Some ecosystems, such as streams running through forests, and mudflats in estuaries, may receive most of their energy second hand from other ecosystems in the form of plant detritus. Therefore ecosystems can exchange energy with one another.

Whereas energy is eventually dissipated as low-grade heat, nutrient elements, such as calcium, nitrogen, potassium and phosphorus, are not irretrievably lost in this way. They are cycled within ecosystems along the same paths as the energy, but are eventually returned to the soil from which they can be reused by plants. Movements of elements between ecosystems occurs, and in terrestrial ecosystems the hydrological cycle (the movement of water from oceans through atmospheric water vapour, through precipitation and back to the oceans via streams and rivers) plays a major role in both delivering and removing elements to and from the ecosystem. Apart from the nutrients arriving in rainfall, the other major source for most ecosystems is the gradual degradation of underlying rocks (weathering) which replenishes elements removed by plants and by water percolating through the soil. The water movement through the soil also leaches away unbound nutrients and takes them from one ecosystem to another, often from a terrestrial ecosystem to an aquatic one. A knowledge of the quantities and flow rates of ecosystems helps us to manage them efficiently. If we wish to crop the ecosystem at any given trophic level (e.g. sheep at the second trophic level), then we must ensure that the rate of removal of nutrients can be compensated for by natural inputs from rainfall and weathering. If this is not so, then we must either reduce the level of exploitation or add those elements in deficit as fertilizers. An element in short supply may limit the rate of ecosystem productivity.

The ecosystem concept therefore involves all the complexity of species interactions within the system, but views these in relation to the processes in which the ecosystem is engaged—productivity, energy flow, nutrient cycling, etc. It is a concept that can be applied at a variety of scales, whether to a pool of water in a rotting log, to the whole forest, or even to the entire earth, that is essentially a closed system for nutrients but still has energy inputs and outputs. The ecosystem concept has proved a very useful one, not only in the assistance it provides in understanding the relationships between organisms and the interactions with the physical environment, but also because it gives us a basis for their rational use as a resource for the support of human populations.

Ecosystems and biodiversity

An important question for biogeographers, ecologists and conservationists is how many species are really necessary to keep an ecosystem functioning? Is there a basic minimum number of species that is needed for any given ecosystem to operate, after which other species are simply excess to requirements? In other words, are some species redundant? If this were the case, we could remove species from an ecosystem without any effect on ecosystem function until all the redundant species had been extracted, after which the ecosystem would begin to suffer with further losses (Fig. 4.4a). A possible alternative model is based on the supposition that all species are important to ecosystem function [7], so the loss of each species renders it a little less efficient (the so-called 'rivet hypothesis'), as shown diagrammatically in Fig. 4.4(b). There is a third option, that the species diversity of an ecosystem may bear no relationship at all to its function, in which case species losses would result in random consequences. Figure 4.4 shows the expected outcome of reducing or increasing the number of species in an ecosystem if either of the main non-random hypotheses were true.

These models have been tested both in the laboratory and in the field, but as yet there does not seem to be a simple answer regarding which is closer to the natural situation. John Lawton and his research

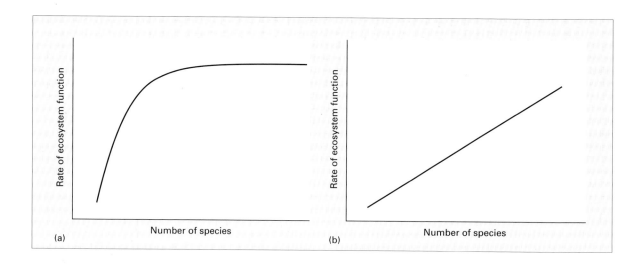

Fig. 4.4 The possible relationships between the number of species in an ecosystem and its rate of function (e.g. primary productivity, decomposition, nutrient turnover, etc.). (a) Redundant species hypothesis, where at high species density some can be lost without affecting ecosystem function. (b) Rivet hypothesis, in which the loss of any species affects ecosystem function. After Lawton [7].

team at Imperial College, London [8], have used miniature ecosystems in controlled-environment chambers ('ecotrons') to test the hypotheses. Keeping the soil conditions and microclimates constant, they have varied the number of animal and plant species between totals of 9 and 31. They found that the species-rich ecosystems were more effective at primary productivity than the species-poor ones (they developed more complex vegetation architecture) and also in the accumulation of inorganic nutrients into plant tissue. This suggests that diversity does improve ecosystem function in a manner similar to that presented in the rivet hypothesis model (Fig. 4.4b). But it is possible that the continued addition of species could eventually bring the system to a saturation point (as in Fig. 4.4a), so the experiment does not settle the question. It does, however, lead us to reject the third possibility, namely that ecosystem function is unrelated, or randomly related, to diversity.

David Tilman and his coworkers at the University of Minnesota [9] have approached the question by experimenting with communities which they have assembled in the field and they have broadly come to similar conclusions to the Lawton group— more species means better ecosystem function, such as productivity and nutrient retention. They go one step further, however, by pointing out that not all species are equally important in an ecosystem (where 'important' means that its removal has a greater impact than might be predicted). Also, how important a species is depends in part on the other species present. For example, plants from the pea family (Fabaceae) often play a significant role in the nutrient cycling of ecosystems because they have symbiotic relationships with nitrogen-fixing bacteria in their roots that ultimately make extra nitrates available to other species as a result of death, decomposition and element recycling. The loss of members of the Fabaceae to an ecosystem therefore can be disproportionately harmful to nutrient cycling and consequently productivity. But the importance of a pea-family species depends upon whether there are other members of the family present; if not, its loss may have very high impact. Effectively, a species of this type can be regarded as a keystone species (see Chapter 3). Therefore, not all species are equal in terms of the impact resulting from their loss, and if a species is the only representative of a particular functional group, such as producer, decomposer, nitrogen-fixer, then clearly its importance is even greater. It is possible that the range of functional groups available in an ecosystem is more important to its efficient functioning than is its biodiversity in taxonomic terms.

A field study of a natural ecosystem in the savanna grasslands of the Serengeti National Park in East Africa by Sam McNaughton and his colleagues from Syracuse University has provided an example of how more animal species could result in better ecosystem function [10]. Comparing sites that have been grazed by the herds of large mammalian herbivores of the tropical grasslands with other sites where these animals had been excluded by the erection of fences, they demonstrated that several nutrients, such as nitrogen and sodium, were cycled more efficiently in the grazed ecosystems. Therefore, grazing animals actually enrich the nutrient availability in the ecosystems they occupy. Again, it is probable that some species are more effective than others in this role, and their loss would be particularly serious for ecosystem function.

A related question, but perhaps even more difficult to answer, is whether diversity has an influence on the stability of an ecosystem. It was the British ecologist Charles Elton who first proposed (in the 1950s) that a more complex and rich ecosystem should also be more stable—meaning that it was less prone to violent fluctuations such as those caused by epidemic disease or pest outbreaks. It seemed a reasonable proposal, since experience showed that such instability was often a feature of simple ecosystems, as in the case of agricultural monocultures. It was also argued that a complex food web could provide a buffer against any perturbation in which certain species might become scarce. With a complex web there are more opportunities for prey-switching, as with the lynx and hare relationship (see p. 68).

But the development of mathematical modelling as an approach to the understanding of populations did not provide the expected results. They showed that a species in a diverse ecosystem was no less subject to fluctuations caused by unfortunate events, such as drought or disease, than is a species in a simple ecosystem. It remains possible, however, that although individual species may still fluctuate, the function of the entire ecosystem could be less vulnerable to such chance events if the system is diverse and complex than if it is species-poor and simple. Thus, individual species may rise and fall in abundance, but the ecosystem survives because many other contenders are available to replace the unfortunate sufferer in the event of catastrophe. David Tilman [11] has illustrated this possibility in his field experiments with natural vegetation plots. Some of these suffered an unplanned natural disturbance in the form of drought, and the plots with higher species numbers suffered lower declines in biomass than species-poor plots. Therefore, although richness does not guarantee the success, or even survival, of individual species, it does provide an ecosystem with a greater capacity to cope with disaster. There are clear lessons here for both conservationists and agriculturalists. On the conservation side, the loss of global biodiversity that we are currently experiencing may well be affecting the functioning of the entire biosphere [12]. On the agricultural side, the use of multicropping systems rather than single-species stands in agriculture should provide advantages both in terms of productivity and stability of the system—a particular concern in marginal areas.

One final point in this debate needs to be clarified, and that is what precisely do we mean by 'stability'? Is a stable ecosystem one which is difficult to deflect from its current composition or function? This approach defines stability in terms of inertia. Or is a stable ecosystem one which rapidly returns to its original state following disturbance? This uses the concept of resilience as a basis for defining stability. Both ideas, of course, are inherent in the concept that most people have of stability. Perhaps the most effective way of combining the two ideas is by employing predictability as a measure of stability [13]. A stable ecosystem should behave in a predictable manner no matter what fate may cast in its path, and biodiversity does appear to render an ecosystem predictable by providing a kind of 'biological insurance' against the failure of certain sensitive species when exposed to particular stresses.

These various experiments are combining the concepts of community and ecosystem and are asking questions regarding the ways in which community composition affect ecosystem operation. These are very important questions in biogeography, because we need to be able to appreciate how the presence or absence of particular species in an area will affect the behaviour and survival of the remaining species. As we begin to realize the extent to which the world is changing, largely as a consequence of the activities of our own species, we need to be able to predict the outcome of biogeographical alterations in species distributions and assemblages.

Biotic assemblages on a global scale

We have seen that we can view assemblages of species of plant, animal and microbe in different ways and at different scales. We may adopt a taxonomic approach, identifying all the organisms that we find in an area, analysing their associations together and possibly defining clusters of species that we can classify and name as communities. We have seen that such definitions are likely to be somewhat artificial because they usually lack sharp boundaries between them, but they may be very useful, enabling biogeographers to map the biota within regions. We have also seen that it is possible to view communities in a different way, depending on how their living and non-living components interact. Using this ecosystem view, the taxonomic identity of species is of lesser interest than their respective roles within the system. Thus, each species falls within a particular functional type, initially determined by its feeding systems, but also by other aspects, such as its nitrogen-fixing capacities, etc.

The application of the community concept on a global scale is complicated by the restriction of species to certain geographical areas. Take, for example, the desert scrub vegetation of northern Iran (Fig. 4.5) [14]. When we select an area of the relatively uniform scrub and sample it at a scale, say, 50 × 50 m, we can make out the individuals of the most important species (a deciduous, woody plant, *Zygophyllum eurypterum*) and some of the smaller dwarf shrubs (e.g. *Artemisian herba-alba*, Plate 4.1a, facing p. 86). Observed at a larger scale, covering hundreds of square kilometres, it is possible to make out landscape patterns related to water movements in the alluvial plain, and disturbance due to human settlements and variations in grazing intensity. The moister regions are occupied by different shrubs (*Haloxylon persicum* (Plate 4.1b, facing p. 86) and *Tamarix* species), the saline flats by salt-tolerant plants (e.g. *Salsola*) and the regions around settlements by grazing-resistant, often toxic plant species (e.g. *Peganum harmala* and *Ephedra* species). At this scale of observation it is therefore possible to pick out 'communities' of species that share an ability to exist under certain types of environmental stress (salinity, drought, grazing, etc.). These communities are characterized by particular species, most of which are confined to this region of the world. Similar desert scrub vegetation in the Mojave Desert of California, for example, closely resembles the Iranian scrub in general appearance, but has different species, such as creosote bush (*Larrea tridentata*—in the same family as *Zygophyllum*) and a ragweed species, *Ambrosia dumosa*, which is in the same family as *Artemisia*.

When considering vegetation on a global scale, a classification by physiognomy (general form and lifestyle) of the vegetation is more helpful than the use of taxonomy in the comparison of similar climatic locations in different parts of the world. The need for a physiognomic approach to the classification and mapping of vegetation is evident if we move up a level of scale on the Iranian map (Fig. 4.5) and observe the northern part of Iran, where the units we observe fall into groups such as temperate deciduous forest, forest–scrub, scrub–steppe and desert. The general nature of the vegetation of any part of the world, is best expressed in terms of its gross structure and appearance, and this is controlled

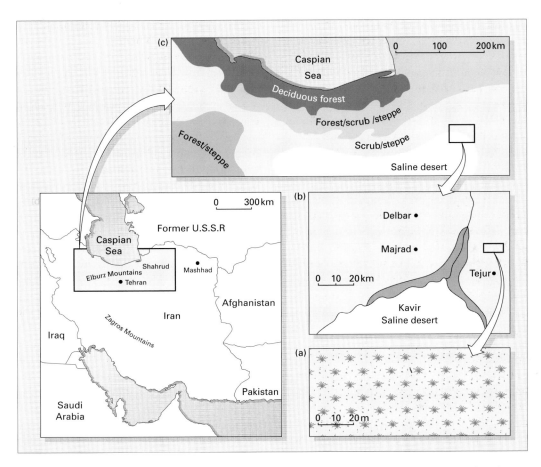

Fig. 4.5 The effect of sampling scale on the appropriateness of different types of vegetation description, illustrated by the vegetation of northern Iran. (a) At a scale of tens of metres, individual shrubs are apparent and their specific identity can be determined. Bushes of *Zygophyllum eurypterum* tend to be evenly spaced and smaller plants of *Artemisia herba-alba* (dots) occur between them. (b) At a scale of kilometres, different groupings of species form distinct assemblages ('communities') having their own constituent species that can cope with various environmental stresses (drought, salinity, grazing, etc.). The saline desert is almost devoid of vegetation except for a halophyte, *Salsola*. The moister alluvial margins and riverside bears a community of *Haloxylon* and *Tamarix* shrubs. Most of the land surface is covered with *Zygophyllum* scrub, except where villages have caused vegetation loss by overgrazing and fire-wood gathering. (c) At a scale of hundreds of kilometres, vegetation is perceived in terms of physiognomic form (deciduous forest, forest–steppe, scrub–steppe, desert, etc.) rather than as species assemblages.

by major factors such as climate and type of soil. Similar patterns recur in different parts of the world, so that a very similar type of vegetation may evolve in each of these areas, as in the case of the desert scrub of Iran and California. This comparability of vegetation form may lead to the evolution of similar types of animal in each area. A general classification of terrestrial, large-scale ecosystems, which are known as 'biomes', has been developed, and which includes the following major types: deserts; the cold tundra of high latitudes and high altitudes; northern coniferous forest or 'taiga';

temperate forest; tropical rainforest; temperate grass-land (known as the 'prairies' in North America, the steppes of Eurasia, the 'pampas' of South America, and the 'veld' in South Africa); and finally the cha-parral of areas with a Mediterranean climate.

Any system of classifying the world's vegetation into units must avoid the simple use of particular groups of species. Although the rainforests of Brazil are comparable in many respects to those of West Africa or South-East Asia, the actual assemblage of species is quite different. Their similarity is mainly due to the fact that they are structurally compara-ble, being dominated by tall trees arranged in a series of layers, most of which are evergreen and broad-leaved. There are other similarities, such as the presence of canopy-dwelling primates, and the fact that pollination is often brought about by birds, and so on. Essentially, we are comparing the com-munities in terms of the functional types of plants and animals present rather than their taxonomic affinities.

The concept of a 'life form' among plants was first put forward by the Danish botanist Christen Raunkiaer in 1903. He observed that the most common or dominant types of plants in a climatic region had a form well suited to survive in the prevailing conditions. Thus, in arctic conditions the most common plants are dwarf shrubs and cushion-forming species that have their buds close to ground level. In this way they survive the winter conditions when wind-borne ice particles have an abrasive effect on any elevated shoots. In warmer climates, buds are usually carried well above the ground and the tree is an efficient life form, but periodic cold or drought may necessitate the loss of foliage and the development of a dormant phase. This has resulted in the evolution of the deciduous habit. More prolonged drought results in a different type of vegetation, with shrubs that have smaller above-ground structures. Some plants of areas with seasonal drought survive the unfavourable period as underground organs (e.g. bulbs or corms), or as dormant seeds. Animals also show distinct life forms adapted to different climates with cold-resistant, seasonal or hibernating forms in cold regions and forms with drought-resistant skins or cuticles in deserts. Nevertheless, animal life forms are usually far less easy to recognize than are those of plants and, consequently, biomes are primarily distin-guished by the plant life forms they contain.

The plant life forms recognized by Raunkiaer consisted of the following.
1 Phanerophytes—woody plants with buds more than 25 cm above ground level. These he subdi-vided into different size classes.
2 Chamaephytes—plants with buds above the soil surface but less than 25 cm, including both woody and herbaceous types, often cushion-shaped in form.
3 Hemicryptophytes—plants with their perennat-ing buds at ground level, either with leafy stems in the growing season, or with a basal rosette of leaves.
4 Geophytes—plants with their buds below the soil surface, having stem tubers, corms, buds, rhizomes, root tubers, etc.
5 Helophytes—marsh plants.
6 Hydrophytes—aquatic plants.
7 Therophytes—plants that survive an unfavour-able season as seeds.

Although Raunkiaer stressed the importance of perennating organs in his classification of plant life forms, he also recognized that leaf size and form were important; large, undivided leaves being par-ticularly associated with the hot, wet conditions of the tropics, and small, thick leaves with tough sur-face layers (sclerophylls) being more characteristic of drier regions. He analysed the flora of different parts of the world into component functional types and found that each region had its own distinctive 'biological spectrum'. Tundra regions were charac-terized by an abundance of chamaephytes and a lack of therophytes; temperate grasslands were typified by hemicryptophytes; tropical rainforests contained a preponderance of phanerophytes; deserts were rich in therophytes and geophytes, and so on.

Using this physiognomic approach to vegetation, it proved possible to define and characterize units, initially referred to as 'plant formations' since they were based on purely botanical criteria. Global maps were constructed by plant geographers that used these plant formations as their basic units, and

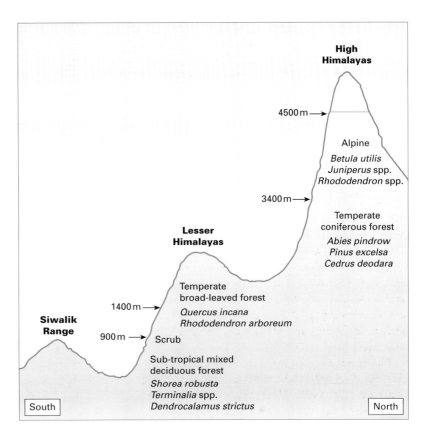

Fig. 4.6 Diagrammatic section of the western Himalayas in northern India, showing the approximate altitudinal limits of the major vegetation types. With increasing altitude one passes through vegetation belts similar to those found on passing from lower into higher latitudes. The scrub zone (900–1400 m) is strongly modified by human deforestation.

these were found to have a broad correspondence with climatic zones. Indeed, in the early days the vegetation of understudied areas of the world was often taken as a basis for predicting climate, so it was inevitable that vegetation maps and climate maps had a general similarity to one another.

There is no real agreement among biogeographers about the number of biomes in the world. This is because it is often difficult to tell whether a particular type of vegetation is really a distinct form or is merely an early stage of development of another, and also because many types of vegetation have been much modified by the activities of human beings. This is very apparent from the global vegetation map produced by Haxeltine and Prentice and reproduced in Plate 4.2a (facing p. 86) (see [15]) Satellite imagery has increased the accuracy with which vegetation can be mapped on a global scale,

but the very considerable impact of humans, especially in the temperate zone and around the edges of the arid regions, means that the expected close relationship between vegetation patterns and climate patterns is no longer clear.

There is, however, an overall latitudinal zonation of biomes apparent as we move from the rainforests of the equatorial regions to the dwarf-shrub tundra of the Arctic, and a similar trend is reflected in altitudinal terms if we examine the changes in biomes with altitude in the Himalayan mountains (Fig. 4.6). Ascending from the northern Indian savanna and thorn scrub, we move up through subtropical monsoon forest, largely occupied by drought-deciduous tree species. A zone of scrub follows, which owes its existence largely to the impact of human activity and that of domesticated animal grazing. Above this is temperate deciduous

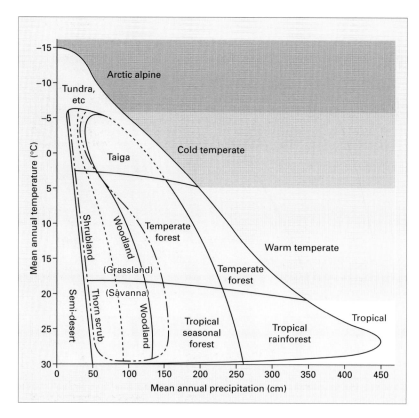

Fig. 4.7 The distribution of the major terrestrial biomes with respect to mean annual precipitation and mean annual temperature. Within regions delimited by the dashed line, a number of factors, including geographical location, seasonality of drought and human land use may affect the biome type which develops. From Whittaker [16].

forest with oak and rhododendron, its general form and appearance being very similar to that found in western Europe or the eastern part of the United States. Coniferous forest lies above the oak zone, dominated mainly by the deodar cedar, and above this is the alpine birch and juniper scrub that leads on up to the tundra and the permanent snows of the high mountains. The altitudinal sequence thus broadly mirrors the latitudinal zones but in a much shorter distance.

The link between biomes and climate becomes even more apparent if we map the biomes against climatic variables, such as precipitation and temperature [16], as has been done in Fig. 4.7. Again, the divisions between biomes are not as sharp as those indicated here, but it is plain that each biome occupies a region where specific climatic requirements are met. To understand the global pattern of biome distribution therefore it is necessary to

appreciate underlying patterns of climates and their causes.

Patterns of climate

The climate of an area is the whole range of weather conditions—temperature, rainfall, evaporation, sunlight and wind—that it experiences through all the seasons of the year. Many factors are involved in the determination of the climate of an area, particularly latitude, altitude and location in relation to seas and land masses. The climate in turn largely determines the species of plants and animals and even the life forms or functional types that can live in an area.

Climate varies with latitude for two reasons. The first reason is that the spherical form of the earth results in an uneven distribution of solar energy with respect to latitude. As the angle of incidence

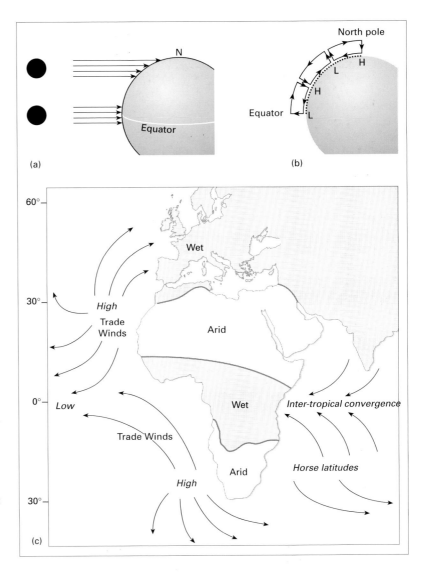

Fig. 4.8 Patterns of climate. (a) Due to the spherical shape of the earth polar regions receive less solar energy per unit area than the equatorial regions. (b) The major patterns of circulating air masses (cells) in the Northern Hemisphere: H, high pressure; L, low pressure. (c) Pressure areas, wind directions and moisture belts around Europe and Africa.

of the sun's rays approaches 90°, the area over which the energy is spread is reduced, so that there is an increased heating effect. In the high latitudes, energy is spread over a wide area; thus, polar climates are cold (Fig. 4.8a). The precise latitude that receives sunlight at 90° at noon varies during the year; it is at the Equator during March and September, at the Tropic of Cancer (23.45°N) during June, and at the Tropic of Capricorn (23.45°S)

during December. These two Tropics mark the limits beyond which the sun is never overhead. The effect of this seasonal fluctuation is more profound in some regions than in others.

Variations in climate also result from the pattern of movement of air masses (Fig. 4.8b). Air is heated over the Equator, and therefore rises (causing a low-pressure area) and moves towards the pole. As it moves towards the pole it gradually cools and increases in density until it descends, where it forms a subtropical region of high pressure, known as the Horse Latitudes (Fig. 4.8c). Air from this high-pressure area either moves towards the Equator, or else moves polewards. The poleward-heading air mass eventually meets cold air currents moving south from the polar region, where air has been cooled and descends (causing a high-pressure area). Where these two air masses meet, a region of unstable low pressure results, in which the weather is changeable.

This idealized picture is complicated by the Coriolis effect (named in honour of the French mathematician Gaspard Coriolis, who analysed it), which results from the west–east rotation of the earth. This force tends to deflect a moving object to the right of its course in the Northern Hemisphere and to the left in the Southern Hemisphere. As a result, the winds moving towards the Equator come to blow from a more easterly direction (Fig. 4.8c). These 'Trade Winds', coming from both the Northern and the Southern Hemispheres, therefore meet at the Equator, and this region is known as the 'intertropical convergence zone'. Where these easterly winds have passed over oceans they become moist, and this moisture is deposited as rain, generally over the easterly portions of the equatorial latitudes of the continents. Similarly, the winds which move polewards from the high-pressure Horse Latitudes come to blow from a more westerly direction, and provide rain along the westerly regions of the higher latitudes of the continents. The Horse Latitudes themselves are regions in which dry air is descending, and arid belts form along these latitudes of the continents.

The distribution of oceans and land masses modifies this simple picture yet further. Because heat is gained or released more slowly by water than by land masses, heat exchange is slower in maritime regions, while at the same time humidities are higher. In summer therefore continental areas tend to develop low-pressure systems as a result of the heating of land masses and the conduction of this heat to the overlying air masses. Conversely, in winter the reverse situation occurs,

continental areas becoming cold faster than the oceans, and high-pressure systems develop over them (Fig. 4.9). One effect of this process is that the continental low-pressure systems draw in moist air from neighbouring seas, for example from the Indian Ocean to East Africa and India, causing summer monsoon rains. The winter of these areas, on the other hand, is usually dry.

The global circulation patterns of the oceans is also of great importance in determining world climate patterns (Fig. 4.10). Warm, surface waters from the equatorial regions of the Atlantic Ocean are carried north-eastwards towards Iceland and Norway, warming this section of the Arctic Ocean, keeping it ice-free through the summer and bringing the whole of western Europe mild conditions. As the water body cools it becomes denser and sinks, reversing its flow pattern as it heads back to the southern Atlantic. From here, the cold, dense water may flow up into the Indian Ocean where it receives new warmth, or it may continue around the southern latitudes and eventually move northwards into the Pacific Ocean and pick up heat there. Once heated, the less dense waters move along the ocean surface and back into the Atlantic. This so-called 'conveyor belt' of oceanic circulation is responsible for dispersing much of the tropical warmth to higher latitudes, particularly in the Atlantic seaboard. Without it, the high latitudes would be much colder. Indeed, one of the main features of the glacial episodes in the earth's recent history has been the shutting down of the global oceanic heat conveyor. Its effect on biome distribution is apparent in that boreal coniferous forest, for example, extends to much higher latitudes in Scandinavia than it does in Alaska.

In addition to the heating and cooling effects of land masses and oceanic currents, climate is also affected by altitude. On average, the air temperature falls by 0.6°C for every 100 m rise in height, but this varies considerably according to prevailing conditions, especially the aspect and steepness of slope and the wind exposure. Because of this tendency for temperature to fall with increasing altitude, the organisms inhabiting high tropical and

(a)

Plate 4.1 (a) Desert scrub vegetation in eastern Iran. The most conspicuous feature is the deciduous shrub, *Zygophyllum eurypterum*, together with smaller plants of *Artemisia herba-alba*. This vegetation is typical of the drier regions of alluvial flats.

(b)

(b) Desert scrub vegetation in eastern Iran in which the main scrub is *Haloxylon persicum*. This vegetation occupies the moister and more saline regions of the alluvial flats.

[*facing p. 86*]

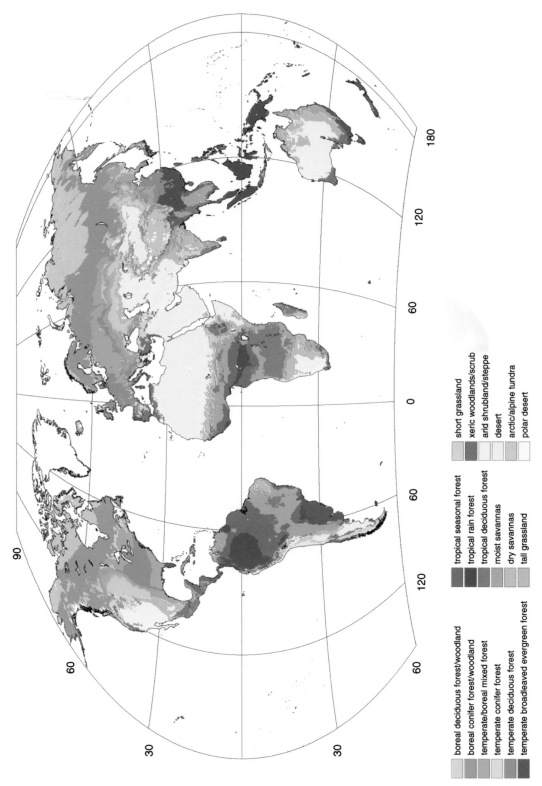

boreal deciduous forest/woodland
boreal conifer forest/woodland
temperate/boreal mixed forest
temperate conifer forest
temperate deciduous forest
temperate broadleaved evergreen forest

tropical seasonal forest
tropical rain forest
tropical deciduous forest
moist savannas
dry savannas
tall grassland

short grassland
xeric woodlands/scrub
arid shrubland/steppe
desert
arctic/alpine tundra
polar desert

Plate 4.2 (a) Global map of the present distribution of vegetation on earth. This is somewhat idealized since it ignores the impact of human beings. It can be regarded as the potential vegetation if climate alone were the operative determinant. Reproduced with permission from Haxeltine and Prentice (Chapter 4, [24]).

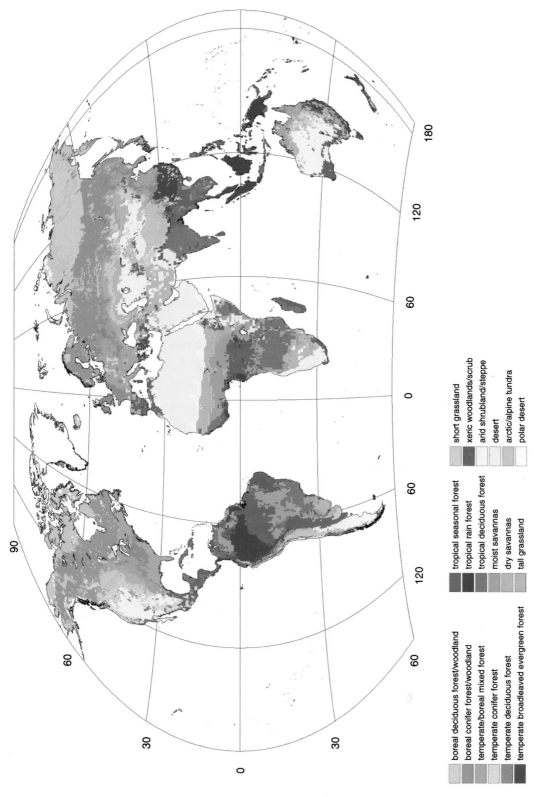

boreal deciduous forest/woodland
boreal conifer forest/woodland
temperate/boreal mixed forest
temperate conifer forest
temperate deciduous forest
temperate broadleaved evergreen forest

tropical seasonal forest
tropical rain forest
tropical deciduous forest
moist savannas
dry savannas
tall grassland

short grassland
xeric woodlands/scrub
arid shrubland/steppe
desert
arctic/alpine tundra
polar desert

Plate 4.2 (b) Global map of vegetation derived from the predictions of a computer model, BIOME 3. As in the case of Plate 4.2 (a), this is a potential vegetation map that is based solely on climatic considerations. It can be seen that the two maps are very similar, which means that this model is a reliable predictor of vegetation under given climatic conditions. This close correspondence suggests that the model could be used to predict reliably the consequences of future climatic change for vegetation distribution patterns. Reproduced with permission from Haxeltine and Prentice [24].

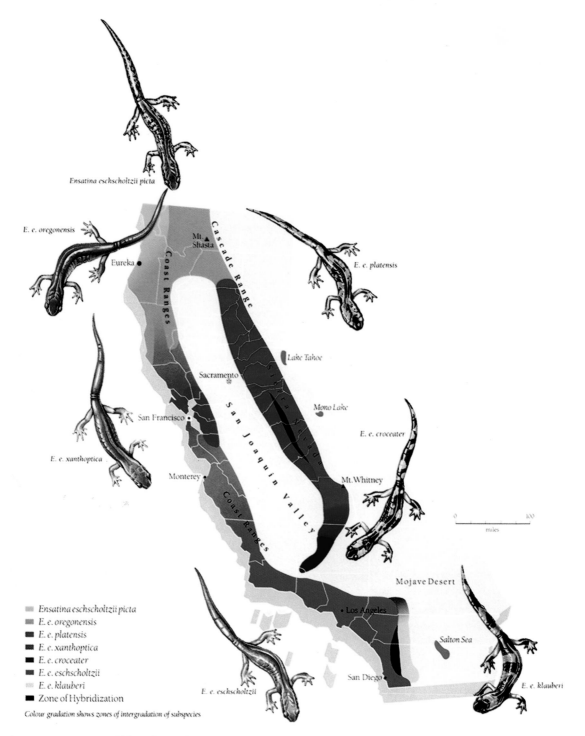

Plate 5.1 Ring species of the salamander *Ensatina* in the western United States. One species (*E. oregonensis*) is found in the north and up into Oregon and Washington. It then devides in northern California and forms a more or less continous ring around the San Joaquin Valley. The salamanders vary in form from place to place, and consequently they have been given a number of taxonomic names. Where the coastal and inland sides of the ring meet in southern California, the salamanders behave as good species at some sites (black zones on the map). Reprinted, by permission of Ten Speed Press, California, from Thelander (Chapter 5, [19]).

Map labels: *Ensatina eschscholtzii picta*; *E. e. oregonensis*; Mt. Shasta; Cascade Range; Coast Ranges; Eureka; *E. e. platensis*; Lake Tahoe; Sacramento; Sierra Nevada; Mono Lake; *E. e. croceater*; San Francisco; San Joaquin Valley; *E. e. xanthoptica*; Monterey; Mt. Whitney; Coast Ranges; Mojave Desert; Los Angeles; *E. e. eschscholtzii*; Salton Sea; San Diego; *E. e. klauberi*

Legend:
- Ensatina eschscholtzii picta
- E. e. oregonensis
- E. e. platensis
- E. e. xanthoptica
- E. e. croceater
- E. e. eschscholtzii
- E. e. klauberi
- Zone of Hybridization

Colour gradation shows zones of intergradation of subspecies

Scale: 0 — 100 miles

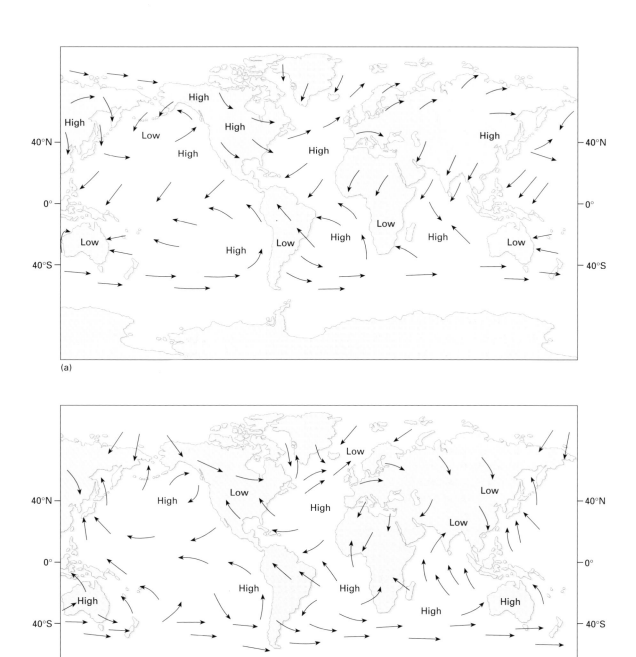

(a)

(b)

Fig. 4.9 General pattern of air movement across the earth's surface and distribution of areas of high and low pressure (a) in January and (b) in July. Note the reversal of winds in the Indian Ocean in July, taking monsoon rains into northern India and East Africa.

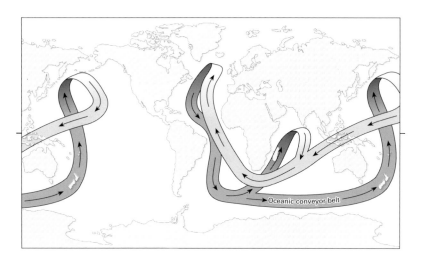

Fig. 4.10 The oceanic conveyor belt carrying warm, low-salinity surface waters from east to west into the North Atlantic and deep, higher salinity, cold waters from west to east into the Pacific.

subtropical mountains, such as the Himalayas in northern India (see Fig. 4.6), may be more like the flora and fauna of colder regions than that of the surrounding lowlands. However, although temperature in general falls as one ascends such mountains, other environmental conditions do not precisely mirror those found at higher latitudes. For example, the seasonal variations in daylength typical of high-latitude tundra areas are not found in the 'alpine' regions of tropical mountains. Also the high degree of insolation resulting from the high angle of the sun produces considerable diurnal fluctuations in temperature that are not found in tundra regions. It is not surprising therefore that the altitudinal zonation of plants and animals should not precisely reflect the global, latitudinal zonation. Also, arctic and alpine races of a single species often differ in their physiological make-up as a consequence of these climatic differences.

Climate diagrams

As we have seen, individual species of plants and animals and also life forms are affected by a whole range of physical factors in their environment, many of these being directly related to climate. Biogeographers have long sought therefore a means of portraying climates in simple, condensed form that would give at a glance an indication of the

main features that might be of critical importance to the organisms of the area. Mean values of temperature and rainfall may be of some use, but one also needs to know something of seasonal variation and of extreme values if the full implications of a particular climatic regime are to be appreciated. It is with this aim in view that Heinrich Walter, of the University of Hohenheim in Germany, devised a form of climate diagram which is now widely used by biogeographers [17]. An explanation of the construction of these diagrams is given in Fig. 4.11, and a selection of climate diagrams displaying the climates of some of the major biomes of the earth is given in Fig. 4.12.

Modelling biomes and climate

The general link between plant formations, defined in terms of life forms, and climate is evident, but modern biogeographers require more robust methods in the study of connections between vegetation and climate, especially if they are to be in a position to predict the future in the event of climate change. Physiognomy, i.e. the form and morphological structure of vegetation, was the only information available to Raunkiaer and those who immediately followed him. He worked from the basic idea that plant formations existed and simply needed to be described and mapped, so his was essentially a

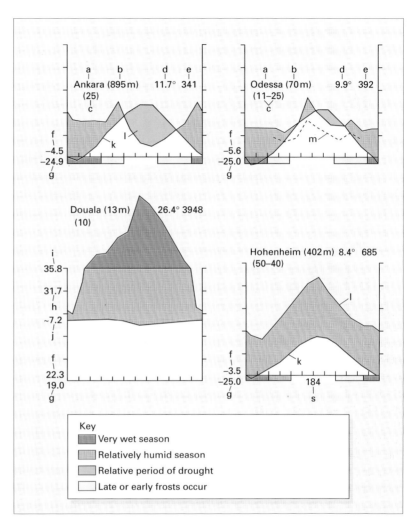

Fig. 4.11 Climate diagrams convey many aspects of seasonal variation in climate at a site in a manner that is easily assimilated visually, and which is of great relevance to the determination of vegetation in the area. Key to the climatic diagrams: Abscissa: months (Northern Hemisphere January–December, Southern Hemisphere June–July); Ordinate: one division = 10°C or 20 mm rain. a = station, b = height above sea level, c = duration of observations in years (of two figures the first indicates temperature, the second precipitation), d = mean annual temperature in °C, e = mean annual precipitation in millimetres, f = mean daily minimum temperature of the coldest month, g = lowest temperature recorded, h = mean daily maximum temperature of the warmest month, i = highest temperature recorded, j = mean daily temperature variations, k = curve of mean monthly temperature, l = curve of mean monthly precipitation, m = reduced supplementary precipitation curve (10°C = 30 mm), s = mean duration of frost-free period in days. Some values are missing, where no data are available for the stations concerned. After Walter [17].

Clementsian view of vegetation. But we now have much more detailed information about the physiology of different plant types, including their tolerance of cold or heat and their ability to cope with drought or flooding. When all of this information is put together, it is possible to define much more precisely the range of plant functional types that can be useful in classifying, mapping and understanding vegetation and its relation to climate. It is also possible to escape the notion that biomes are fixed in their make-up; instead we can admit the possibility of regional variations and transitions

between biomes. In other words, we can adopt a more Gleasonian approach to global vegetation.

One of the most elaborate attempts is that of E.O. Box [18], who compiled a list of 90 plant functional types, on the basis of which vegetation could be described on a global scale. He was able to assess the climatic tolerances and requirements of his functional types, involving temperature, precipitation and the seasonal variation in these. Precipitation alone, however, can be misleading because other conditions, such as high temperature and intense solar radiation, can lead to the rapid evaporation of

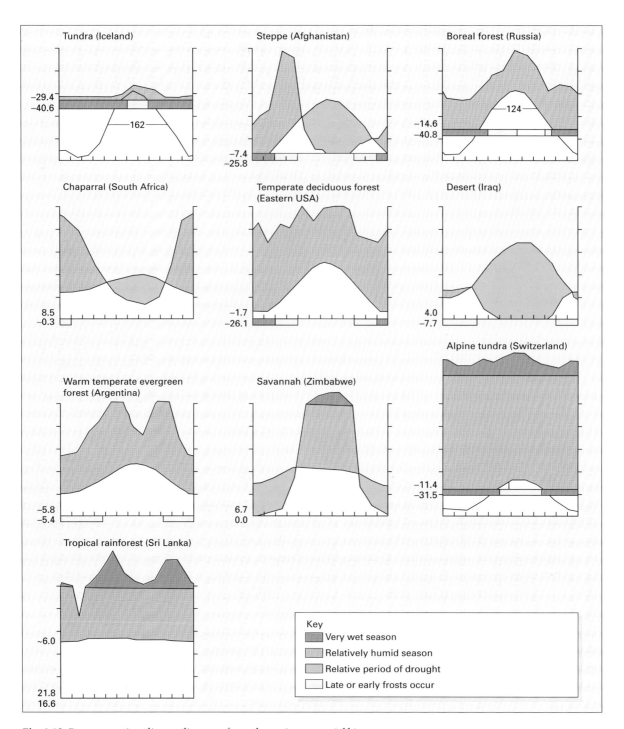

Fig. 4.12 Representative climate diagrams from the major terrestrial biomes.

water; therefore a moisture index expressing the ratio of precipitation to potential evapotranspiration (the combination of evaporation of water from surfaces and the upward movement and loss of water from plants) is required. Once these data are available, it is possible to inspect a set of climatic conditions and predict the plant functional types that would be found, together with their relative abundance and importance. The predictions from such a model [19] can then be checked against observations in the field to see how closely the model matches reality. It must be borne in mind, of course, that the predictions on this model will show potential vegetation on the assumption that climate is the determining factor; it makes no allowance for human modification of the natural habitats.

One of the most effective and widely used models relating global vegetation to climate is that constructed by Colin Prentice from the University of Lund, Sweden, and his colleagues [15]. They have sought to maintain a simple approach in their efforts to build a computer model of world vegetation, and they use just 13 functional types based not simply on morphology but also on physiology, especially the temperature tolerances of the plants. They regard minimum temperature as particularly important, as this seems to determine woody plant distribution, as Raunkiaer correctly asserted. Most broad-leaved evergreen tropical trees, for example, are killed if exposed to frost. Temperate deciduous trees, such as the oaks, are damaged by temperatures below –40°C, while many of the boreal evergreen, needle-leaved conifers, such as spruce and fir, can cope with temperatures between –45 and –60°C. Some of the needle-leaved deciduous conifers, such as larch, are able to tolerate even lower temperatures.

On the other hand, many temperate trees need to be chilled during the winter if they are to bud effectively and produce flowers the following spring, so this requirement needs to be built into the model. It means that some of the needle-leaved evergreens will demand winters in which the coldest month has an average temperature of below –2°C. Warmth

in the growing season, however, is also needed, and even cold-dwelling trees need about 5°C for adequate photosynthetic activity. Lower-growing plants, on the other hand, such as dwarf shrubs and cushion chamaephytes, may survive effectively where the average temperature of the warmest month is only 0°C. To allow for the wetness of an area, Prentice and his colleagues used a moisture index of the type developed by Box, building in an allowance for seasonal variation in wetness.

Their model works by taking into account all of these, and several other, climatic variables, and determining which of the functional types could survive and then assessing which is likely to be dominant. In this way the potential biome for the area is deduced. The outcome of the model is shown in Plate 4.2b (facing p. 86), which depict the potential biome distribution in both the Old and the New World. The model is essentially Gleasonian in that it begins from the requirements of the plant functional types, and assembles groupings on the basis of what could tolerate given climatic conditions in different parts of the world. There is no initial assumption about how many biomes exist or how they should be comprised. In fact, 17 terrestrial biomes emerge, as can be seen from the maps. The model predictions of the distribution of these biomes is in very reasonable agreement with the actual information available (see Plate 4.2a), put together by Olson. We have to bear in mind that the information is incomplete (hence the gaps in the real world maps), and often reflects the consequences of human cultural activities rather than potential vegetation.

The importance of this approach to the global classification and mapping of biomes is that it provides us with a robust model on which we can project new circumstances and observe their possible outcome. Climatic scenarios of different types can be imposed, and the consequences for biome distribution can be estimated. Since we know the relationships of certain types of agriculture to these biomes, we can also examine the impact of climate change on agricultural potential of different parts of the world.

Biomes in a changing world

It is now generally accepted that the climate is changing, largely as a consequence of human activities in the biosphere [20], so the boundaries we observe around the biomes (Plate 4.2a) can be expected to change in response to rising global temperature. Theoretical studies based on predictive models of rising carbon dioxide and the development of a 'greenhouse effect' have been extensively used to estimate the future trends in global temperature, but the models are by their very nature complex and many variables are still uncertain. It is now firmly established, however, that global temperatures have been rising for the last 150 years, having increased by about 1°C in that time, and there is some evidence that the rate of rise is itself increasing, with about 0.5°C gained in the last 50 years (see Fig. 11.1, p. 225). One major problem facing biogeographers is the modelling of climate and biome distribution under new climatic regimes.

This is a very complicated process involving many different stages. As we have seen, it is possible to model present-day vegetation in relation to climate and to produce quite a close fit. The next stage is to manipulate the climate in the model to mimic conditions that we think existed in the relatively recent past (say 6000–21 000 years ago (e.g. [21])). When very different sets of conditions were in operation (see Chapter 11) and an ice age was in the process of departing. If the model can produce vegetation maps that correspond with what we know of former vegetation from the fossil record, then it should prove robust enough to project into the future.

But the problems facing modellers should not be underestimated. One of the most significant difficulties is determining the extent of feedback mechanisms [22]. If, for example, the global vegetation became more productive as a result of raising atmospheric carbon dioxide levels, how would this affect the greenhouse system? If vegetation becomes more sparse in areas subject to drought, will this result in more energy being reflected back into space? Will certain biomes become more prone to fire? If so, how will they change? If the rise in temperature of the earth influences the flow of the global oceanic conveyor system, how would this affect future climates?

One oceanic complication has already become apparent, namely the phenomenon known as El Niño in the Pacific Ocean. Under 'normal' conditions, strong winds blow westwards across the Pacific from South America towards Indonesia, and result in the bringing of rain to the western Pacific islands [20]. Deep ocean currents flow in the opposite direction, replacing water displaced by the surface winds, and these currents bring nutrients to the west coast of South America, resulting in high plankton productivity and rich fisheries. During an El Niño episode, the winds are reduced, rain systems move eastwards and cause storms along the west coast of South and Central America, and the western Pacific (Indonesia, Borneo, the Malay Peninsula, Northern Australia, even Hawaii) experiences drought. The consequences of El Niño are also felt in many other parts of the world, involving drought in Amazonia and South Africa, a lower level of flooding in the Nile, and a weak summer monsoon in India. Recently, El Niño events have been occurring more frequently, and have been stronger in their effects, which may well be a consequence of rising global temperature and its impact on the atmospheric/oceanic circulation patterns of the Pacific Ocean. We may well expect alterations in biome patterns as a consequence of such changes.

Perhaps our whole attitude to the world's biomes is too static for the present situation of rapid change. The picture of the relationship between biomes and climate depicted in Fig. 4.7 is comfortable and reassuring, but it is also based on the idea of equilibrium. It assumes that there is time for vegetation to settle down in its climatic mould. In the coming century this is unlikely to be the case, and we need to develop more dynamic attitudes and models to cope with the understanding of the consequences of our activities on the world's biotic assemblages [23].

If, as may reasonably be expected, biome boundaries do shift with climatic change, it becomes extremely important to ensure that examples of the different biomes are carefully conserved and that their biodiversity is not lost because of an inability of the community of organisms to change their ranges in response to climatic shifts. This lack of mobility could occur as a result of fragmentation of habitats and the erection of artificial barriers (such as agricultural land) to species movements. With this in mind, the World Wildlife Fund, New York, has begun on the task of selecting the best examples of all the world's biome types. So far they have selected 136 terrestrial 'ecoregions' (together with 35 freshwater and 61 marine ecoregions). These are effectively an extension of the biodiversity hotspots discussed in Chapter 2, but emphasize the importance of having all biomes represented, even the less diverse ones. The emphasis of conservation should not always rest on the individual species, but also upon whole communities of organisms and their physical habitats.

Summary

1 The idea of a 'community' of organisms that occurs in discrete units and is predictable in its species composition is attractive and useful to biogeographers, but nature often exhibits a continuous change in species assemblages depending on the individual requirements of species.

2 If the landscape consists of a fragmented mosaic, communities are more likely to be recognizable in nature.

3 The ecosystem is a useful way of considering biotic (animal and plant) assemblages in relation to the non-living world. It is a concept based on the ideas of energy flow through a series of feeding (trophic) levels and the circulation of elements between organisms and the physical world.

4 The use of the ecosystem concept and the notion of functional types of organisms (producers, decomposers, nitrogen-fixers, etc.) within the community provides a way of investigating the implications of biodiversity for natural systems. It allows us to ask the question, are all species really necessary?

5 Global ecosystems, biomes, are best defined in terms of functional types, either morphological or, better, physiological. Models relating biome distribution to climate can then be developed.

6 Climate–biome models will provide a means of predicting the outcome of climate change on the earth's biogeography and will have implications both in conservation and agriculture.

Further reading

Archibold OW. *Ecology of World Vegetation*. London: Chapman & Hall, 1995.

Crawley MJ. *Plant Ecology*, 2nd edn. Oxford: Blackwell Science, 1997.

Forman RTT. *Land Mosaics: The Ecology of Landscapes and Regions*. Cambridge: Cambridge University Press, 1995.

Smith TM, Shugart HH, Woodward FI. *Plant Functional Types: Their Relevance to Ecosystem Properties and Global Change*. Cambridge: Cambridge University Press, 1997.

References

1 Ellenberg H. *Vegetation Ecology of Central Europe*, 4th (English) edn. Cambridge: Cambridge University Press, 1988.

2 Rodwell JS. *British Plant Communities*, Vol. I. *Woodlands and Scrub*. Cambridge: Cambridge University Press, 1991.

3 Callaway RM. Positive interactions among plants. *Bot Rev* 1995; 61: 306–49.

4 Callaway RM, D'Antonio CM. Shrub facilitation of coast live oak establishment in central California. *Madroño* 1991; 38: 158–69.

5 Forman RTT. *Land Mosaics: The Ecology of Landscapes and Regions*. Cambridge: Cambridge University Press, 1995.

6 Reagan DP, Waide RB. *The Food Web of a Tropical Rain Forest*. Chicago: University of Chicago Press, 1996.

7 Lawton JH. What do species do in ecosystems? *Oikos* 1994; 71: 367–74.

8 Naeem S, Thompson LJ, Lawler SP, Lawton JH, Woodfin RM. Empirical evidence that declining species diversity may alter the performance of terrestrial ecosystems. *Phil Trans R Soc London B* 1995; 347: 249–62.

9 Symstad AJ, Tilman D, Wilson J, Knops MH. Species loss and ecosystem functioning: effects of species identity and community composition. *Oikos* 1998; 81: 389–97.

10 McNaughton SJ, Banyikwa FF, McNaughton MM. Promotion of the cycling of diet-enhancing nutrients by African grazers. *Science* 1997; 278: 1798–800.

11 Moffat AS. Biodiversity is a boon to ecosystems not species. *Science* 1996; 271: 1497.

12 Chapin FS, Walker BH, Hobbs RJ, Hooper DU, Lawton JH, Sala OE, Tilman D. Biotic control over the functioning of ecosystems. *Science* 1997; 277: 500–4.

13 Naeem S, Li S. Biodiversity enhances ecosystem reliability. *Nature* 1997; 390: 507–9.

14 Moore PD, Bhadresa R. Population structure, biomass and pattern in a semi-desert shrub, *Zygophyllum eurypterum* Bois. and Buhse, in the Turan Biosphere Reserve of north-eastern Iran. *J Appl Ecol* 1978; 15: 837–45.

15 Prentice IC, Cramer W, Harrison SP, Leemans R, Monserud RA, Solomon AM. A global biome model based on plant physiology and dominance, soil properties and climate. *J Biogeogr* 1992; 19: 117–34.

16 Whittaker RH. *Communities and Ecosystems*, 2nd edn. New York: Macmillan, 1975.

17 Walter H. *Vegetation of the Earth*, 2nd edn. Heidelberg: Springer-Verlag, 1979.

18 Box EO. *Macroclimate and Plant Forms: An Introduction to Predictive Modeling in Phytogeography*. The Hague: Dr W. Junk, 1981.

19 Cramer W, Leemans R. Assessing impacts of climate change on vegetation using climate classification systems. In: Solomon AM, Shugart HH, eds. *Vegetation Dynamics and Global Change*. London: Chapman & Hall, 1992: 190–217.

20 Moore PD, Chaloner WG, Stott P. *Global Environmental Change*. Oxford: Blackwell Science, 1996.

21 Claussen M, Brovkin V, Ganopolski A, Kubatzki C, Petoukhov V. Modelling global terrestrial vegetation–climate interaction. *Phil Trans R Soc London B* 1998; 353: 53–63.

22 Woodward FI, Lomas MR, Betts RA. Vegetation–climate feedbacks in a greenhouse world. *Phil Trans R Soc London B* 1998; 353: 29–39.

23 Woodward FI, Beerling DJ. The dynamics of vegetation change: health warnings for equilibrium 'dodo' models. *Global Ecol Biogeogr Lett* 1997; 6: 413–18.

24 Haxeltine A, Prentice IC. BIOME 3: An equilibrium terrestrial biosphere model based on ecophysiological constraints, resource availability, and competition among plant functional types. *Global Biogeoche Cyc* 1996; 10: 693–709.

CHAPTER 5: *The source of novelty*

Some insects are protected from detection by predators by having an almost perfect resemblance to a leaf or a twig. This is perhaps the most dramatic example of the intricate way in which an organism is adapted to its environment. Other adaptations are just as complex and thorough, although not always so obvious. Every aspect of the environment makes its demands upon the structure or the physiology of the organism: the average state of the physical conditions, together with their daily and annual ranges of variation; the changing patterns of supply and abundance of food; the occasional increased losses due to disease, predators or increased competition from other organisms. Every species must be adapted to all this, so that it can tolerate and survive the hostile aspects of its environment, and yet take advantage of opportunities it offers.

In the nineteenth century, it was accepted that each species had always existed precisely as we now see it. God was thought to have created each one, with all its detailed adaptations, and these had remained unchanged. Fossils were considered to be merely the remains of other types of animal, each equally unchanging during its span of existence, that God had destroyed in a catastrophe (or a number of catastrophes) such as the biblical Flood.

In his journey round the world in the ship HMS Beagle from 1831 to 1836, Darwin saw two phenomena that eventually led him to consider alternative explanations. On the Galápagos Islands in the Pacific, isolated from South America by 960 km (600 miles) of sea, different birds were well adapted to feeding on different diets. Some, with heavy beaks, used them to crack open nuts or seeds; some, with smaller beaks, fed on fruit and flowers; others, with fine, narrow beaks, fed on insects. On the mainland these different niches are occupied by quite different, unrelated types of bird, for example toucans,

parrots and flycatchers. The remarkable fact was that, on the Galápagos Islands, each of these varied niches was instead filled by a different species of one type of bird, the finch. It looked very much as though finches had managed to colonize the Galápagos Islands before other types of bird and then, free from their competition, had been able to adapt to diets and ways of life that were normally not available to them. This logical explanation, however, ran directly against the current idea of the fixity of characteristics. Equally disturbing were the fossils that Darwin had found in South America. The sloth, armadillo and guanaco (the wild ancestor of the domesticated llama) were each represented by fossils that were larger than the living forms, but were clearly very similar to them. Again, the idea that the living species were descended from the fossil species was a straightforward explanation, but one that contradicted the view that each species was a special creation and had no blood relationship with any other species.

Natural selection

The explanation that Darwin eventually deduced and published in 1858 is now an almost universally accepted part of the basic philosophy of biological science. Darwin realized that any pair of animals or plants produces far more offspring than would be needed simply to replace that pair: there must therefore be competition for survival amongst the offspring. Furthermore, these offspring are not identical to one another, but vary slightly in their characteristics. Inevitably, some of these variations will prove to be better suited to the mode of life of the organism than others. The offspring that have these favourable characteristics will then have a natural advantage in the competition of life, and will tend to survive at the expense of their less fortunate relatives. By their survival, and eventual mating, this process of natural selection will lead to the persistence of these favourable characteristics into the next generation.

Evolution is therefore possible because of competition between individuals that differ slightly from one another. But why should these differences exist, and why should each species not be able to evolve a single, perfect answer to the demands that the environment makes upon it? All the flowers of a particular species of plant would then, for example, be of exactly the same colour, and every sparrow would have a beak of precisely the same size and shape. Such a simple solution is not possible, because the demands of the environment are neither stable nor uniform. Conditions vary from place to place, from day to day, from season to season. No single type can be the best possible adaptation to all these varying conditions. Instead, one particular size of beak might be the best for the winter diet of a sparrow, while another might be better adapted to its summer food. Since, during the lifetimes of two sparrows differing in this way, each type of beak is slightly better adapted at one time and slightly worse adapted at another, natural selection will not favour one at the expense of the other. Both types will therefore continue to exist in the population as a whole, as in the case of the oystercatchers described on p. 64.

Because we do not normally examine sparrows very closely, we are not aware of the many ways in which the individual birds may differ from one another. In reality, of course, they vary in as many ways as do different individual human beings. In our own species we are accustomed to the multitude of trivial variations that make each individual recognizably unique: the precise shape and size of the nose, ears, eyes, chin, mouth, teeth, the colour of the eyes and hair, and the type of complexion, the texture and waviness of the hair, the height and build, and the pitch of voice. We know of other, less obvious characteristics in which individuals also differ, such as their fingerprints, their degree of resistance to different diseases, and their blood

group. All of these variations are, then, the material upon which natural selection can act. In each generation, those individuals that carry the greatest number of less advantageous characteristics would be least likely to live long enough to have children who would perpetuate these traits, while those with a large number of more advantageous characteristics would be more likely to survive and breed.

Darwin's theory and Darwin's finches

Until recently, it seemed as though evolution took place too slowly for it to be detectable over the time scale of scientific studies of living organisms, so that it could only be detected in the fossil record. However, it is now clear that this perception was wrong. One of the most detailed and fruitful studies of evolution in action has been carried out precisely where Darwin first noted evidence of it—in the Galápagos Islands, and on the birds now known as 'Darwin's finches'. These all belong to the genus *Geospiza*, found only in the Islands. It includes 13 species, six of which feed on the ground, eating the nuts and seeds of some two dozen species of plant. Three of these species of ground finch are found together on the tiny island called Daphne Major, only 34 ha (84 acres) in area and 8 km (5 miles) from the nearest large island. The largest of them is *G. magnirostris* and the smallest is *G. fuliginosa*, while *G. fortis* is of intermediate size (Fig. 5.1). Their beak size is the only difference in the appearance of these three species, and it is also the vital aspect of their lives, for it controls what seeds they feed on. Both the size of the finch's body and that of its beak are strongly dependent on those of its parents, i.e. there is a strong inherited factor in their size.

For many years, these three species on Daphne Major have been studied by the 'Finch Unit' led by the British workers Peter and Rosemary Grant, whose work has been vividly described in Jonathan Weiner's excellent book, *The Beak of the Finch* (see Further reading). Since 1973, they have measured and weighed some 19 000 of these birds, belonging to 24 generations. Their populations have ranged

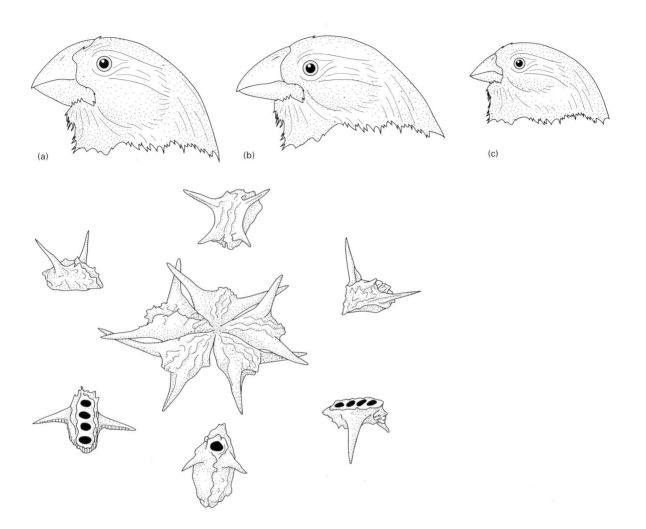

Fig. 5.1 The three species of ground finch: (a) *Geospiza magnirostris*, (b) *G. fortis* and (c) *G. fuliginosa*. In the bottom part of the figure is the fruit of *Tribulus*, to show how it breaks up into mericarps, some of which have been opened by finches and the seeds removed, leaving holes.

from fewer than 300 birds in the hardest, driest year, to approximately 1000 in the best year. Since 1977, the Finch Unit has recognized, measured and ringed every single bird, and recorded with which others it has mated, how many offspring they had, how many survived, in turn mated, etc. They have recorded the abundance of each type of food plant and its seeds, the hardness of the seeds, and the pattern of temperature and rainfall. Analysis of their results showed with startling clarity the extent of the differences in the environment from year to year, and the immediacy of the impact of these changes upon the populations of the finches.

When food is plentiful, all three species prefer to eat the softest types of seed. In drier seasons, there are fewer of these favourite seeds, because they are produced by smaller plants, which tend to wither and die during a drought. As a result, each species has then to become more specialized, spending more of its time feeding on those seeds to which its beak size is best adapted. For example, *G. magnirostris* has the biggest and strongest beak, and is the best at cracking the hardest seeds, which belong to the shrub *Tribulus cistoides*. The fruit of this (Fig. 5.1) breaks up into hard, spiny 'mericarps', each of which contains up to six large seeds. *G. magnirostris* can crack two mericarps in less than a minute, and extract at least four seeds, while *G. fortis* obtains only three seeds in over 1.5 min. Therefore, in this particular, tiny example of competition, *G. magnirostris* is gathering food at 2.5 times the rate of *G. fortis*. Furthermore, not all the *G. fortis* can even try to compete for this food, for only those with beaks at least 11 mm long can crack a mericarp; those whose beak is only 10.5 mm long cannot do so. Meanwhile, little *G. fuliginosa* has to feed on smaller, softer seeds. This provides a very precise example of the nature and results of the link between morphology (beak size) and ecology (the availability of different types of food).

But evolution is not merely about relative ease of existence; natural selection is about life and death. The Finch Unit was fortunate that their studies covered a period when the adaptations of the finches were tested to the full, for it included two of the most extreme years in the century—both the driest year, 1977, and the wettest year, 1983. This climatic switchback had immense effects on the finches of the island.

The drought of 1977 (Fig. 5.2a) first affected the amount of food available for all the ground finches. During the wet season of 1976, there had been more than 10 g of seeds per square metre of ground. Through 1977, as the drought struck harder and harder, the plants failed to flower and set the new season's seeds, and the finches increasingly consumed the seeds scattered in 1976: by June, there were only 6 g/m² of seeds, and by December only 3 g/m². The food shortage hit first the new generation of finches, because of the shortage of the small seeds that they all feed on (Fig. 5.2c). As a result, all the fledglings of 1977 died before they were 3 months old. But the drought also hit the adults. In June 1977 there had been 1300 ground

finches on Daphne Major; in December, there were fewer than 300—only about a quarter of the population had survived. But death hit harder at the finches that could only eat softer seeds. *G. fortis* lost 85% of its population, from 1200 birds down to 180 (Fig. 5.2b), but little *G. fuliginosa* suffered worst, for its population dropped from a dozen to only a single bird [1], so that the population could only recover by immigration from the neighbouring larger island of Santa Cruz. Finally, the drought had been kindest to birds with big beaks. The average size of the *G. fortis* that survived was 5.6% greater than the average size of the 1976 population, and their beaks were correspondingly longer (and stronger)—11.07 mm long and 9.96 mm deep, compared with 1976 averages of 10.68 and 9.42 mm.

As if that opportunity to watch and document the harsh workings of natural selection were not enough, 1983 allowed the Finch Unit to see a total reversal in the demands of the environment of Daphne Major. That was the year of the strongest El Niño event (see p. 92) so far in the twentieth century; the rainfall was 10 times the previously known maximum. The island was drenched, and its plants grew rampantly. By June, the total mass of seeds was almost 12 times greater than in the previous year, but small, soft seeds predominated—they formed up to 80% of the total mass of seeds, up to 10 times more than the previous maximum. That was partly because the smaller plants grew luxuriously and produced many more seeds, and partly because the growth of *Tribulus* had been hampered by smothering vines.

As a result of these lush conditions, the finch population spiralled upwards: by June, there were more than 2000 ground finches on the island, the numbers of *G. fortis* having increased by more than four times. But in 1984 there was only 53 mm of rain, and in 1985 only 4 mm. Now there was a new episode of drastic selection, but in the opposite direction from that which had followed the drought of 1973. Now it was selecting smaller birds, with smaller beaks, more suited to eating the plentiful smaller seeds [2].

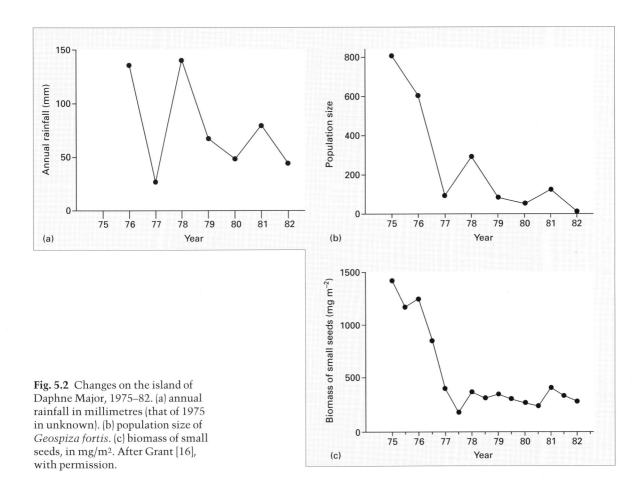

Fig. 5.2 Changes on the island of Daphne Major, 1975–82. (a) annual rainfall in millimetres (that of 1975 in unknown). (b) population size of *Geospiza fortis*. (c) biomass of small seeds, in mg/m². After Grant [16], with permission.

The moral of this second part of the finch story is that conditions, and selection, can oscillate violently. The Finch Unit had witnessed, and documented, a total reversal of the selection pressures within 6 years—a time span that would have been totally invisible in the fossil record, had there been one.

But there was still another aspect of the action of evolution on Daphne Major to be found in the Finch Unit's records. Normally, each species has its own niche, and hybridization between species is therefore disadvantageous, for the hybrid young are less well adapted than either of their parents, and therefore cannot compete successfully with either of them. That had indeed been true on Daphne Major before the flood year. Interspecies mating was rare, and the occasional hybrid was unsuccessful and did not find a mate. But, when there is so much food available that competition has for the moment disappeared, hybridization is no longer necessarily a disadvantage—and that was the situation on Daphne Major in the Year of Abundance, 1983. After that, hybrids between *G. fortis* and *G. fuliginosa* in fact did better than pure-bred members of either species and, by 1993, about 10% of the finches of Daphne Major were hybrids [3]. (This does not imply that the different species present will disappear; at the present rate of hybridization, and if the environment remains unchanging and favourable to hybridization, it would take about a century for *G. fortis* and *G. fuliginosa* to merge completely—and, of course, the environment *will* change.)

The Finch Unit has therefore seen evolution in action. Their records show precisely what Darwin's theory had predicted, and what biologists have long accepted. Each species is adapted to its existing environment, within a pattern of the availability of food and of competition with other species. But that adaptation is not immutable: it cannot afford to be, for the environment itself is fluid and changeable. So, as from year to year the environment makes new demands and provides new opportunities, so the adaptations of the species will change in harmony with those demands and opportunities. It is changes of this kind in the characteristics of a species that have led, over the last few decades, to the evolution of strains of bacteria that are resistant to commonly used antibiotics, to strains of insects that are resistant to DDT and to other chemicals designed to control pests, and of strains of malarial parasite that are resistant to modern drugs. Adaptation by evolution is never far behind our efforts to control the biological world.

It is clear, then, that the characteristics of the members of a species can alter if there is a change in the selective forces that act upon them. But this raises two questions. Firstly, what mechanism within the living organism controls its characteristics and, secondly, how did the species arise in the first place? These two questions will now be dealt with in turn.

The controlling force within the organism

The force within the organism lies in the system that is responsible for the transmission of the characters of the parents to the next generation. Within each cell lies a rather opaque object called the nucleus, inside which is a number of thread-like bodies called chromosomes. Each chromosome consists of a chain of large, complex molecules known as genes. It is the biochemical activity of these genes that is responsible for the characteristics of every cell of an individual, and thus for the characteristics

of the organism as a whole. There might, then, be a particular gene that determined the colour of an individual's hair, while another might be responsible for the texture of the hair and another for its waviness. Each gene exists in a number of slightly different versions, or alleles. Taking the gene responsible for hair colour as an example, one allele might cause the hair to be brown while another might cause it to be red. Many different alleles of each gene may exist, and this is the main reason for much of the variation in structure that Darwin noted.

An individual of course inherits characteristics from both of its parents. This is because each cell carries, not one set of these gene-bearing chromosomes, but two: one set derived from the individual's mother, and the other derived from its father. A double dose of each gene is therefore present, one inherited from the mother, the other from the father. Both parents may possess exactly the same allele of a particular gene. For example, both may have the allele for brown hair, in which case their offspring would also have brown hair. But very often they may hand down different alleles to their offspring; for example, one might provide a brown-hair allele, while the other provided a red-hair allele. In such a case, the result is not a mixing or blurring of the action of the two alleles to produce an intermediate such as reddish-brown hair. Instead, only one of the two alleles goes into action, and the other appears to remain inert. The active allele is known as the dominant allele and the inert one as the recessive allele. Which allele is dominant and which recessive is normally firmly fixed and unvarying; in the above example, the brown-hair allele might be dominant, and the red-hair allele might be recessive.

The genes themselves are highly complex in their biochemical structure. Although normally each is precisely and accurately duplicated each time a cell divides, it is not surprising that from time to time—due to the incredible complexity of the molecules involved—there is a slight error in this process. This may happen in the cell divisions that lead to the production of the sexual gametes (the male sperm or pollen, the female ovum or egg). If so, the individual resulting from that sexual union may show a completely new character, unlike those of either of its parents. In the example given above, such an individual might have completely colourless hair. Such sudden alterations in the genes are known as mutations.

The genetic system outlined above can lead to changes in the characteristics of an isolated population, in two ways. First, new mutations may appear and, if they are advantageous, spread through the population. Secondly, since each individual carries several thousand genes, and each may be present in any one of its several different alleles, no two individuals carry exactly the same genetic constitution, or genotype—unless they are identical twins, developed from the splitting of a single original developing egg. Inevitably, therefore, the two isolated populations will differ somewhat in their initial genetic content, some alleles being rarer in one population than in the other or, perhaps, being absent altogether. As mating goes on in the two populations, new combinations of alleles will appear haphazardly in each, and this will lead to further differences between them.

Whether they are new mutations, or merely new recombinations of existing alleles, new characteristics will therefore appear within an isolated population. Any of these which confer an advantage on the organism are likely to spread gradually through the population, and so change its genetic constitution. However, it is important to realize that chance, as well as its genetic constitution, plays a role in determining whether a particular individual survives and breeds. Even if a new, favourable genetic change appears in a particular individual, it may by chance die before it reproduces, or all its offspring may similarly die, so that the new mutation or recombination disappears again. Nevertheless, however rare each genetic change may be, each is likely to reappear in a certain percentage of the population as a whole. In a larger population, each mutation or recombination will therefore reappear sufficiently often that the effects of random chance are reduced. The underlying advantages or disad-

vantages that they confer will then eventually show themselves as increased or decreased reproductive success. For this reason, it is the population, and not the individual, that is the real unit of evolutionary change. In smaller populations, however, chance will play a greater role in controlling whether a particular allele becomes common or rare or disappears; this effect is known as 'genetic drift', because it is not directed by selective pressures. Smaller populations therefore contain less genetic variability, and are less closely adapted to their environment; they are therefore more likely to become extinct than larger populations.

The way in which the genotype is expressed, as the morphology, physiology, behaviour, etc. of the organism, is known as the phenotype. This is somewhat variable and can be modified by the environment. Thus, identical twins (sharing therefore an identical genotype) will come to differ from one another if they are brought up in areas with, for example, differing amounts of sunlight or of available food. This slight plasticity of the genotype is valuable from an evolutionary point of view, for it makes it possible for a single genotype to survive in slightly different habitats.

This, then, is the genetic system that provides both the stability that ensures that the complex systems of the organism will function and be adapted to the demands of the environment, and the plasticity that allows it to respond to changes in that environment. The final, and central, question in the story of evolution is how a single, well-adapted species of this kind can split into two species, adapted to two slightly different environments.

From populations to species

Before explaining its origin, we must first of all define the meaning of the term 'species'. Why do biologists consider that a sparrow and a robin are separate species, but that two dogs such as a German shepherd dog and a greyhound (which appear just as different from one another) are both members of the same species? To biologists the essential difference is that, under normal conditions in the wild, a sparrow and a robin do not mate together, while a German shepherd dog and a greyhound will (the great difference in the appearance of the two dogs is due to artificial selection by humans).

However great may be the area of land (or water) within which a particular species is to be found, it is not found everywhere within that area. Any area is a patchwork quilt of differing environments, of meadow, pond and woodland, of dense forest or of regrown forest—and even the meadow or woodland is itself made up of a myriad of differing habitats, as we saw in Chapter 3. As a result, the species is broken up into many individual populations that are separate from one another. Furthermore, no two patches of woodland, no two freshwater ponds, will be absolutely identical, even if they lie in the same area of country. They may differ in the precise nature of their soil or water, in their range of temperature, or their average temperature, or in the particular species of animal or plant that may become unusually rare or unusually common in that locality. Each population independently responds to the particular environmental changes that take place in its own location, and the response of each population is also dependent on the particular pattern of new mutations and of new genetic combinations that have taken place within it. Each population will therefore gradually come to differ from the others.

Once populations have started to diverge in their genetic adaptations in this way, the foundations for the appearance of a new species have been laid. If two divergent populations should meet again when the process has not gone very far, they may simply interbreed and merge with one another. The further, vital step towards the appearance of a new species is when hybrids between the two independent populations are found only along a narrow zone where the two populations meet. Such a situation suggests that, though continued interbreeding within this zone can produce a population of hybrids, these hybrids cannot compete elsewhere with either of the pure parent populations.

The length of time that it would take for this process of gradual adaptive genetic change to lead to the appearance of a new species in nature is normally greater than the length of time over which organisms have been studied in detail by biologists. However, where a species has been extending its range and, in the process, has had to adapt to new environments, we can see the resulting pattern of evolutionary change laid out on the landscape as a pattern of biogeography. In a few cases, the end-products of that process have come into contact and demonstrated the extent of the genetic change by refusing to mate with one another: they have evolved into separate species, known as 'ring species'.

One of the best examples of a ring species is the pattern of distribution of the salamander *Ensatina* in the western United States [4]. The story appears to have begun with the species *E. oregonensis* living in Washington State and Oregon, which spread into northern California, where it formed the new species *E. eschscholtzii*. As this species continued to spread southwards, encircling the hot lowlands of the San Joaquin Valley, it developed populations with different the genetic constitutions. One result of this was that the populations had different colour patterns (Plate 5.1, facing p. 86): the populations on the western side of the valley became lightly pigmented, while those on the eastern side developed a more blotchy pattern. As a result, biologists gave them different names, but because these populations did not appear to interbreed with one another, they were not recognized as full species, but only as separate subspecies of the single species *E. eschscholtzii* (*E. e. picta*, *E. e. platensis*, etc.).

However, at two points, these two sets of subspecies came into contact. At some time in the past, the subspecies that lives in the San Francisco area, *E. e. xanthoptica*, colonized the eastern side of the valley, where it met one of the 'blotched' subspecies, *E. e. platensis*. Here, the amount of genetic divergence between these two subspecies is not enough to prevent them from interbreeding (Plate 5.1). The other meeting point is below the southern end of the San Joaquin Valley, in southern California, where *E. e. eschscholtzii* meets *E. e. croceater* and *E. e. klauberi*. (There is nowadays a gap in the ring of subspecies in this area, perhaps because of climatic change, but populations of *E. e.*

croceater are found both north-west and east of Los Angeles, showing that the chain was once complete [5].) Over much of this area of overlap, the two subspecies hybridize to a limited extent (approximately 8% of the salamanders of the area are hybrids). However, at the extreme southerly end of the chain, the length of time since the two types of salamander diverged from their common northern ancestors is at its maximum. As a result, the genetic differences between them are so great that they there behave as completely separate species, *E. eschscholtzii* and *E. klauberi*.

There can be no general rule as to the length of time that it will take for the descendants of one original species to diverge so far from one another in their genetic constitution that they have become separate species. The most important factor in determining the rate of genetic change is the speed at which the environment changes. If it changes rapidly, the organism must also change rapidly, or else become liable to extinction. But the rate at which an organism can respond is also dependent on population size. In a small population, the random effect of genetic drift may by chance produce a new mixture of genetic characteristics that match the new requirements of the environment. This is less likely to happen in a larger population, where the sheer size of the gene pool makes rapid evolutionary change of this kind less likely.

The fastest well-documented example of speciation is that of some species of cichlid fish in a small lake in Africa. This became separated from the large Lake Victoria by a strip of land that has been dated by radiocarbon analysis at 4000 years old [6], and the species of fish in the small lake therefore cannot be older than that. Many of the species of the fruitfly *Drosophila* that are found in the narrow valleys between the lava flows on the slopes of the volcanic mountains in the Hawaiian Islands (see p. 284) are probably also only a few thousand years old.

Barriers to interbreeding

One can distinguish several different degrees of separation between species. In the first, the two species will not mate under any circumstances. In the second, they will mate, but they will have no offspring—the mating is sterile. In the third, they will mate and have offspring, but these descendants are themselves sterile. An example of this is the case of the horse and the ass; although these are separate species, they do sometimes breed together, but the resulting mule (male) or hinny (female) is sterile. Finally, the two species may breed together and have offspring, but the fertility of the offspring is reduced, so that the hybrids soon become extinct in competition with the more fertile results of mating within each species. As we have seen in the case of the Californian salamanders, these different degrees of biological separateness reflect the length of time for which the two species have been separate from one another, and therefore the time available for each to become independently, and differently, adapted to its environment.

Once independent evolution in isolation has produced a situation in which the hybrids are less well adapted than either of their parents, then natural selection will favour individuals that do not help to perpetuate this more poorly adapted hybrid population. This may be either because they cannot, or will not, mate with individuals from the other group, or because such unions are infertile. The barrier to hybridization is known as an isolating mechanism, and it may take many forms. In animals such as birds and insects that have a complicated courtship and mating behaviour, small differences in these rituals may in themselves effectively prevent interbreeding. Sometimes the preference for the mating site may differ slightly. For example, the North American toads *Bufo fowleri* and *B. americanus* live in the same areas, but breed in different places [7]. *B. fowleri* breeds in large, still bodies of water such as ponds, large rainpools and quiet streams, whereas *B. americanus* prefers shallow puddles or brook pools. Interbreeding between species is also hindered by the fact that *B. americanus* breeds in early spring and *B. fowleri* in the late spring, although there is some mid-spring overlap.

Many flowering plants are pollinated by animals that are attracted to the flowers by their nectar or

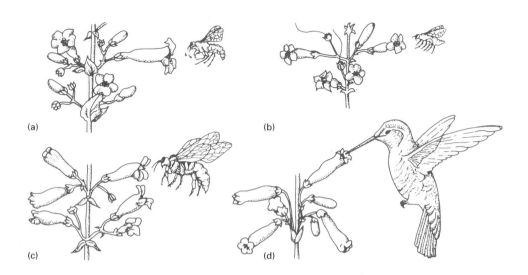

Fig. 5.3 Four species of the beard-tongue (*Pentstemon*) found in California, together with their pollinators. Species (a) and (b) are pollinated by solitary wasps, species (c) by carpenter bees, and species (d) by humming birds. After Stebbins [17].

pollen. Hybridization may then be prevented by the adaptation of the flowers to different pollinators. For example, difference in size, shape and colour of the flowers of related species of the North American beard-tongue (*Pentstemon*) adapt them to pollination by different insects, or, in one case, by a hummingbird (Fig. 5.3). In other plants, related species have come to differ in the time at which they shed their pollen, thus making hybridization impossible. Even if pollen of another species does reach the stigma of a flower, in many cases it is unable even to form a pollen tube, because the biochemical environment in which it finds itself is too alien. It cannot therefore grow down to fertilize the ovum. Similarly, in many animals alien spermatozoa cause an allergy reaction in the walls of the female genital passage, and the spermatozoa die before fertilization.

Other isolating mechanisms may not prevent mating and fertilization taking place, but instead ensure that the union is sterile. These may be genetic isolating mechanisms, the structure and

arrangement of the genes on the chromosomes being so different that the normal processes of chromosome splitting and pairing that accompany cell division are disrupted. These differences may make themselves felt at any stage from the time of fertilization of the ovum, through all the steps in development, to the time at which the sexual gametes of the hybrid itself are produced. Whenever the effects are felt, the result is the same; the hybrid mating is sterile or, if offspring are produced, these are themselves sterile, or of reduced fertility.

Polyploids

Another method by which new species can appear is by polyploidy—the doubling of the whole set of chromosomes in the nucleus of a developing egg or seed, so that each chromosome automatically has an identical partner. This may occur in the development of a hybrid individual (in which case it can overcome any genetic isolating mechanisms), or in the development of an otherwise normal offspring of parents from a single species. In either case, the new polyploid individual will be unlikely to find another similar individual with which to mate, and the origin of new species by polyploidy has therefore been important only in groups in which self-fertilization is common. Only a few animal groups fall into this category (e.g. turbellarians, lumbricid earthworms and weevils), but in these groups an appreciable proportion of the species probably arose in this way. In plants, however, in which self-fertilization is common, polyploidy is an important mechanism of speciation. More than one-third of all plant species have probably arisen in this way, including many valuable crop plants such as wheat (see p. 208), oats, cotton, potatoes, bananas, coffee and sugar cane. Polyploid species are often larger than the original parent type, and also more hardy and vigorous; many weeds are polyploids.

An example of a pest species resulting from such polyploidy is the cord-grass, a robust, rhizomatous plant of coastal mudflats around the world. There are several species of this plant, but none of them was a serious pest until two species were brought into contact with one another in the waters around the major port of Southampton, England, in the latter half of the last century. An American species of cord-grass, *Spartina alterniflora*, was brought into the area, probably carried in mud on a boat, and was able to hybridize with the native English species, *S. maritima*. The hybrid was first found in the area in 1870, and was named *Spartina* x *townsendii*. It contained 62 chromosomes in its nucleus but, because these chromosomes were derived from two different parent species, they were unable to join together in compatible pairs before gamete formation, and so the hybrid did not produce fertile pollen grains or egg cells. It was nevertheless able to reproduce vegetatively, and it is still found along the coasts of western Europe. But in 1892 a new fertile cord-grass appeared near Southampton, and was named *S. anglica*. This has 124 chromosomes, as a result of a doubling of the number found in the sterile hybrid, so that the chromosomes could once again form compatible pairs and fertile gametes could be produced. This new species, formed by natural polyploidy, has been extremely successful and has spread around the world, often creating problems for shipping by forming mats of vegetation within which sediments are deposited, and so contributing to the silting up of estuaries.

Polyploidy can also be artificially produced, for example by the use of colchicum, an extract of the meadow saffron plant, *Colchicum autumnale*. Techniques of this kind have been used to provide new strains of commercially valuable plants, such as cereals, sugar-beet, tomatoes and roses.

Clines and 'rules'

Environmental conditions usually vary in complex fashion, so that it is difficult to distinguish any regular changes that may result from them. There are, however, exceptions to this. Such conditions as average temperature or rainfall may change gradually and regularly according to either latitude or altitude. Some aspects of the organism, such as its size or height, may then also vary gradually and continuously across the area; this regular change is known as a cline. For example, the yarrow *Achillea lanulosa* grows in North America from the Pacific coast to the 4000 m crest of the Sierra Nevada Mountains. The higher the altitude at which it is found, the lower is the average height of the plant. This is a genetically controlled adaptive feature, for seeds from specimens from different altitudes retain their relative height characteristics even if grown together.

Some similar changes have been noted in warm-blooded animals, and have been called 'rules'. For example, warm-blooded animals lose both heat and moisture through their body surface. Because a larger animal has a smaller surface area compared with its volume than does a smaller animal, warm-blooded animals tend to be larger in cooler, drier environments than in hotter, more humid environments, thus conserving heat and moisture; this is known as 'Bergmann's rule' [8]. Similarly, 'Allen's rule' notes that such projecting parts of the body as the ears and tail tend to be shorter in colder, drier regions, for the same reasons. However, although these 'rules' are of some historic interest, there are many other characteristics, such as brood size or frequency, which vary in this way and which are no less important. It must also be stressed that these 'rules' apply primarily to variations within a species: the differences between species are more complex, so that simple effects of this kind are often concealed by other results of their differing ways of life.

Competition for life

However successful their adaptations to their physical environment, organisms must also adapt to the demands of the biological world around them, either to avoid being eaten, or to compete for space or food supply with other organisms. There can be no final solution to any of these problems for, as quickly as new adaptations appear that reduce predation or allow more successful competition with other species, the predator or competitor will in its turn adapt, in the biological version of an 'arms race'. The herbivorous group that becomes able to run faster and escape from its predators, itself provides the stimulus that leads to the evolution of faster predators. The plant that evolves spines or unpleasant-tasting biochemicals to avoid being eaten by herbivores similarly stimulates the appearance of herbivores insensitive to these defences. For example, milkweeds are rich in poisonous glycosides and are avoided by most caterpillars. The caterpillars of the monarch butterfly, however, not only feed on these plants, but also in turn use the biochemical to make the adult butterfly unpalatable to birds. Other biochemicals commonly used by plants as feeding deterrents include alkaloids, flavoids, quinones and raphides (crystals of calcium oxalate) [9].

One method by which the problem of competition can be at least reduced is for the two competing groups gradually to become specialized to different ways of life; they may then be able to exist together in the same area without competing with one another. Dolph Schluter, of the University of British Columbia in Vancouver, has studied this aspect of evolution in sticklebacks (*Gasterosteus*) that became isolated in small lakes in British Columbia at the end of the last Ice Age, about 12 000 years ago. Where there is only one species in a lake, it is a generalist, feeding both on the bottom and in the open water of the lake. But, where there are two species in the lake, one has become specialized to feed on the bottom and has a larger, more bulky body, a wide mouth and a coarse food-filtering system in the gills, while the other species feeds only in the open waters and has a more slender body, a narrower mouth and a finer filter system. Schluter designed an experiment to check on the interaction between the two types of fish, placing generalists together with open-water feeders in an artificial pond; he found that those generalists that were better adapted to feeding on the bottom grew more rapidly than those that were more similar to the open-water feeders, with which they were having to compete [10]. Again, evolution could be seen in action.

Finally, a part of the adaptation of any population is to ensure that its numbers are approximately adjusted to the food supply of the area. The territorial behaviour of some birds, such as the Scottish red grouse (*Lagopus lagopus scoticus*), does this very effectively [11]. Each male takes possession of an area of heather moor large enough to provide an adequate food supply for its family, and defends it against other members of its species. In a year when food is scarce, the territory claimed is larger. The males compete for these territories by display, and this system therefore not only ensures that it is the weakest birds that are excluded from the moor (and are frequently killed by predators or starvation), but also ensures that an adequate food supply is available for the successful birds. This type of social competition is a close parallel to that in human societies. In both, as a result of social competition those which are successful receive a variety of advantages, sexual, nutritive and environmental. The red grouse society no longer contains a group of moderately successful males, sometimes adequately fed and at other times weakened by malnutrition. Instead, it is permanently divided into the 'haves', assured of the necessities of life and of the opportunity, by reproduction, to transmit their characteristics to the next generation, and the 'have-nots' of whom about 60% die during the winter.

Controversies and evolutionary theory

Although there is now a vast amount of evidence for the theory of evolution by natural selection, controversy still exists about some details of the circumstances in which new species evolve or the rate at which this happens. For example, some biologists believe that evolutionary change normally takes place at a steady, gradual rate. Others instead believe that, even if genetic alterations gradually accumulate within a population, this may not be reflected in detectable morphological or physiological changes until they are so numerous as to shift the balance of the whole genotype. At this point a comparatively large number of changes are seen to take place at the same time; this is known as the 'punctuated equilibrium' model of evolutionary change. Each group of theorists provides examples that may support their view, and, at times, the same example is interpreted by each as supporting their own view.

It is also difficult to isolate such underlying patterns from the more direct effects of the environment. For example, a study of the fossil shells of gastropod molluscs from Pliocene and Pleistocene deposits (see Fig. 6.2, p. 114) in northern Kenya shows long periods during which their structure and size remained unchanged, interrupted by shorter periods (5000–50 000 years) during which they changed rapidly (Fig. 5.4). This was interpreted as an example of punctuated equilibrium [12]. However, the fact that the periods of change took place in several lineages at about the same time suggests that they were the result of external events that affected all of them, rather than resulting from some inherent evolutionary mechanism.

The main difficulty with such studies is that, in general, the fossil record is not sufficiently detailed for us to be able to be certain whether gradualistic evolution or punctuated evolution was involved. In any case, we have no reason to believe that either style of evolution systematically prevails over the other. The real point of interest should instead therefore be to identify the circumstances under which one or the other type of evolution would be more likely to take place.

Another argument that concerns evolutionary biologists is whether new species always arise in isolation, separated from the area in which the ancestral, related species is to be found. This is known as 'allopatric' speciation, and is the situation most frequently found, as, for example, in the case of the separate islands within which the distinct species of finch recognized by Darwin had evolved. But some biologists believe that a new species can also arise 'sympatrically', within the area of distribution of the ancestral species. For example, two different species of lacewing insect are found in the cold-temperate and boreal regions of North America. *Chrysopa carnea* is found in grasslands, meadows and on deciduous trees, but only rarely on conifer trees; it is light green in colour. *C. downesi* is found only on conifer trees, and is a very dark green. Apart from their different colours, the two species are nearly indistinguishable morphologically and, although they will interbreed in the laboratory, they do not in nature. It was originally suggested [13] that the common ancestor of these two living species lived in both habitats. A simple genetic mutation could then have led to the appearance of the dark green colouration in some individuals. This colouration gave a selective advantage to these individuals in the coniferous habitat, but a corresponding selective disadvantage in the grassland–meadow deciduous tree habitat. These selective forces would then have led to the appearance of two populations, each living in only one of these habitats. The consequent genetic isolation of these populations would have permitted independent evolutionary change to take place in each. It has subsequently been shown [14] that the two species also differ in their complex mating calls, and it has been suggested that this, rather than their colouration, was the original basis for the separation of the two species. However, it is equally possible that this reproductive difference evolved subsequently as a barrier to hybridization between the two populations as they became closely adapted to the two different habitats. The difficulty is that there is no evidence of these past genetic events, and it is also possible that *C. downesi* evolved

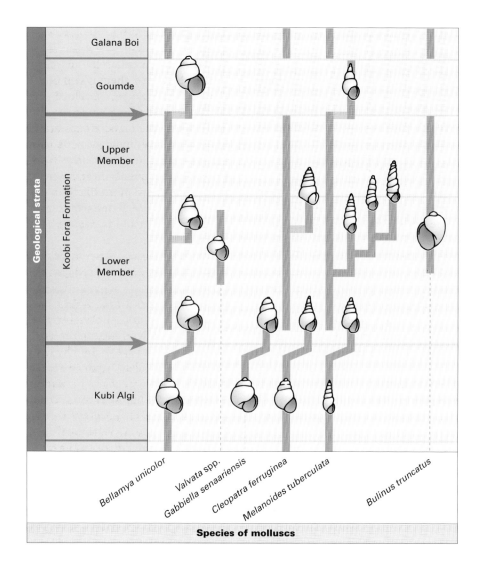

Fig. 5.4 Evolutionary changes in fossil gastropod molluscs in northern Kenya. The arrows indicate the levels at which sudden evolutionary changes took place simultaneously in several different species. From Dowdeswell [18].

allopatrically in an area in which the meadow habitat had disappeared, and then spread back into this part of the environment in areas that were still inhabited by the light-green type of *C. carnea*, which it then displaced from the conifer habitat.

It is possible that the two types of oystercatcher described in Chapter 3 (p. 64), with differing thicknesses of bill, are undergoing sympatric speciation.

These controversies, and others like them, are common in any area of science, as new observations provoke new theories or suggest modifications of existing theories. But the protagonists in these disputes are only arguing about details of the theory of evolution: all accept that the theory itself is correct, and is indeed the only one that makes sense of the phenomena of the living world. It is particularly important to realize this, as some groups within society, such as the 'Creation Science' organization, are opposed to the idea of evolution and try to present these academic controversies as symptoms of widespread and fundamental scepticism of the validity of the theory of evolution. They are, of course, nothing of the kind.

Evolution and the human race(s)

As one travels around our planet, it is obvious that the types of men and women that have long inhabited the different continents differ from one another in systematic ways. They differ most obviously in physical characteristics of the skin, eyes, hair and build, but also in less obvious ways, such as aspects of their physiology and biochemistry. Should they therefore be regarded as separate species? The answer to that is clear. As we have seen, the distinguishing feature of members of the same species is that they can interbreed with one another without any marked reduction in fertility of the succeeding generations. Because there is no evidence whatever of any decrease in fertility resulting from matings from even the most geographically distant or apparently different types of human being, there can be no doubt that they are all merely different races of one species, *Homo sapiens*. In any case, these racial differences account for very little of the genetic diversity of our species. Nearly all of that diversity (85%) is made up of the genetic differences between the individuals within a single population (such as the peoples of Spain or of a single African tribe). Another 8% is made up of genetic differences between populations of this kind, and only 7% of all the diversity is contributed by differences between one race and another.

Like any other species, ours has been subject to natural selection, and the racial differences are an aspect of this process. Each race has its own particular assemblage of many alleles. To some extent the process by which the alleles that come to dominate or be common in each race is haphazard, for the selective value of some of them is still uncertain. For example, the blood group Rh is found in

30% of all Caucasians and is rare in the peoples of the Far East and in American Indians, and the AB blood group is totally absent from American Indians. However, there is still no clear correlation between these aspects of blood immunology and any selective advantage or disadvantage. It is known that some of the blood groups are associated with a higher or lower incidence of a particular disease. For example, individuals with blood group O are more likely to suffer from stomach ulcers, and those with group A are more likely to develop stomach cancer. But these illnesses normally only appear late in life, so that they are unlikely to have had a significant selective effect.

Despite the uncertainty over the selective reasons for their different blood group characteristics, the advantages of many of the other differences between the races are quite clear. For example, skin coloration is probably related to the fact that vitamin D (which is needed for calcium fixation and bone growth) is produced in the skin by a reaction between ultraviolet light and various precursor biochemicals. The light skin of northern Caucasians probably evolved in order to allow more of the weak northern sunlight to penetrate the skin, so as to produce enough vitamin D, while the very dark skin of more tropical races may have prevented an overproduction of the vitamin, which could have been toxic. The tightly curled hair of African races helps to prevent excessive evaporation, and heat stroke, in the tropical sun. The narrower eye slit of the Asian race may help to protect their eyes from grit and glare in the deserts, and their padding of fat may help them to retain heat during the cold nights and winters. The smaller nostrils and longer noses of peoples that live at higher altitudes may help to warm and humidify the air before it passes down into the lungs.

A good example of the recent action of natural selection upon the human species is the well-known history of the human blood condition known as sickle-cell anaemia, which causes anaemia and malfunctions of the liver and blood system. The gene for this condition was common in West Africa, probably occurring in over 20% of the population, because it also provided resistance to malaria. In

West Africans who were transported as slaves to North America, where malaria does not occur, this compensating advantage of the sickle-cell anaemia gene was no longer relevant, and the frequency of the gene has now dropped to below 5%. In Central America, where malaria still persists, the gene is still found in 20% of the population.

Another example of adaptive genetic change in our species is provided by the distribution of the ability of adults to digest milk. Most adults cannot do this, because they lack the enzyme lactase, which is responsible for the biochemical utilization of the lactose sugar found in milk; if they drink milk they suffer from nausea, vomiting and abdominal pains. It is interesting to find that lactose is retained into adult life in just those groups (the Europeans and some northern African tribes) that are, or have been, pastoral nomads and for whom the milk of domesticated animals was therefore a readily available additional source of nourishment. The possession of the gene that led to the retention of lactose in adults must have provided a high selective advantage, and this has led to the frequency of the gene rising from perhaps 0.001% some 9000 years ago, when the domestication of livestock commenced, to its present 75% in these populations [15].

Although, via natural selection, the environment controlled the evolution of our own species until comparatively recently, we have now largely turned the tables on Nature. At first we merely modified our immediate environment to make it more congenial, by the use of fire, tools, clothing and dwellings. But now we have gone further, to modify the animals, plants and total environment of much of the planet to provide the food and energy sources we require. The human population of the earth has risen so dramatically over the last 2000 years that the impact of natural selection in determining which individual will survive to reproduce, and which will not, has been greatly diminished. To a large extent we now control the environment, rather than being selected by it. Understanding the problems that arise from this fact is the subject of much of this book.

Summary

1 The genetic mechanism of inheritance involves the continual production of slight variations in the characteristics of a species, due to the recombination of existing characteristics and to the appearance of new features by mutation.
2 In the short term, the natural selection of the individuals with the greatest number of favourable characters ensures that the species is adapted to its existing environment. In the longer term, when the environment changes, natural selection acts on the population to change its characteristics, so that it remains adapted.
3 Modern observations of populations in the wild, aided by computer analysis, have shown how quickly these respond to changes in their environment, and have shown clearly the action of natural selection in bringing this about.
4 Natural selection takes place independently in each population of a species, adapting it to local conditions. Continued isolation between these populations can lead to them gradually becoming so different from one another that they have become new species.

Further reading

Ridley M. *Evolution*, 2nd edn. Boston: Blackwell Science, 1996.
Weiner J. *The Beak of the Finch*. London: Vintage Books, 1994.
Desmond A, Moore J. *Darwin (A Biography of Charles Darwin)*. London: Michael Joseph, 1991.

References

1 Grant PR, Boag PT. Rainfall on the Galápagos and the demography of Darwin's finches. *Auk* 1980; 97: 227–44.
2 Gibbs HL, Grant PR. Ecological consequences of an exceptionally strong El Niño event on Darwin's finches. *Ecology* 1987; 68: 1735–46.
3 Grant PR. Hybridization of Darwin's finches on Isla Daphne Major, Galápagos. *Phil Trans R Soc London B* 1993; 340: 127–39.
4 Wake DB, Yanev KP, Brown CW. Intraspecific sympatry in allozymes in a 'ring species', the

plethodontid salamander *Ensatina eschscholtzii*, in southern California. *Evolution* 1986; 40: 866–8.

5 Jackman TR, Wake DB. Evolutionary and historical analysis of protein variation in the blotched forms of salamanders of the *Ensatina* complex (Amphibia: Plethodontidae). *Evolution* 1994; 48: 876–97.

6 Fryer G, Iles TD. *The Cichlid Fishes of the Great Lakes of Africa; their Biology and Evolution.* Edinburgh: Oliver & Boyd, 1972.

7 Blair AP. Isolating mechanisms in a complex of four species of toads. *Biol Symp* 1942; 6: 235–49.

8 James FC. Geographic size variations in birds and its relationship to climate. *Ecology* 1970; 51: 365–90.

9 Ehrlich PR, Raven PH. Butterflies and plants: a study in evolution. *Evolution* 1964; 18: 586–08.

10 Schluter D, McPhail JD. Ecological character displacement and speciation in sticklebacks. *Am Naturalist* 1992; 140: 85–108.

11 Watson A. Population limitation and the adaptive value of territorial behaviour in the Scottish red grouse, *Lagopus l. scoticus.* In: Stebbins B, Perrins C, eds. *Evolutionary Ecology.* London: Macmillan, 1977: 19–26.

12 Williamson PG. Palaeontological documentation of speciation in Cenozoic molluscs from Turkana Basin. *Nature* 1981; 293: 437–43.

13 Tauber CA, Tauber MJ. A genetic model for sympatric speciation through habitat diversification and seasonal isolation. *Nature* 1977; 268: 702–5.

14 Henry TS. Sibling species, call differences, and speciation in green lacewings (Neuroptera: Chrysopidae: *Chrysoperla*). *Evolution* 1985; 39: 965–84.

15 Bodmer WF, Cavalli-Sforza LL. *Genetics, Evolution and Man.* San Francisco: Freeman, 1976.

16 Grant PR. *Ecology and Evolution of Darwin's Finches.* Princeton: Princeton University Press, 1986.

17 Stebbins GL. *Variation and Evolution in Plants.* New York: Columbia University Press, 1950.

18 Dowdeswell WH. *Evolution: a Modern Synthesis.* London: Heinemann, 1984.

19 Thelander CG. *Life on the Edge.* Berkeley, California: Biosystems Books, 1994, 233.

CHAPTER 6: *Patterns in the past*

It is interesting and instructive to attempt to explain the smaller scale patterns of distribution by using the ecological approach that has been dealt with in Chapters 1–3. But biogeographers also wish to explain the larger patterns of life on the planet. Can we identify major regions that appear to differ fundamentally from one another in the types of plant or animal to be found there and, if so, how did this come about? That is the type of problem that is analysed in the historical approach to biogeography.

Although very few groups have precisely the same pattern of geographical distribution, there are some zones that mark the limits of distribution of many groups. This is because these zones are barrier regions, where conditions are so inhospitable to most organisms that few of them can live there. For terrestrial animals, any stretch of sea or ocean proves to be a barrier of this kind—except for flying animals whose distribution is for this reason obviously wider than that of solely terrestrial forms. Extremes of temperature, such as exist in deserts or in high mountains, constitute similar (though less effective) barriers to the spread of plants and animals.

These three types of barrier—oceans, mountain chains and large deserts—therefore provide the major discontinuities in the patterns of the spread of organisms around the world. Oceans completely surround Australia. They also virtually isolate South America and North America from each other and completely separate them from other continents. Seas, and the extensive deserts of North Africa and the Middle East, effectively isolate Africa from Eurasia. India and South-East Asia are similarly isolated from the rest of Asia by the vast, high Tibetan Plateau, of which the Himalayas are the southern fringe, together with the Asian deserts which lie to the north.

Each of these land areas, together with any nearby islands to which its fauna or flora has been able to spread, is therefore comparatively isolated. It is not surprising to find that the patterns of distribution of both the faunas (faunal provinces or zoogeographical regions) and the floras (floral regions) largely reflect this pattern of geographical barriers. Before the detailed composition of these faunal provinces and floral regions can be understood fully, it is first necessary to follow the ways in which today's patterns of geography, climate and distribution of life came into existence. From what has been discussed in earlier chapters, it is clear that the differences between the faunas and floras of different areas might be due to a number of factors. Firstly, any new group of organisms will appear first in one particular area. If it competes with another, previously established group in that area, the expansion in the range of distribution of the new group may be accompanied by contraction in that of the old. However, once it has spread to the limits of its province or region, whether or not it is able to spread into the next will depend initially on whether it is able to surmount the geographical ocean or mountain barrier, or to adapt to the different climatic conditions to be found there. (Although, even if it is able to cross to the next province or region, it may be unable to establish itself because of the presence there of another group which is better adapted to that particular environment.) Of course, changes in the climate or geographical pattern could lead to changes in the patterns of distribution of life. For example, gradual climatic changes, affecting the whole world, could cause the gradual northward or southward migrations of floras and faunas, because these extended into newly favourable areas and died out in areas where the climate was no longer hospitable. Similarly, the possibilities of migration

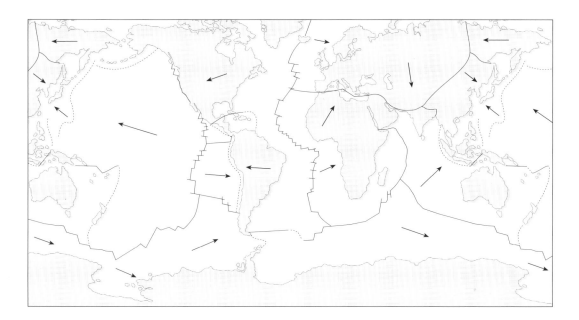

Fig. 6.1 The major tectonic plates today. Lines within the oceans show the positions of spreading ridges: dotted lines show the positions of trenches. Lines within the continents show the divisions between the different plates. Arrows show the directions and proportionate speeds of movement of the plates. The Antarctic plate is moving clockwise.

between different areas could change if vital links between them became broken by the appearance of new barriers, or if new links appeared.

Plate tectonics

Overwhelmingly the most important factor in causing major, long-term changes in the patterns of distribution of organisms has been the slow alteration in the geography of the world that used to be called continental drift, but is now known as plate tectonics. This not only affects the distributional patterns directly, by the splitting and collision of land masses and their movement across the latitudinal bands of climate; it also affects them indirectly, as new mountains, oceans or land barriers deflect the atmospheric and oceanic circulations, so changing the climatic patterns upon the land masses.

The motive force for plate tectonics is sea-floor spreading, caused by great convection currents that bring heated material to the surface from the hot interior of the earth. Where these upward currents reach the surface at the floors of the oceans, chains of underwater volcanic mountains form. These are known as 'spreading ridges', because new sea floor is formed there and is moved progressively away from the spreading ridge as more new material is steadily produced there. The age of the ocean floor is therefore progressively greater as one moves, to one side or the other, further away from the spreading ridges. Old ocean floor is consumed at the deep troughs or 'trenches' around the edges of the Pacific Ocean, where it disappears back into the earth (Fig. 6.1). Where these lie adjacent to the edge of a continent, the material of which is too light to be drawn back into the earth, the volcanism caused by the inwards movement of old ocean floor raises mountain chains such as the Andes of today.

Where spreading ridges lie within the oceans, their activity will cause continents to move apart, and ultimately may cause them to collide with one another, the collision raising mountains such as the Himalayas and Urals. A ridge may also gradually

Era	Period	Epoch	Approximate duration in millions of years	Approximate date of commencement in millions of years BP	
Cenozoic	Quaternary	Pleistocene	2.4	2.4	
	Tertiary	Pliocene	2.6	5	Millions of years ago
		Miocene	18	23	
		Oligocene	12	35	
		Eocene	21	56	50
		Paleocene	9	65	
Mesozoic	Cretaceous		81		100
				146	150
	Jurassic		62		
				208	200
	Triassic		37		250
				245	
Palaeozoic	Permian		45		
				290	300
	Carboniferous		72		350
				362	
	Devonian		46		400
	Silurian		32	408	
				440	450
	Ordovician		75		
				510	500
	Cambrian		60		550
				570	
Proterozoic			4000		
					4600

Formation of the Earth's crust about 4600 million years ago

Fig. 6.2 The geological time scale.

elongate and extend under a continent; its activity will then cause the gradual rifting apart of those regions of the continent that lie on either side of the spreading ridge. As these move apart, they become separated by a new, widening ocean, the floor of which is similarly moving to one side or the other, away from the spreading ridge. The moving units at the surface of the earth are therefore areas that may contain continental masses, or that may consist only of ocean floor. These units are known as 'tectonic plates', and earth scientists today refer to 'plate tectonics' rather than to 'continental drift'.

The movements of the continents are quite slow —only about 5–10 cm per year. These movements must have affected life in several ways, even though the changes must have been incredibly gradual, and noticeable only over a period of millions of years. The most obvious change would have resulted directly from the movements of the continents relative to the poles of the earth, and to the Equator. As they moved, so the different areas of land would have come to lie in cold polar regions, in cool, damp temperate regions, in dry subtropical regions, or in the hot, wet equatorial regions.

The true edges of the continents are marked, not by their coastlines, but by the edge of the continental shelf. Between the continental shelves, the deep oceans separate the continental plates. Sometimes the whole of these plates has been above sea level. At other times comparatively shallow 'epicontinental' seas have covered the edges of the continents (for example the North Sea today) or formed seas within the continents (such as Hudson's Bay today). Although the extents of these shallow seas have varied over geological time (for geological time scales see Fig. 6.2), they must have formed a barrier to the spread of organisms, and they are shown in the palaeogeographical maps (Figs 6.3, 6.4). They were particularly extensive in the Jurassic and the Cretaceous Periods of geological time (Fig. 6.2).

Drift also affected the climates of the continents in other, less direct ways. The climate of any area is controlled by its distance from the sea, which is the ultimate source of rainfall. The central part of great supercontinental land masses, such as Eurasia today or Gondwana in the past, is therefore inevitably dry, and the climate of such areas is monsoonal (see p. 86), experiencing great daily and seasonal changes of temperature. The break-up of a supercontinent, or the spread of shallow epicontinental seas into the interior of continents, would have brought moister, less extreme climates to these regions. The extent of these seas may also have been at least partly the result of plate tectonic activity. There appears to be a correlation between the total length of the system of oceanic spreading ridges and the extent of the transgression of seas over the continents. Because the ridges are formed of high underwater mountain chains, an increase in their length will decrease the capacity of the ocean basins and cause an increase in the areas of the continents that are covered by epicontinental seas.

The pattern of ocean spreading ridges, as well as the resulting splitting and movement of the continents themselves, will also have affected their climates by altering the patterns of water circulation in the oceans. For example, the cold Benguela Current up the western side of southern Africa, the warm clockwise Gulf Stream of the North Atlantic, and the southward movement of cold, deep water from the Arctic Ocean into the North Atlantic, all result from the present pattern of continents, and all affect the climates of the adjacent continents.

The different continental patterns of the past would therefore have resulted in different patterns of ocean circulation, and different climates.

Furthermore, the distribution of climate within the continental masses must also have been affected by the appearance of new mountain ranges as a result of plate tectonics. These would have had particularly great effects on the climate of the continents if they arose across the paths of the prevailing moisture-bearing winds, since areas in the lee of the mountains would then become desert. These can be seen today in the Andes, to the east of the mountain chain in southern Argentina, and to the west along the coast from northern Argentina to Peru; the winds in these two regions blow in opposite directions. A huge area of desert, including the arid wastes of the Gobi Desert of outer Mongolia, has also formed in central Asia, far from seas from which winds could gain moisture to fall as rain.

The evidence for past geographies

The simple pattern of age banding of the ocean floor that results from the process of sea-floor spreading provides direct evidence of the positions of the continents over the last 180 million years, to the middle of the Jurassic Period. By removing from the map any ocean floor younger than, for example, the end of the Cretaceous Period, one can return the continents to their positions at that time. However, all ocean floor older than 180 million years has been consumed, and the positions of the continents before that time therefore have to be deduced from a different line of evidence—palaeomagnetism. This relies on the fact that many rocks contain particles of iron-containing minerals. As the rock cooled after its deposition, these particles became aligned along the lines of the then-prevailing direction of the earth's magnetic field, like tiny compass needles. It is therefore possible to calculate how far from the magnetic pole the rocks lay when they were deposited, and in which direction the pole lay. This indicates the orientations of the continents and their north–south positions (but not their

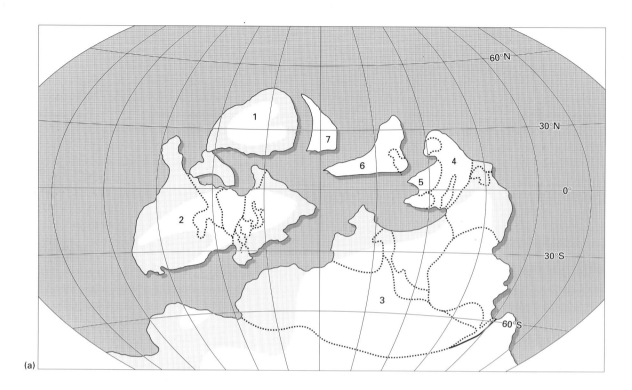

(a)

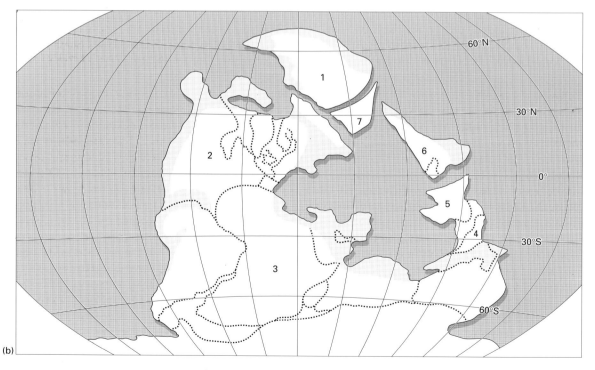

(b)

(c)

Fig. 6.3 (*Above and facing page.*) World geography in:
(a) the Early Devonian, about 400 million years ago;
(b) the Late Carboniferous; (c) the Late Permian.
Continental positions after Cambridge Palaeomap
Services [43]. Tripel–Winkel projection. Dark tint
indicates ocean, light tint indicates epicontinental seas.
Dotted lines indicate the future coastlines of the modern
continents; outlines of land areas after Stanley [44].
1, Siberia; 2, Euramerica; 3, Gondwana; 4, South-East Asia;
5, southern China; 6, northern China; 7, Kazakhstan.

east–west, longitudinal positions). Other evidence
for the past positions of the continents can be gained
from the types of rocks that were laid down within
them at that time (e.g. desert sandstones or glacial
deposits). The former patterns of union of continents can also be deduced from matching the
sequences of rock types or rock ages, and from
dating the times of the rise of mountain chains that
mark their collision.

But the evidence for past geographies does not
only come from physical geology, for palaeobiogeography has always played an important role,
and was particularly important in the formulation of the original theory of continental drift.
The realization that the distributions of Permo-
Carboniferous plants in the Southern Hemisphere,
and of various Mesozoic reptiles on either side

of the South Atlantic, did not make simple sense
in the context of modern geography helped to con-
vince the German scientist Arthur Wegener that
the continents must have moved [1]. Many palaeo-
biogeographers believed in the correctness of
Wegener's theory (first published in 1912), long
before the new sea floor and palaeomagnetic data of
the 1960s convinced the geologists. Nevertheless,
the degree of detail provided by the geophysical
data is now on the whole greater than that of the
palaeobiogeographical evidence, so that the latter
more often has only a confirmatory role. The geophysical data, by establishing the times at which
land masses split or united, have also identified
the units of time and geography within which it
is appropriate to make palaeobiogeographical analyses [2]. Until then, such analyses often made

little sense, for they frequently combined units of time within which major changes of geography had taken place, or combined geographical areas in inappropriate patterns.

Today, we can see that patterns of distribution of plants in the Permo-Carboniferous, and of vertebrates in the Cretaceous conform closely to the patterns of land and sea that are deduced by geologists when the continents are placed in the locations suggested by the geophysical data, and the outlines of shallow seas are added (see below). Fossil marine faunas provide evidence on the development of faunal provinces on either side of widening oceans (p. 254). The shapes of the fossil leaves of flowering plants indicate the climatic regime of the areas they inhabited, as do the types of plant themselves (see below), and their pollen also provides information on climatic change during the Ice Ages (p. 179).

Although palaeobiogeography therefore often plays a subordinate role, there are also situations in which its data provide more direct, detailed evidence than the geophysical record. For example, both the fossil marine faunas on either side of the Panama isthmus (p. 243), and the fossil mammals of North and South America (p. 152), provide clearer and more reliable evidence of the date of the final linking of those two continents than does any geophysical data from the area. Because sea-floor spreading does not provide data on the presence or spread of epicontinental seas, palaeobiogeography also provides crucial evidence on the times during which these seas separated areas of land.

Changing patterns of continents

A major feature of the geography of the Late Palaeozoic and the Mesozoic (Fig. 6.3) was the great supercontinent we call Gondwana. It has long been known that this included the land areas that later became South America, Africa, Antarctica, Australia and India. Asia, as we know it today, did not exist. It is a great jigsaw that gradually came together, between 400 and 200 million years ago, from two major land masses (Siberia and Kazakhstan), that

were joined by a number of smaller fragments (including North China, South China, Tibet, Indo-China and South-East Asia) which split off from the northern edge of Gondwana and moved northwards.

By the Silurian, 435–410 million years ago, when complex living organisms first colonized the land, North America and northern Europe had formed a second major land mass, Euramerica (see Fig. 6.3a). Euramerica, Gondwana, Siberia, Kazakhstan and Tibet joined together in the Late Carboniferous to Early Permian, 300–270 million years ago (see Fig. 6.3b), and Indo-China and the two portions of China joined this enormous land mass in the Late Permian, 260 million years ago, to form the single world continent we call Pangaea (see Fig. 6.3c). Throughout this period of time, Gondwana had gradually been moving across the South Pole.

However, it was not long before Pangaea started to become divided. Gondwana separated from the more northern land mass, which comprised North America and Eurasia, and which is called Laurasia. While Gondwana started to break up into separate continents, shallow epicontinental seas penetrated across Laurasia. Both these processes started in the Jurassic and continued throughout the Cretaceous Period (Fig. 6.4). In the Late Cretaceous (Fig. 6.4c) Europe was still separated from Asia by a sea known as the Obik Sea, and the Mid-Continental Seaway completely bisected North America into eastern and western land areas. The western part was connected to Asia via Alaska and Siberia to form a single 'Asiamerican' land mass, and the eastern portion was connected to Europe via Greenland to form a 'Euramerican' land mass. South America finally became separated from Africa in the middle Cretaceous. India separated from the rest of Gondwana in the Early Cretaceous.

It was not until the Early Cenozoic that the geography of the world became similar to that we see today (Fig. 6.4d). In the Eocene, Australia finally separated from Antarctica, and India became united with Asia. At about the same time, Europe became separated from Greenland, and the drying of the Obik Sea finally made Europe continuous with Asia. Although the continents of Africa and Eurasia had for long been close together, shallow seas separated their land areas from one another until the Miocene. The final link between South America and North America did not form until the Late Pliocene, about three million years ago.

Early land life on the moving continents

Our understanding of the movements of the continents and of the timing of the different episodes of continental fragmentation or union is now fairly detailed, and the distribution of fossil organisms correlates very well with the varying patterns of land.

The earliest time at which there is enough evidence to discern patterns of life is the Early Devonian, 380 million years ago (see Fig. 6.3a), by which time separate floras and fish faunas can be distinguished in the northerly placed Siberian continent, the equatorially placed Euramerican continent and the eastern region of Gondwana [3,4]. Early amphibians are found in near-equatorial positions in both Euramerica and Australia [5,6], where the climate would have encouraged a rich growth both of plants and of the invertebrates that the earliest terrestrial vertebrates would have fed upon. The fossil record suggests that all the early amphibian and reptile groups were largely confined to a narrow warm, humid equatorial zone, bordered by dry subtropical belts, until the Late Permian [7].

The great expansion of the land plants began in the Devonian, and this may well itself have had an effect on the world's climate. Today, vegetated surfaces decrease the albedo (or reflectivity) of an area by 10–15%, while plants recycle much of the rainfall (up to 50% in the Amazon Basin). Theory suggests that the photosynthetic activities of the plants would have reduced the carbon dioxide content of the atmosphere, and there is also both botanical evidence [8] and geological evidence [9] that this took place. This would have led to an 'icehouse' effect (the reverse of the 'greenhouse' effect of higher temperatures, that results from *increased* carbon dioxide). Therefore, this enormous build-up of the world's vegetation was the cause of the next

significant event, which was the onset of global cooling.

This cooling began in the middle of the Carboniferous Period, and led to the appearance of ice sheets around the South Pole, similar to those of Antarctica today. As Gondwana moved across the South Pole, the glaciated area moved across its surface. Although the whole South Polar area must have been icy cold, the proportion that was actually glaciated probably varied according to the position of the Pole itself. When this was near to the edge of the supercontinent, the adjacent ocean would have provided enough moisture to create the heavy snowfalls that would have formed extensive continental ice sheets. But when the Pole lay further inland, away from the ocean, it is likely that the inland areas, although cold, would not have been heavily glaciated.

This polar glaciation caused the latitudinal ranges of the Carboniferous floras to be compressed towards the Equator. In the equatorial region there was a great swampy tropical rainforest, rather like that of the Amazon Basin today. This lay across Euramerica and was fed by rains from the warm westward equatorial ocean current that would have washed the eastern shore of that continent. Surprisingly, the distribution of different types of rock laid down at this time shows no sign of the presence of monsoonal, dry conditions in the interior of the great supercontinent, and the climatic pattern seems to have been essentially latitudinal [10]. The equatorial wet belt was bordered to the south by a subtropical desert that stretched across northern South America and northern Africa; beyond this lay the cold lands around the glaciated South Pole. Another desert belt covered northern North America and northeastern Europe, but Siberia (still a separate island continent) lay in a wetter zone further to the north.

The absence of dormant buds and of annual growth rings in the fossil remains of the equatorial coal-swamp flora indicates that it grew in an unvarying, seasonless climate. The flora was dominated by great trees belonging to several quite distinct groups. *Lepidodendron*, 40 m tall, and *Sigillaria*, 30 m tall, were enormous types of lycopod (related to the tiny living club-moss, *Lycopodium*). Equally tall *Cordaites* was a member of the group from which the conifer trees evolved, and *Calamites*, up to 15 m high, was a sphenopsid, related to the living horsetail, *Equisetum*. Tree ferns such as *Psaronius* grew up to 12 m high, and seed ferns such as *Neuropteris* were among the most common smaller plants living around the trunks of all these great trees. In the eastern United States, and in parts of Britain and central Europe, the land covered by this swamp forest was gradually sinking. As it sank, the basins that formed became filled with the accumulated remains of these ancient trees. Compressed by the overlying sediments, dried and hardened, the plant remains have become the coal deposits of these regions.

Far to the south of the equatorial coal-swamp flora, the lands around the growing South Polar ice sheets bore a different flora, lacking many of the northern trees and with fewer ferns and seed ferns. It is known as the *Glossopteris* flora, after a genus of woody shrub that is found only in that region. Surprisingly, this flora is found as close as 5° to the then South Pole, in a region that must have been not only cold but also with a very short winter daylength. The very marked seasonal growth rings of plants from these regions are therefore not surprising, although the size of these is unexpectedly great from so high a latitude.

After the Carboniferous gave way to the Permian, the coal swamps of southern Euramerica disappeared and, by the Late Permian, deserts lay in their place. This was partly because these regions had moved northwards, away from the Equator, and partly because the mountain ranges of northern Africa and eastern North America had extended and risen higher, blocking the moist winds from the ocean that lay to the east.

Four different Permian floras can be distinguished [11]. To the north of the northern subtropical desert lay the temperate 'Angara' flora, with *Cordaites*-like conifers, and also herbaceous horsetail plants, ferns and seed ferns. The flora is richest towards the eastern coast, and becomes less diverse towards the colder north of Siberia. The second flora is the rich, varied, ever-wet tropical 'Cathaysian' rainforest flora, with sphenopsid and *Lepidodendron* trees, *Gigantopteris* lianas and many types of seed fern. Conifers and *Cordaites* were rare in this flora, which is found along the eastern margins of Euramerica and in the land masses that were moving from Gondwana towards Asia, across the tropical ocean. The seasonal tropical Euramerican flora occupied the rest of Euramerica. Finally, the Gondwana flora, descendant from the earlier Late Carboniferous *Glossopteris* flora, occupied the whole of that great supercontinent. It was almost certainly divided up into subsidiary floras according to the varying climates of the different parts of that great region, but as yet too little is known of it for us to be able to recognize these.

In the Late Permian the equatorial belt narrowed and the subtropics expanded, while the polar ice caps disappeared, so that warmer environments spread towards the poles. These environmental changes may have been the stimulus to the faunal changes that took place at this time. Many new groups of land vertebrate evolved and spread widely through the world, while many of the original equatorial groups became extinct [7]. Not surprisingly, land vertebrates did not reach Siberia or China until after those land masses had joined the world supercontinent in the Late Permian.

In the Late Permian, as in the Late Carboniferous, evidence for the presence of arid areas in the centre of the mid-latitudinal regions of the supercontinent is lacking. In fact, rich faunas of Late Permian fossil reptiles have been found in regions of southern South America and Africa that, according to climatic modelling experiments, would have had annual temperature changes of 40–50°C—similar to those of Central Asia today, and not very congenial to reptiles [12]. However, many of the rivers drained internally into great lakes, whose waters would have had a cooling, stabilizing effect on the climate.

One world—for a while

The coalescence of the different continental fragments to form Pangaea had geographical effects and also consequential climatic effects. In general, the world became steadily warmer and drier during the Triassic. The disappearance of the oceans that had previously separated the continents, and the formation of the enormous supercontinent, had left vast tracts of land now far from the oceans and the moist winds that originated there. Furthermore, the new, lofty mountain ranges that still marked the regions where Euramerica, Siberia and China had collided provided physical and climatic barriers to the dispersal of their faunas and floras. Although one would in any case have expected the Permian floras described above to change through time, these barriers made it more likely that any changes would take place independently in each, rather than these floras gradually blending. Therefore, we can identify a general evolutionary change, in which older types of tree, such as those belonging to the lycopods and sphenopsids, and *Calamites*, disappeared. They were replaced either by the radiation of existing types of tree, such as caytonias and ginkgos, or by new groups such as cycads and bennettitaleans. A new type of fern, the osmundas, also appeared. This floral change was complete by the end of the Triassic Period. In addition, we can also identify changes within the floras. This is most clearly seen in the Gondwana flora, where in the Early Triassic *Glossopteris* itself was replaced by the seed fern *Dicroidium*.

In the Jurassic and Early Cretaceous, floras seem to have gradually become more similar to one another, approaching the modern pattern in which there are gradual latitudinal changes governed by climate, and these manifest themselves as changing patterns in the dominance of different groups as one moves from lower latitudes to higher latitudes. For example, in the Jurassic and Early Cretaceous, two floral provinces can be distinguished in the northern land mass, Laurasia [13] (see Fig. 6.4a). The Northern Laurasian (or 'Siberian-Canadian') province had a lower floral diversity, with more deciduous plants such as conifers, and with evidence of annual growth rings, while the Southern Laurasian (or 'Indo-European') province does not show these features. From the mid-Jurassic onwards there was a floral change that was more marked in the Southern Laurasian province, and which may have been linked to increasing aridity. Several types of fern and of sphenopsid became extinct, and the cycads became rare. Among the conifers the Cheirolepidiaciae became important, and the araucarias were numerous in the moister, low to middle latitudes.

The southern boundary between the Northern and the Southern Laurasian floral provinces in Asia shifted northwards between the Late Jurassic and the Early Cretaceous. This seems to have been caused by a climatic change, the southern part of Asia becoming increasingly hot and arid. The change is sometimes quoted as evidence for a general increase in global warmth. However, there is no evidence for this elsewhere, and it seems instead to reflect a more local phenomenon, perhaps related to changes in the temperatures of the ocean currents adjacent to southern Asia, and therefore to the rainfall of this area. These Asian Mesozoic floras extended to high (about 70°) northern and southern latitudes, to areas that, although clearly warm, must have had seasonal very brief periods of daylight [14].

To outline now the biogeographical history of the vertebrate land animals (amphibians and reptiles) of Pangaea, one must return to the Permian and Triassic. By the middle of the Permian, these animals appear to have been quite competent at dispersing through regions of different climate, and Pangaea soon came to contain a fairly uniform fauna, with little sign of distinct faunal regions [15].

Great changes took place in the worldwide fauna during the Triassic [16]. The bulk of the Permian faunas had been made up of mammal-like reptiles and other older types of reptile. These disappeared during the Early and Middle Triassic, and were at first replaced by a radiation of early reptiles known as archosaurs. These in turn were soon replaced (in the Late Triassic) by their own descendants, the dinosaurs, which came to dominate the world throughout the Jurassic and Cretaceous. Although comparatively little is known about the Jurassic dinosaurs, it is enough to show that they were

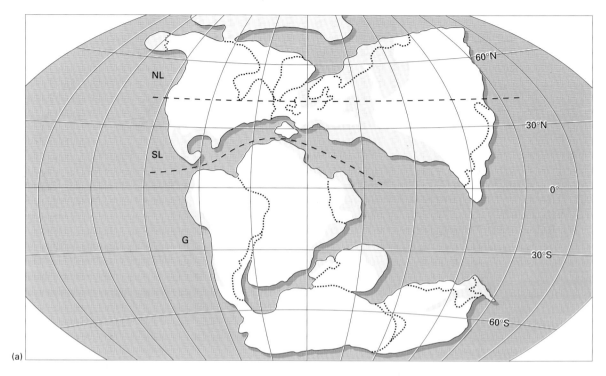

(a)

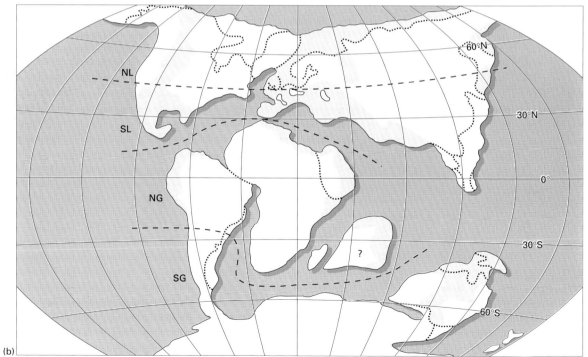

(b)

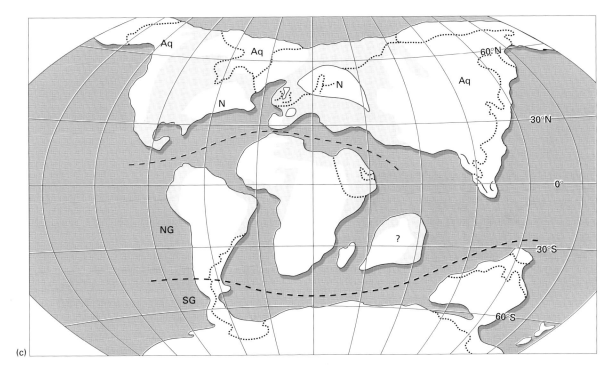

(c)

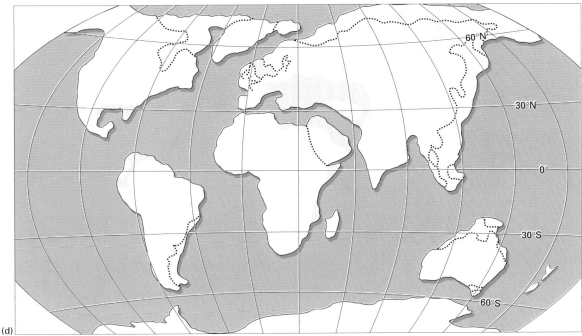

(d)

Fig. 6.4 (*Above and facing page.*) World geography in (a) the Early Cretaceous, about 140 million years ago; (b) the middle Cretaceous, about 105 million years ago; (c) the Late Cretaceous, about 90 million years ago; (d) the Eocene, about 40 million years ago. Tripel–Winkel projection; outlines of epicontinental seas (light shading) after Smith *et al.* [45]. Dotted lines indicate modern coastlines. Continental positions after Cambridge Palaeomap Services [41] Floral provinces in (a)–(c) after Crane [26]. Aq, *Aquilapollenites*; G, Gondwanan; N, *Normapolles*; NG, Northern Gondwanan; NL, Northern Laurasian; SG, Southern Gondwanan; SL, Southern Laurasian? refers to the fact that the affinities of the Cretaceous flora of India are unknown.

able to spread throughout the world. Their route between North America and Asia must have been via Alaska and Siberia, which still had quite mild climates (see below). That the dinosaurs were also able to reach Gondwana is shown by the similarities between the Jurassic dinosaur faunas of North America and East Africa. In these two areas are found not only the same families of dinosaur, but also in some cases the same genera, for example the sauropod dinosaurs *Brachiosaurus*, *Bothriospondylus* and *Barosaurus*, and the ornithopod *Dryosaurus*. Although the position of the land connection between Gondwana and the north is unknown, it seems likely that it was via South America.

The distribution of the new types of dinosaur that evolved during the Cretaceous mirrors the palaeogeographical maps very closely. Most of those that evolved during the Early Cretaceous (ostrich dinosaurs, dome-headed dinosaurs, dromaeosaurs and primitive duck-billed dinosaurs) dispersed throughout the Northern Hemisphere, which was still undivided by seaways (see Fig. 6.4a). Those types that evolved later, near the end of the Cretaceous, however, showed a more restricted pattern of distribution: tyrannosaurs, protoceratopsids and advanced duck-billed dinosaurs are all found only in Asiamerica (Asia plus eastern North America), the new north–south seaways having prevented them from reaching Euramerica (Europe plus western North America) [2] (see Fig. 6.4c). The rhinoceros-like ceratopsians, which only evolved in the very latest Cretaceous in North America, did not even reach Asia, which suggests that a marine barrier had by then separated the two continents.

There is little evidence for the Late Cretaceous groups in Gondwana, although this is partly because much less has so far been discovered of the Cretaceous record of the Southern Hemisphere. In the Late Cretaceous, duck-billed dinosaurs, sauropods and a possible ceratopsian are known from South America. As we shall see in the next chapter, there does appear to have been a number of islands between North and South America in the Late Cretaceous, and these must have provided a filter route for the exchange of organisms between the two continents. Although the exchange also included some mammals, as we shall see below, the link cannot have been a full link similar to the Panama Isthmus today, as there was no wholesale exchange of faunas.

The early spread of mammals

The first, most primitive, mammals appeared in the Triassic Period, not long after the first dinosaurs, but they almost certainly laid eggs, as do the living monotremes (the platypus and spiny anteater) of Australia. The two major modern groups, the marsupials and the placentals, did not appear until the Late Cretaceous. In the marsupial type of mammal the young leave the uterus at a very early stage and then complete their development in the mother's pouch, whereas in the placental mammals the whole period of embryonic development takes place in the uterus. Although the placentals spread to all parts of the world except Australia (see below), the marsupials were only successful in the Southern Hemisphere. Unfortunately, our understanding of the patterns of distribution of these two groups is hampered by our incomplete knowledge of the timing of some crucial plate-tectonic events and of the composition of some early faunas.

The sea-floor spreading record provides reliable data on the timing of some plate-tectonic events. For example, we know that the last land contact between South America and Africa finally broke at the end of the Early Cretaceous, about 110 million years ago. Similarly, although sea-floor spreading between Antarctica and Australia commenced in the Early Cretaceous, final separation between the two continents was not until the Late Eocene, 40 million years ago. India had parted company from Gondwana much earlier, in the Early Cretaceous, 118 million years ago. In some areas, however, complex plate-tectonic events were taking place and clear sea-floor spreading data are absent. This includes the regions that link South America northwards to North America and southwards to Antarctica, so that the exact timing of the separations of these continents is uncertain. A further plate-tectonic complication is that Antarctica was at that time made up of two separate plates. West Antarctica, which lay adjacent to South America, was itself made up of a series of microplates, and East Antarctica was still joined to Australia. The fossil record, too, is exasperatingly silent on the mammal faunas of Australia before the Oligocene, while little is known of the Cretaceous floras of India. We therefore have to try to piece together a plausible biogeographical history of the mammalian groups that is compatible with what is known, while acknowledging the underlying uncertainties.

The earliest marsupials and placentals are known from the Late Cretaceous of Asia, and some of them, like the contemporary Asian dinosaurs, spread into western North America. From there, the ancestors of the living groups of marsupials reached South America (presumably over the same route used by the Late Cretaceous invasion of dinosaurs from North America to South America), and thence via Antarctica (which was still forested and ice-free at that time) to Australia [17]. Placentals, too, reached South America, where they coexisted with the marsupials—but neither of these groups were able to reach Africa or India in the Cretaceous, for both those continents had already separated from the rest of Gondwana. (Marsupials reached Europe via Greenland (see p. 156) in the Eocene and, from there, northern Africa, but became extinct on both those continents.)

There is still a major uncertainty over whether placental mammals ever reached Australia and, if not, why not. They were certainly absent from that continent in the Late Tertiary and the Quaternary (see Fig. 6.1), until rats and bats entered from South-East Asia in the last few million years, and humans introduced the dingo and other domestic animals quite recently. Unfortunately, the Early Tertiary deposits of Australia are almost entirely lacking in fossils, so that they do not provide any evidence of the Australian mammal fauna at that time. The resulting lack of any evidence of placentals in the early history of Australia has led many biogeographers to assume that placentals reached South America later than the marsupials and that, by that time, some kind of sea barrier had arisen that prevented placentals from reaching Australia.

Such a barrier would have had to have existed from some time in the Palaeocene, when placentals are known in South America, until the Eocene, when Australia separated from Antarctica. It might have lain in the southernmost part of South America (for the Andes had not yet started to rise), or in West Antarctica, which at that time was made up of a series of microplates which might have formed individual islands separated by sea. A few mammalian fragments have been found in the Late Cretaceous or Early Tertiary of Australia during the 1990s [18,19], and hailed as early placentals. However, all of them may belong to early types of mammal, that were neither placental nor marsupial, and that were widespread in the world in the Cretaceous. Therefore, until more convincing evidence is found, it seems best to continue to accept the above theory, that placentals were originally absent from Australia, despite all its uncertainties.

The great Cretaceous extinction event

The diversification of both the marsupials and the placentals was only made possible by the sudden disappearance of the dinosaurs, which had dominated the world since the Late Triassic, over 140 million years before. But the dinosaurs were not the only group to disappear at that time, for they were accompanied into extinction by other reptile groups, such as the flying pterosaurs and the great marine reptiles (the plesiosaurs and mosasaurs). The ocean biota also lost the ammonoids, which had been in existence for 250 million years, as well as many types of marine plankton and of such invertebrate groups as reef-building corals and the bivalved, mollusc-like brachiopods. It has been estimated that 75% of all the species, and 25% of all families, of animal became extinct at that time. In some groups, such as the plankton and brachiopods, there was a later wave of replacement by new types, but other groups had disappeared for ever. What could have caused the simultaneous extinction of vertebrates and invertebrates (but not all of either group) on land, in the air and in the oceans?

Although many explanations have been put forward as to the nature of this sudden event, more and more evidence is now accumulating that it was the result of the impact of an enormous comet or meteor, about 10 km (6 miles) in diameter. The first evidence of this was the discovery by father and son Walter and Luis Alvarez, of Berkeley University, California, of unusually high levels of 'rare-earth' minerals, such as iridium and osmium, in a thin layer of clay laid down at the K/T boundary at Gubbio in Italy [20]. Significantly, these minerals were also present in precisely the proportions in which they are found in some types of meteorite. Later research has shown that this thin clay layer, together with the minerals, is found in over 100 locations as far apart as North America, New Zealand, the South Atlantic and the North Pacific, so it was clearly deposited worldwide, rather than being merely a local event. The Alvarezes suggested that these deposits were the result of the impact of an enormous meteorite or comet, and this was eventually confirmed by the discovery of a great crater at Chicxulub in northern Yucatan, Mexico, which was made 64.5 million years ago—exactly the time of the extinctions. The crater is 200–300 km (125–190 miles) in diameter, and was made by a meteorite at least 10 km (6 miles) in diameter, and which impacted at approximately 32 000 kph (20 000 mph).

The shock wave from the meteorite impact would have had a devastating influence on the forests and larger animals over a very wide area. A fireball, many hundreds of kilometres across, would have formed, and burning hot winds would have swept around the world. This, together with the red-hot ejecta from the crater, would have started forest fires in many areas. The heat would have produced quantities of nitric acid in the atmosphere, which would have produced acid rain, defoliated many plants, and reduced the ozone content of the atmosphere and so allowed more of the damaging ultraviolet radiation to penetrate to the earth's surface. The K/T boundary clay also contains large quantities of soot—so much that it has been suggested that 90% of the world's forests may have burned, releasing quantities of pyrotoxins into the atmosphere.

Studies of the sediments at the Chicxulub crater suggest that between 1000 and 4000 km³ of rock were thrown up by the impact. The minerals of the rocks where the comet struck contained carbonates and sulphates; their heating in the impact would have released great quantities of carbon dioxide and sulphur dioxide into the atmosphere, as well as water vapour from the heated ocean. These would have had profound and complicated effects upon the world's weather [21].

The sulphur dioxide and water vapour would have combined in the atmosphere to form a sulphuric-acid aerosol. Together with the smoke from the forest fires, this would have darkened the skies for about 6 months, so interfering with plant photosynthesis and disrupting the world's ecosystems, and also causing an initial global cooling of about 10°C. The resulting 'ice-house' climate of the earth probably lasted for 1 or 2 years. But, after this, the particles of limestone rock that had been thrown into the atmosphere would have increased the amount of carbon dioxide there, which would have led to a greenhouse effect, increasing the temperature by 3–10°C. These conditions may have lasted for several tens of years, until the carbon dioxide could be absorbed by the oceans.

These deductions are supported by botanical studies that show a sudden reduction in the amount of flowering plant pollen at the K/T boundary in North America, from 65 species below it to only eight above (although, of course, the effects may have been less dramatic in other parts of the globe, further from the impact site). Above the soot–rich boundary clay in North America lies a layer that suggests the presence of rotting or burnt vegetation, with no pollen, and then a sudden, short-lived increase in the amount of fern spores [22]. This pattern of change is characteristic of the sequence of plant communities that in turn colonize an area after a forest fire. In the middle northern latitudes there was a particularly high rate of extinction

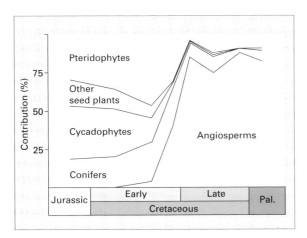

Fig. 6.5 The percentage contribution of the major plant groups to Jurassic, Cretaceous and Palaeocene (Pal) floras. Pteridophytes include ferns and lycopods; cycadophytes include cycads, bennettitaleans and pinnate-leaved seed ferns. After Crane & Lidgard [24]. Copyright 1989 by the AAAS.

in warmth-loving, broad-leaved, evergreen forest taxa, together with the survival and expansion of deciduous northern taxa—a pattern suggesting the extinction of taxa that were vulnerable to darkness or to cold, or to both. This abrupt climatic change was followed by a new, stable climatic regime that continued well into the Palaeocene.

The marine plankton record, too, shows signs of a temperature decrease at the K/T boundary (see p. 243), followed by climatic instability, while many types of plankton died out at this time, and marine productivity remained low for 0.5–1.0 million years [8]. Like the terrestrial plants, they would have been affected by the long darkness. Therefore, both on land and in the oceans, many types of plant died out, unable to adapt to the sudden climatic change, and the effects of this would have risen through the food chain. As plant food became scarce, larger herbivores would have become extinct, followed by the larger carnivores that preyed upon them. The death of the plant plankton would similarly have been followed by that of the animal plankton that fed upon them. On land, only the smaller, cold-blooded reptiles and the little mammals that

could feed on carcasses and invertebrates would have found it easy to survive the K/T catastrophe.

The rise of the flowering plants

Angiosperms are first known in the Early Cretaceous, 135 million years ago. Many modern angiosperm families are recognizable in the Northern Hemisphere by 95 million years ago, at the Early/Late Cretaceous boundary, so differentiation of the flowering plants seems to have been fairly rapid. Although our knowledge of their early distribution is heavily dependent on records from the lowlands of the Northern Hemisphere, the pattern consistently suggests that they first diversified and became dominant in low latitudes, up to 20° from the Equator [23–25]. The first angiosperms may have been early successional herbs or shrubs in areas recently disturbed by erosion and deposition, only later diversifying ecologically to occupy streamside and aquatic habitats, the forest understorey and early successional thickets [26]. Even at the end of the Cretaceous, although angiosperms formed 60–80% of the low-latitude floras, they comprised only 30–50% of those in high latitudes. The rise of the angiosperms was paralleled by a corresponding reduction in the numbers and variety of mosses, club mosses, horsetails, ferns and cycads, but there was less change in the overall diversity of conifers (see Fig. 6.5).

Understanding the dispersal of the flowering plants presents fewer problems than understanding that of mammals, for it commenced earlier, before the break-up of the supercontinents had progressed very far. If they first evolved and diversified in low, near-equatorial latitudes, the tropical flowering plants would have been able to spread eastwards through Africa and thence eventually into southern Eurasia. They would also have been able to spread northwards into North America and from there, once they had colonized the higher latitudes of the north, across the still dry and warm land bridges from North America to Europe via Greenland and Scandinavia and to Siberia via Alaska. Even in the Southern Hemisphere, those angiosperms that adapted to the high south-latitude cool-temperate region were able to disperse through the South America–Antarctica–Australia land mass before this became divided up [27].

In the middle Cretaceous, it is clear from pollen records that in Gondwana, as in Laurasia, separate Northern and Southern floral provinces had differentiated, so that broadly latitudinal floras now existed in both hemispheres (see Fig. 6.4b) [13]. These comprised a humid temperate Northern Laurasian flora dominated by the endemic conifer family Pinaceae and other conifers, with some ferns; a rather similar Southern Laurasian flora with more ferns and with conifers other than the Pinaceae; a tropical Northern Gondwana flora with many cycads and ephedras and few ferns; and a humid Southern Gondwana flora with many podocarp and araucarian conifers and many ferns. (Whether or not this last flora extended into southern Africa or India, already isolated from the rest of Gondwana by ocean, is unknown.)

This was the pattern of the floral provinces into which and through which the flowering plants spread and diversified, apparently beginning in the tropical regions of the Southern Laurasian and Northern Gondwana provinces. In the Southern Hemisphere, a unique Southern Gondwana angiosperm flora evolved, including mainly evergreen trees (such as the southern beech tree, *Nothofagus*), and various shrubs and herbs, which lived alongside the existing conifers and ferns. The total flora was very like that of New Zealand today, and covered the southern parts of South America plus Antarctica, Australia and New Zealand, but it is not known from southern Africa, by then separated from the rest of Gondwana by a wide stretch of ocean.

By the Late Cretaceous, the Northern Gondwana province contained a diversity of pollen of types now found in palms, but the floras of tropical South America and West Africa were becoming progressively more different as the two continents moved apart; they had become distinctly different by the Eocene. A more complicated pattern appeared in the Northern Hemisphere, where the latitudinal pattern of Northern Laurasian vs. Southern Laurasian floras was affected by the longitudinal division of the land mass into Asiamerica and Euramerica (see Fig. 6.4c). Pollen belonging to the group known as *Aquilapollenites* is found throughout the Asiamerican region, between the Obik Sea at the western end of Siberia and the Mid-Continental Seaway that bisected North America, but it is also found in the northern part of Euramerica. Pollen that belongs to the group known as Normapolles (which may be related to the hamamelid group of flowering plants, which includes the witch hazel) has a more restricted distribution, being found only in the warmer southeastern parts of North America and the scattered islands that then made up southern Europe. This flora contains a greater abundance of flowering plants than the *Aquilapollenites* flora.

Late Cretaceous and Cenozoic climatic changes

There is much evidence that the climate of the world in the Mesozoic was very different from that of today. Oxygen-isotope measurements of the composition of the fossil shells of marine Cretaceous plankton (see p. 176) show that the intermediate to deep waters of these oceans were 1.5°C warmer than those of today. On land the presence of plants, dinosaurs and early mammals in high latitudes, and their spread through high-latitude routes such as the Bering region and Antarctica, support these observations. But from the Late Cretaceous onwards the earth's climate changed in two distinguishable, but linked, ways: it became cooler, changing from a generally warm world to one with polar ice sheets, and it became more seasonal. Several different factors can be identified that have influenced these changes, but we still do not fully understand precisely how these factors combined to produce changes of the magnitude and pattern that our analyses reveal [21].

An important factor has been the changing patterns of land and water, in which both the comparatively shallow epicontinental seas and the deep oceans had effects on the climate. Water absorbs more heat than does the land, and releases it more gradually, thus slowing the intensity of any developing seasonal cycles. When, in the Early Cenozoic, the seaways that had subdivided North America and Asia withdrew, the climates of the central parts of those continents must have become less equable and more seasonal, with hotter, drier summers and colder, wetter winters. The deeper oceans not only affect the climates of adjacent land areas directly, but also affect climatic patterns more generally, for the ocean currents redistribute heated or cooled water masses, transporting them to other parts of the globe.

Both these oceanic influences are affected by the sizes and positions of the continents. For example, the widening of the gap between Australia and Antarctica which commenced 46 million years ago allowed the development of a great Antarctic circumpolar current. This current's persistently cold water has had a great influence on the origin and development of Antarctic ice and on the increasing aridity of Australia (see p. 149), because it has prevented the warmer waters of the South Pacific from penetrating further southwards. In contrast, although cold, deep Arctic water has been able to enter the North Atlantic since the opening of the Norwegian Sea between Greenland and Scandinavia in the Late Eocene, the lands surrounding the North Atlantic have been affected by the warm surface waters of the Gulf Stream, channelled northwards by the east coast of North America. The Kuroshio current in the western North Pacific produces a similar effect, warming Japan.

The changing distribution of land and ocean as Africa and India moved northwards towards and into Eurasia, shrinking and finally obliterating the oceans that had previously lain between them, is also likely to have affected the amount of solar radiation that the earth absorbed. This radiation is more effectively absorbed at low latitudes (where it is more vertical to the surface) than at high latitudes, and is more efficiently absorbed by water than by land. Therefore, the fact that, over the last 60 million years, the area of land in low latitudes has increased very greatly, especially in the Northern Hemisphere, is likely to have had a perceptible influence on temperature. Combined with the greater seasonality of the northern continents after the epicontinental seas withdrew, this may have been an important factor in causing the declining temperatures that culminated in the appearance and spread of ice sheets at high latitudes and altitudes in the Late Cenozoic.

Changes in the epicontinental seas will also have had more local effects. For example, the Late Miocene saw a gradual reduction in the size of a large sea that had previously covered much of western Central Asia, and which had brought milder climates to much of northern Eurasia. Its shrinkage led to increased summer temperatures there, and to increased seasonality in the Indian rainfall, with consequent replacement of tropical vegetation by savanna in parts of India [28]. This factor may have been more important than the rise of the Himalayas in affecting the climate of the region.

Another important factor was the rise of more mountain ranges during the Cenozoic. Mountains that rise 2–3 km above sea level not only themselves gather a covering of ice and snow, but also act as barriers to wind flow. They therefore cause the appearance of rain shadows in their lee, and also act as a dam to the movement of air that has been cooled in mid-latitude interiors in the winter. It has been suggested that the rise of the Tibetan Plateau, which covers over a million square kilometres, after the collision of India with Asia in the Eocene, may have been a major cause of the global cooling that took place at the Eocene/Oligocene boundary [29]. This cooling (Fig. 6.6a) led eventually to the appearance of polar ice sheets, commencing 35 million years ago in the Antarctic, and with a major increase at 30 million years ago; Northern Hemisphere glaciation does not seem to have commenced until within the last 10 million years. Ice reflects the sun's rays back into space, so the appearance of these ice caps will in itself have contributed to further climatic deterioration. The removal of this amount of water from the sea led to falling sea levels—in the mid-Oligocene, 30 million

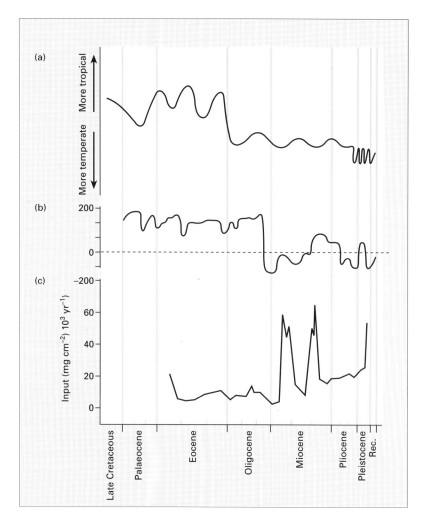

Fig. 6.6 The climatic record of the Late Cretaceous and Cenozoic. (a) Temperature, as suggested by floras from middle latitudes, after Wolfe [32]. (b) Generalized history of sea-level changes, relative to present day. (c) Amount of wind-borne dust in deep-sea sediments. (b) and (c) after Tallis [33].

years ago, sea levels dropped by at least 120 m (Fig. 6.6b). This, together with the cooling of the ocean itself, with consequently decreased evaporation to provide rain, also led to increased aridity of the continents, especially in mid-latitudes, which can be detected from the amount of wind-blown dust deposited in deep-sea sediments (Fig. 6.6c).

In North America, the Rockies started to rise in the latest Cretaceous to Middle Eocene, and reached half their present height 10–15 million years ago. The Sierra Nevada range began to rise in the latest Oligocene, but most of its rise took place over the last 10 million years. The Cascade and Coast ranges only rose over the last six million years. All these mountains will have increased the seasonality of the climate of North America, and it has also been suggested that, by diverting the westerly winds into a more northerly track, they may have played a part in initiating the climatic cooling that eventually led to the Ice Ages. The effects of the rise of the Andes in South America during the Late Cenozoic were less severe, because the continent as a whole is narrower and is mainly at a lower latitude.

Despite our recognition of all the above factors, and our ability nowadays to produce complex computer-generated models of the effects of alterations in their magnitude or location, it is still not possible to construct models that accurately match the high temperatures of the Cretaceous and Early Cenozoic, especially in the continental interiors. This at present can only be explained by making additional *ad hoc* assumptions, for example, higher solar output of radiation or higher levels of atmospheric carbon dioxide in these earlier periods, although rather limited evidence of these phenomena has yet been identified.

Late Cretaceous and Cenozoic floral changes

The most comprehensively-studied floras of this period of time are those of the North American continent, which are described in detail by the American continent, which are described in detail by the American palaeobotanist Alan Graham (see Further reading).

Although, as already noted, there is a variety of different sources of evidence as to the world's past climatic regimes, the fossilized leaves of flowering plants provide an additional source after they became common and diverse, in the middle Cretaceous. In areas of high mean annual temperature and rainfall the leaves have 'entire' margins, not subdivided into lobes or teeth; they are large and leathery, and are often heart-shaped, with tapering pointed tips, and a joint at the base of the leaf. These features are less common in floras from areas of low mean annual temperature and rainfall. Further information can be gained from the climate preferred by the living representatives of the families of plant found in each flora [30], though it would be unwise to assume that living species could never have existed in beyond the range of environmental conditions within which they are found today. They may in the past have lived in environments from which they later became excluded by competitive interaction with other, newly-evolved, types. In any case, many of the Cretaceous and Early Eocene flowering plants belonged to groups that later became extinct, or that were only transitional to modern genera, so that the climatic implications of their presence is

unclear. Finally, the fossil data are often biased towards near-coastal areas of deposition, whose climates are usually milder and less seasonal than those of more inland areas. At the community level, Alan Graham remarks that most of the modern plant biomes appeared comparatively recently, and that the fossil record shows that different combinations of ecologically compatible species can result from each successive climatic change.

The cooling of the world's climate began in the middle Cretaceous, and this is clearly shown in a series of floras, ranging over 30 million years of that time, from about 70°N in Alaska [31]. The earliest of these floras contains the remains of a forest dominated by ferns and by gymnosperms such as cycads, ginkgos and conifers. The nearest living relatives of this flora are found in forests at moderate heights in warm-temperate areas at about 25–30°N, for example in South-East Asia. By the time of the last of these middle Cretaceous Alaskan floras, the flora had changed in two ways. First, the angiosperms had by this time diversified to such an extent that they dominated the flora. Secondly, this forest contains the remains of forest similar to that found today at a latitude of 35–40°N in the region of North China and Korea—much further north than the living relatives of the earlier flora. The differences between these successive fossil floras therefore suggest that the climate of northern Alaska was already becoming cooler in the middle Cretaceous. Other floras from the same area show that, by the end of the Cretaceous, the mean temperature had dropped by about 5°C, and that the diversity of the flowering plants had dropped very greatly [32]. However, the climate of the main part of the North American continent appears to have changed less during this time [33]. In the eastern, Normapolles, region (see p. 128) there was tropical forest, especially in the coastal regions and northwards along the Atlantic margin. The floras of the western, *Aquilapollenites*, region ranged from subtropical forest in the south, via broad-leaved evergreen forest (including types now found exclusively in the Southern Hemisphere, such as *Araucaria*, evergreen Taxodiaceae, and *Gunnera*) in north-western

USA and south-western Canada, to broad-leaved deciduous forest in the polar regions. At the end of the Cretaceous there seems to have been a brief period of lower temperatures, perhaps associated with the extinction event at that time, followed by a long-lasting increase in rainfall in North America. All the known Early Cenozoic floras were dominated by woody trees and lianas, whose nearest living relatives mostly live in tropical, subtropical or deciduous temperate communities. Forests extended close to the poles in this warm, ice-free world [34]. It is therefore not surprising to find that taxa that are today restricted to particular latitudes were, in the Eocene, found much further from the Equator, or existed over a much wider band of latitudes. Tropical or subtropical floras are found in northern North America and Eurasia, extending to latitudes of 45–50°N. Most of the Early Cenozoic flowering plants were therefore what are today called megatherms and mesotherms (preferring, respectively, mean annual temperatures of above 20°C and between 20°C and 13°C); there were few microtherms (those preferring mean annual temperatures of below 13°C).

The climate improved again 50 million years ago, in the Early Eocene. This appears to have been caused by a sudden change in the temperature of the deep sea, for the American oceanographers James Kennett and Lowell Stott [35] have found that, around Antarctica, this increased at that time from 10°C to 18°C in less than 6000 years. They suggest that this change may have been due to a reversal of the normal pattern of oceanic circulation, in which cold water sinks at high latitudes and returns to the surface at low latitudes. But the possible cause of such a reversal is unclear, and it is also difficult to explain the fact that such a fundamental change in oceanic circulation lasted for only 30 000 years. An example of the Early Eocene warming is southern England, then at a latitude of about 45°N. The fossil seeds and fruits of 350 species of plant, belonging to over 150 genera, are preserved in the London Clay deposits near the mouth of the River Thames [36]. The flora includes magnolias, vines, dogwood, laurel, bay and cinnamon, as well as the palmtrees

Nipa and *Sabal* and the conifer *Sequoia*. Some of these plants are now found in the tropics, especially in Malaysia and Indonesia, while others live in temperate conditions like those of east central China today. The closest analogue today would be a subtropical rainforest but, its composition is not identical to that of any single modern flora. The accompanying fauna, which includes crocodiles and turtles, is similar to that found today in the tropics.

From the Middle Eocene to the Early Miocene, climates became cooler, drier and more seasonal. The plant biomes became more like those of today, both in their systematics and in their structural composition. For example, the warm-temperate deciduous forests of the lowlands of North America, and the coniferous forests of its highlands were similar to those of today. But there were also exceptions: for example, the tropical rainforest of south-eastern North America was replaced by a semi-deciduous tropical dry forest unlike anything that can be seen in the world today. Similarly, the forest of such high northern latitudes as Ellesmere Island in the Canadian Arctic (81°N) was unlike any modern forests. It included elements of more southern temperate broad-leaved deciduous forest, but also characteristically northern needle-leaved conifers such as pine, larch and spruce. The diverse fauna included mammals, snakes, lizards and tortoises [37,38]. Furthermore, it is now clear that some of the tropical elements found in the temperate deciduous forests of high latitudes in North America originated in the Old World tropics, and spread to North America via the Bering or Greenland connection. The American palaeobotanist Jack Wolfe has named this flora the 'boreotropical flora' [39].

In the Southern Hemisphere, the floras of South America, Antarctica and Australia still contained, as they do today, descendants of the old Late Cretaceous Southern Gondwana angiosperm flora, with such families as the protects, myrtles, *Nothofagus* and the 'southern conifers' such as *Araucaria*, *Podocarpus* and *Dacrydium*. These forests covered at least the periphery of Antarctica, but the inland flora of that continent is unknown.

After the Early Eocene climatic improvement, studies of the leaf characteristics of the Northern Hemisphere floras of the Cenozoic show that there were two episodes of cooling, with the appearance of some glaciers in Antarctica, and each followed by warmer conditions (although not to the warmth of the Early Eocene). A much more dramatic cooling took place in the Early Oligocene (see Fig. 6.6a), when significant areas of Antarctica became glaciated. The climatic change involved both a decrease in the mean annual temperature, and an increase in the mean annual range (i.e. a decrease in equability). For example, within one or two million years the mean annual temperature of the Pacific north-west of North America dropped from about 22°C to about 12°C, while the mean annual temperature range increased from about 7°C to nearly 24°C, and annual rainfall almost halved. In this short space of time the band of broad-leaved evergreen forests shifted southwards, from 40 to 60°N to 20–35°N, and were replaced by temperate broad-leaved deciduous forests and woodlands. It is not surprising to find that the floral diversity became much lower at this time, for many families must have become extinct, in these areas at least. The degree of cooling at this time in Europe was lower than in North America, and there is less evidence of drying, largely because most of the European floras of that time are known from areas not far from the coasts.

One result of the increasing seasonality of the climate was that the old closed evergreen forests became thinner and deciduous, providing an evolutionary opportunity for the appearance of new types of non-woody flowering plant; it is only in the Late Cenozoic that herb-dominated flowering plant communities became important. Another opportunity arose with the spread of arid environments. In the Late Eocene, the warm seas had easily evaporated to provide moist, rain-bearing winds, and there were not yet regional zones of arid and semi-arid climate in the subtropics, where today the high-pressure areas of descending dry air produce deserts (see p. 86), although there were deserts in the continental interiors. Even in the later Eocene,

the cooler drier climates had led to the first appearance of such biomes as chaparral-woodland-savanna in North America. After the Early Oligocene cooling, the colder seas provided less rain, and there was a great expansion of arid-land vegetation [40]. In Eurasia, this climatic change is accompanied by a marked change in the mammal faunas, to smaller types such as rodents and rabbits, and the old forest types become extinct [41].

The earliest dry, savanna vegetation appeared in low latitudes in the Early to Middle Miocene, followed by the first grasslands in mid-latitudes. The spread of silica-rich C_4 grasses in North America may have been the cause of the great wave of extinction of herbivorous mammals there, as those that did not have high-crowned teeth could not withstand the high rates of tooth-wear caused by such a diet. Temperatures dropped again in the late Middle Miocene, but high latitudes remained relatively warm, and mixed coniferous forest did not appear in the Arctic until the Late Miocene, five to seven million years ago, when there seems to have been a considerable increase in the area of high-latitude ice sheets. Tundra appeared in the Antarctic at that time, but it may not have appeared in the Arctic until later, in the Pliocene. However, it is still difficult to be sure of the timing and course of lowland glaciation in the Northern Hemisphere from seven to three million years ago, and it may well have varied from area to area [42]. But there is still no evidence for extensive tundra, grassland or desert at that time.

Summary

1 The patterns of distribution of animals and plants are mainly controlled by the patterns of geography—by the positions of oceans, shallow epicontinental seas, mountains and deserts. Because these were different in the past, mainly because of the movement of continents due to plate tectonics, the patterns of distribution of life in the past were different also.

2 Climate changes have also had a great effect on the patterns of distribution of living organisms. In the Late Cretaceous and Early Cenozoic, the world's climate was warmer than it is today. Many organisms were then able to spread via high-latitude routes that are now closed both by climatic changes and by the separation of continents.

3 These climatic changes also led to changes in the nature of the floras, as increasing aridity led to the replacement of forests by woodland and grasslands. These in their turn caused changes in the nature of the mammal faunas that lived in them.

Further reading

Prothero DR. *The Eocene–Oligocene Transition. Paradise Lost.* New York: Columbia University Press, 1994.

Culver SJ, Rawson PS (eds) *Biotic Response to Global Change: The Last 145 Million Years.* London: Natural History Museum, 1999.

Graham A. *Late Cretaceous and Cenozoic History of North American Vegetation.* New York & Oxford: Oxford University Press, 1999.

References

1 Wegener A. *The Origin of Continents and Oceans* (1966 English Translation of 4th edn.) London: Methuen, 1929.

2 Cox CB. Vertebrate palaeodistributional patterns and continental drift. *J Biogeogr* 1974; 1: 75–94.

3 Edwards D. Constraints on Silurian and Early Devonian phytogeographic analysis based on megafossils. In: McKerrow WS, Scotese CR, eds. *Palaeozoic Palaeogeography and Biogeography.* Geological Society Memoir no. 12, 1990: 233–42.

4 Young GC. Devonian vertebrate distribution patterns and cladistic analysis of palaeogeographic hypotheses. In: McKerrow WS, Scotese CR, eds. *Palaeozoic Palaeogeography and Biogeography.* Geological Society Memoir no. 12, 1990: 243–55.

5 Milner AR. Biogeography of Palaeozoic tetrapods. In: Long JA, ed. *Palaeozoic Vertebrate Biostratigraphy and Biogeography.* London: Belhaven, 1993: 324–53.

6 Thulborn T, Warren A, Turner S, Hamley T. Early Carboniferous tetrapods in Australia. *Nature* 1996; 381: 777–80.

7 Berman DS, Sumida SS, Lombard RE. Biogeography of primitive vertebrates. In: Sumida SS, Martin KLM, eds. *Amniote Origins: Completing the Translation to Land.* London: Academic Press, 1997: 85–139.

8 Chaloner WG, McElwain J. The fossil plant record and global climatic change. *Rev Palaeobotany Palynol* 1997; 95: 73–82.

9 Mora CI, Driese SG, Colarusso LA. Middle to Late Paleozoic atmospheric CO_2 levels from soil carbonate and organic matter. *Science* 1996; 271: 1105–7.

10 Witzke BJ. Palaeoclimatic constraints for Palaeozoic palaeolatitudes of Laurentia and Euramerica. In: McKerrow WS, Scotese CR, eds. *Palaeozoic Palaeogeography and Biogeography.* Geological Society Memoir no. 12, 1990: 57–73.

11 Ziegler AM. Phytogeographic patterns and continental configurations during the Permian Period. In: McKerrow WS, Scotese CR, eds. *Palaeozoic Palaeogeography and Biogeography.* Geological Society Memoir no. 12, 1990: 367–79.

12 Crowley TJ, Hyde WT, Short DA. Seasonal cycle variations on the supercontinent of Pangaea. *Geology* 1989; 17: 457–60.

13 Batten DJ. Palynology, climate and the development of Late Cretaceous floral provinces in the Northern Hemisphere: a review. In: Brenchley P, ed. *Fossils and Climate.* Geological Journal, Special Issue no. 11. London: John Wiley, 1984: 127–64.

14 Chaloner WG, Lacey WS. The distribution of Late Palaeozoic floras. *Spec Papers Palaeontol* 1973; 12: 271–89.

15 Cox CB. Triassic tetrapods. In: Hallam A, ed. *Atlas of Palaeobiogeography.* Amsterdam: Elsevier, 1973: 213–23.

16 Cox CB. Changes in terrestrial vertebrate faunas during the Mesozoic. In: Harland WB, ed. *The Fossil Record.* London: Geological Society, 1967: 71–89.

17 Rougier GW, Wible JR, Novacek MJ. Implications of *Deltatheridium* specimens for early marsupial history. *Nature* 1999; 396: 459–63.

18 Godthelp H, Archer M, Cifelli R, Hand SJ, Gilkeson CF. Earliest known Australian Tertiary mammal fauna. *Nature* 1992; 356: 514–6.

19 Rich TR, Vickers-Rich P, Constantine A, Flannery JF, Kool J, van Klaveren N. A tribosphenic mammal from the mesozoic of Australia. *Science* 1997; 278: 1438.

20 Alvarez LW, Alvarez W, Asato F, Michel HV. Extraterrestrial cause for the Cretaceous—Tertiary extinctions. *Science* 1980; 208: 1095–108.

21 Crowley TJ, North GR. Palaeoclimatology. *Oxford Monogr Geol Geophysics* 1991; 8: 1–330.

22 Spicer RA. Plants at the Cretaceous–Tertiary boundary. *Phil Trans R Soc London B* 1989; 325: 291–305.

23 Crane PR, Lidgard S. Angiosperm diversification and paleolatitudinal gradients in Cretaceous floristic diversity. *Science* 1989; 246: 675–8.

24 Crane PR, Lidgard S. Angiosperm radiation and patterns of Cretaceous palynological diversity. In: Taylor PD, Larwood GP, eds. *Major Evolutionary Radiations*. Systematics Association Special Volume no. 42, 1990: 377–407.

25 Lidgard S, Crane PR. Angiosperm diversification and Cretaceous floristic trends: a comparison of palynofloras and leaf macrofloras. *Paleobiology* 1990; 16: 77–93.

26 Crane PR. Vegetational consequences of the angiosperm diversification. In: Friis EM, Chaloner WG, Crane PR, eds. *The Origins of Angiosperms and Their Biological Consequences.* Cambridge: University Press, 1987: 107–44.

27 Drinnan AN, Crane PR. Cretaceous paleobotany and its bearing on the biogeography of austral angiosperms. In: Taylor TN, Taylor EL, eds. *Antarctic Paleobiology*. New York: Springer-Verlag, 1990: 192–219 .

28 Ramstein G, Flateau F, Besse J, Joussaume S. Effect of orogeny, plate motion and land-sea area distribution on Eurasian climate change over the past 30 million years. *Nature* 1997; 386: 788–95.

29 Ruddiman WF, Kutzbach JE. Forcing of Late Cenozoic Northern Hemisphere climate by plateau uplift in southern Asia and the American West. *J Geophys Res* 1989; 94D: 18405–27.

30 Collinson ME. Cenozoic evolution of modern plant communities and vegetation In: Culver SJ & Rawson PS, eds. *Biotic Response to Global Change: The Last 145 Million Years*. London: Natural History Museum, 1999.

31 Smiley CJ. Cretaceous floras from Kuk River area, Alaska; stratigraphic and climatic interpretations. *Bull Geol Soc Am* 1966; 77: 1–14.

32 Parrish JT, Spicer RA. Late Cretaceous vegetation: a near-polar temperature curve. *Geology* 1988; 16: 22–5.

33 Wolfe JA, Upchurch GR. North American nonmarine climates and vegetation during the Late Cretaceous. *Palaeogeogr Palaeoclimatol Palaeoecol* 1987; 61: 33–77.

34 Tallis JH. *Plant Community History: Long-Term Changes in Plant Distribution and Diversity.* London: Chapman & Hall, 1991.

35 Kennett JP, Stott LD. Abrupt deep-sea warming, palaeoceanographic changes and benthic extinctions at the end of the Palaeocene. *Nature* 1991; 353: 225–9.

36 Collinson ME. *Fossil Plants of the London Clay.* London: Palaeontological Association, 1983.

37 McKenna MC. Eocene paleolatitude, climate, and mammals of Ellesmere Island. *Palaeogeogr Palaeoclimatol Palaeoecol* 1980; 30: 349–62.

38 Estes R, Hutchinson JH. Eocene lower vertebrates from Ellesmere Island, Canadian Arctic Archipelago. *Palaeogeogr Palaeoclimatol Palaeoecol* 1980; 30: 325–47.

39 Wolfe JA. Some aspects of plant geography of the Northern Hemisphere during the Late Cretaceous and Tertiary. *Ann Missouri Bot Garden* 1975; 62: 264–79.

40 Singh G. History of aridland vegetation and climate: a global perspective. *Biol Rev* 1988; 63: 159–95.

41 Meng J, McKenna MC. Faunal turnovers of Paleogene mammals from the Mongolian Plateau. *Nature* 1998; 394: 364–7.

42 Frakes LA, Francis JE, Syktus JI. *Climate Modes of the Phanerozoic.* Cambridge: Cambridge University Press, 1992.

43 Cambridge Paleomap Services. ATLAS version 3.3. Cambridge: Cambridge Paleomap Services, 1993.

44 Stanley S. *Earth and Life through Time.* New York: W.H. Freeman, 1986.

45 Smith AG, Smith DG, Funnell BM. *Atlas of Mesozoic and Cenozoic Coastlines.* Cambridge: Cambridge University Press, 1994.

CHAPTER 7: *Patterns of life today*

The currently accepted systems of biogeographical regions have their roots in the nineteenth century, when our growing knowledge of the world allowed biologists to realize that its surface could be divided into separate areas that differed in their endemic animals and plants. Naturally enough, these divisions were based upon the distribution of dominant, easily visible groups. Thus, de Candolle in 1820, followed by Engler in 1879, used the patterns of distribution of flowering plants as the basis for a system of floral regions, while Sclater in 1858, working on birds, and Wallace in 1860–1876, working on mammals, defined the system of zoogeographical regions. Apart from some modifications to the botanical scheme by Good [1] and Takhtajan [2], and quite minor changes to the zoogeographical scheme by Darlington [3], these nineteenth-century interpretations have come down to the present day almost unchanged.

As we now know, the nature of the biota of mammals and of flowering plants that eventually developed on each continent was the result of an interaction between their early history of origin and diversification, the gradual fragmentation of the continents (especially in the Southern Hemisphere), and the climatic changes that took place during the Cenozoic. In trying to understand the geographical origins of particular elements in the biota of these faunal and floral regions, it is important to realize that they may in the past have dispersed via now-broken land connections that also had climates far more congenial than those found there today. In particular, the warmer climates of the Early Cenozoic allowed animals and plants to spread and disperse via high-latitude routes that are now too cold to allow this. These routes were the Bering land bridge between Alaska and Siberia, and the land connections between eastern North America and Europe via Greenland (see p. 156). During the period of greatest warmth in the Cenozoic, the Late Palaeocene/Early Eocene, the average temperature of the coldest month in these southern Arctic regions was 10–12°C, and frost was absent or rare [4]. Therefore these are probably the routes taken by Old World tropical genera that are found today also in the tropical flora of Mexico and the Panama Isthmus [5]. Although climates had cooled somewhat by the Miocene, in the Late Cenozoic, it was still warm enough for some mammals that we now view as typically African (such as lions, hyaenas, giraffids and macaques) to spread into southern Europe and live there alongside typically Eurasian genera [6].

This last point also illustrates the fact that in the past there would normally have been a transitional zone between neighbouring regions. The clear distinctions that can be made today between the different zoogeographical regions and the different floral kingdoms, with sharp lines drawn between them on the world maps, are the results of coincidences of current geography and recent climatic changes. As we shall see, the Ice Ages of the last couple of million years progressively reduced the northwards extent of the range of the more warmth-loving flowering plants and mammals. But, as they were pushed southwards, they found their way blocked in North America by the narrow filter of the Panama Isthmus, and blocked in Eurasia by the combination of the Mediterranean Sea, the deserts of North Africa, the Middle East and southern Asia, and the Himalayan mountain range. As a result, many of these families became extinct north of those barriers. When the climate recovered during the interglacial periods, those same barriers prevented them from returning northwards. Therefore, instead of there being broad zones of gradual transition between neighbouring faunal or floral regions, they are separated by areas within which the climate forms a barrier between very different biotas.

In fact, these coincidences of geographical and climatic barriers are probably the only reasons why we can identify separate biogeographical regions. If, instead, South America, Africa and India were broadly connected, without barriers, to the continents to the north, we should merely be describing the patterns of latitudinal gradients of diversity, with endemic families in the equatorial tropical belt, and others in the temperate and cold-temperate zones to the north and south.

We can now turn to outline the global patterns of biogeographical regions of the mammals and flowering plants, and to examine how the early history of these groups created those patterns. This is followed by a detailed account of what took place in each region.

Mammals: the final patterns

Figure 7.1 shows the pattern of zoogeographical regions that is recognized here. This follows Wallace's 1860–1876 system, except that the boundaries of the Oriental and Australian regions in the East Indies follow the edges of the Asian and Australian continental shelves, rather than the two regions meeting along a line within the East Indies pattern of islands (see p. 150).

Because both South America and Africa were largely isolated from other continents in the Early

Cenozoic, each developed a characteristic mammalian fauna. In the Late Cenozoic, India and South-East Asia had a fauna similar to that of Africa. But their different climatic histories, and the appearance of desert and mountains separating the two areas, led to the divergence of their mammalian faunas. As a result, a separate Oriental zoogeographic region is recognized. The Australian region, with its unique marsupials, forms another distinct fauna. The two land masses of the Northern Hemisphere have mammalian faunas that differ somewhat from one another, although both are similar in having been greatly impoverished by the climatic change of the Pleistocene Ice Ages.

The pattern of distribution of the orders of mammal in the Late Cenozoic Miocene/Pliocene Epochs is shown in Table 7.1. The final pattern found today is slightly different from this, because elephants became extinct in the Palaearctic and Nearctic during the Pleistocene, and because edentates and marsupials dispersed to the Nearctic via the Panama land bridge. The final total of orders for each region in Table 7.1 takes account of these changes. The last line of Table 7.1 also shows the total number of terrestrial families of mammal in each region. These figures therefore exclude whales, sirenians (dugongs and manatees), pinnipedes (seals etc.) and bats, and also humans and the mammals they took with them in their travels (such as the dingo and rabbit in Australia).

The individual families within the orders of mammal show considerable variations in their success at dispersal. A few have been extremely successful. Nine families have dispersed to all the regions except the Australian: soricids (shrews), sciurids (squirrels, chipmunks, marmots), cricetids (hamsters, lemmings, voles, field mice), leporids (hares and rabbits), cervids (deer), ursids (bears), canids (dogs), felids (cats), and mustelids (weasels, badgers, skunks, etc.). In addition, the bovids (cattle, sheep, impala, eland, etc.) have dispersed to all the regions except the Neotropical and Australian, and the murids (typical rats and mice) have dispersed to all except the Neotropical and Nearctic regions. This group of 11 families can conveniently be called

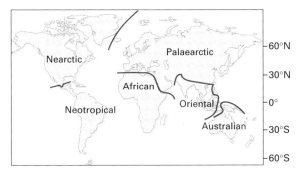

Fig. 7.1 Zoogeographical regions of the world today, based on the distribution of mammals.

Table 7.1 The distribution of the orders of terrestrial mammals during the Late Cenozoic (Miocene–Pliocene). The final total of orders also takes account of Quaternary extinctions and dispersals (see text).

	Africa	Orient	Pal.	Nearc.	Neotrop.	Austr.
Rodents	×	×	×	×	×	×
Insectivores, carnivores, lagomorphs	×	×	×	×	×	
Perissodactyls, artiodactyls, elephants	×	×	×	×	×	
Primates	×	×	×		×	
Pangolins	×	×				
Conies, elephant-shrews, aardvarks	×					
Edentates					×	
Marsupials					×	×
Monotremes						×
Total number of orders today	12	9	7	8	9	3
Total number of terrestrial families today	44	31	29	23	32	11

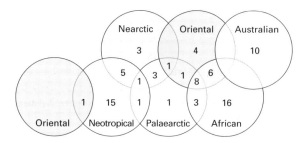

Fig. 7.2 Venn diagram showing the interrelationships of the families of terrestrial mammal of the six zoogeographic regions today, excluding the 11 'wandering' families.

Table 7.2 The degree of endemicity of the families of terrestrial mammal: number of endemic families × 100 ÷ total number of families.

Region	Endemicity
Australian	$10 \times 100 \div 11 = 91\%$
Neotropical	$15 \times 100 \div 32 = 47\%$
African	$16 \times 100 \div 44 = 36\%$
Holarctic	$7 \times 100 \div 37 = 19\%$
Nearctic	$3 \times 100 \div 23 = 13\%$
Oriental	$4 \times 100 \div 31 = 13\%$
Palaearctic	$1 \times 100 \div 30 = 3\%$

'the wanderers'. Their inclusion in any analysis of the patterns of distribution of the living families of terrestrial mammal tends to blur the underlying patterns of relationship of these zoogeographic regions. These 'wanderers' have therefore been excluded from Fig. 7.2, which shows the distribution of the remaining 79 families. The Oriental region is shown twice, so that the single family shared with the Neotropical region (the relict distribution of camelids) can be shown.

As can be seen from Fig. 7.2, the majority of all the families of terrestrial mammal (51 out of 90, i.e. 57%) are endemic to one region or another. The degree of endemicity of the mammals in each of the different regions is calculated in Table 7.2. (It is worth noting that the rodents, the most successful

of all the orders of mammal, contribute 19 of these endemic families: two Nearctic, one Palaearctic, 10 Neotropical and six African.) It is clear from Fig. 7.2 and Table 7.2 that the degree of distinctiveness of the six zoogeographic regions of today, if judged by the endemicity of their mammals, varies greatly. These figures are the resultant of three main factors: isolation, climate and ecological diversity.

The results of the long isolation of the Australian and Neotropical regions are obvious. The other four regions were all interconnected during the Middle Cenozoic. The Nearctic region was connected to Eurasia via the high-latitude Bering region, and many mammal groups were able to disperse across this region during those warmer times. Nevertheless, these groups did not include the tropical and subtropical groups that then ranged throughout Africa and southern Eurasia—including parts of Eurasia well north of the present limits of the Oriental region, which was therefore not recognizable as a separate zoogeographic region at that time. The Pleistocene glaciations of the Northern Hemisphere then decimated the mammal faunas of both the Nearctic and the Palaearctic. These two regions therefore contain few mammal families, and these families are mostly also still found in the adjoining southern regions. As a result, the Nearctic and Palaearctic contain few endemic mammal families. Tropical and subtropical Old World mammal families are therefore now found only in the Oriental and African regions. Without the three families found only in Madagascar, the degree of endemicity of the mammals of the African region would be slightly lower $(13 \times 100 \div 41 = 32\%)$, but it would still be significantly higher than that of the Oriental region. It is interesting that this difference is found also in birds: 13 families are endemic to the African region, but only one is endemic to the Oriental region. These differences are probably mainly the result of the greater latitudinal spread of Africa, which extends from the equator northwards through the Sahara Desert to approximately 30°N, and southwards to 35°S, and therefore includes a wider range of environments than India, which only extends between 14°N and 35°N. But the greater area of Africa must also have played a part in this, providing more space for the evolution of novel groups.

Eisenberg [7] has made an interesting analysis of the extent to which the different ecological niches have been filled in the different zoogeographical regions. Because East Africa has an extensive area of grasslands and savanna, 21% of the mammal genera of Africa are browsers or grazers, compared with only 9% of those in South America. Similarly, because great areas of South America are covered by rainforests, 22% of the mammal genera of northern South America are fruit-eating or omnivorous, compared with only 11% in southern Africa. For the same reason, northern South America also has a high proportion of mammals (bats) that feed on insects, either in the air or on leaves. Surprisingly, only two genera of South-East Asian bat feed on insects living on leaves, compared with nine in South America.

Comparison of data of this kind provokes interesting questions. Australia and northern South America contain similar proportions of arboreal genera, although Australia today has a much smaller area of forests. Eisenberg suggests that this may be an inheritance from the past, when Australia was more heavily forested, so that many marsupials became arboreal. Similarly, he speculates on the possible reasons for the fact that there is a comparatively small number of fruit-eaters/omnivores in Australia. This may be because there are fewer fruit trees in Australia, or because parrot-type birds there have been more successful in that niche than have the mammals. It is also possible that the Australian fruit trees, which are less taxonomically diverse than those of South America and live in a more seasonal climate, may be more seasonal than those of South America, so that fruit is not a reliable, year-round source of food. This may also be the reason why there are also many more bats in northern South America (see p. 14).

Another interesting aspect of comparative mammalian biogeography is that Australia has very few large carnivores. Flannery [8] points out that there may be several reasons for this. Much of that continent is unproductive desert or semidesert, and the vegetation in general is of low productivity because of Australia's impoverished soils (see p. 149). In addition, Australia's annual rainfall is also extremely variable, because it is highly affected by the El Niño events in the Pacific Ocean (see p. 92). All this leads to wide fluctuations in the populations of Australian herbivores. As a result, the carnivores that prey on them, and whose population sizes are always lower than those of their prey, must be very vulnerable to extinction. Flannery notes that Australia has an unusually high number of reptilian predators, such as pythons and varanid lizards, and had giant members of these groups in the Pleistocene Period, as well as a large land crocodile. He suggests that the reptiles, being cold-blooded, may have been better able to survive periods of starvation and so to retain a higher, safer level of population.

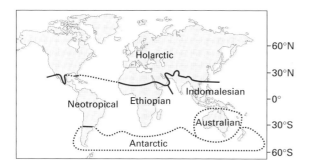

Fig. 7.3 Floral kingdoms of the world today. Modified, after Takhtajan [2].

The distribution of flowering plants today

Figure 7.3 shows the pattern of floral regions that is recognized in this book. In the main, this follows the system laid down by the Russian botanist Armen Takhtajan [2], who defined six kingdoms, 12 subkingdoms and 37 floristic regions. However, it differs from Takhtajan's system in two respects. First, for reasons that are explained on pp. 148–9, the flora of the tiny Cape region of South Africa is no longer recognized as a distinct floral kingdom. Secondly, the flora of Africa south of the Sahara, and that of India and South-East Asia, are recognized as separate Ethiopian and Indomalesian floral kingdoms, instead of being united as a single Palaeotropical floral kingdom. This reflects the fact that the floras of these two regions are as different from one another as, for example, the African flora is different from the flora of South America. Although there is, at the family level, much similarity between the floras of the three areas (cf. Figure 7.5, below), the African flora has been greatly changed by the Cenozoic uplifting, and consequent comparative aridity, of much of southern and eastern Africa; this now lies at a height of over 1000 m, rather than at the 400–500 m level it occupied previously. This led to much floral extinction and impoverishment, and may be the cause of disjunct distributions such as that of the primitive family Winteraceae, which is found in South America, Australia and Madagascar, but not in Africa.

The patterns of distribution of the families of flowering plant are much more difficult to interpret than those of the families of mammal. There are several reasons for this. First, there are many more flowering plants than there are mammals—some 300 living families and 12 500 genera have been described, compared with only 100 families and 1000 genera of living mammal.

Another problem is that the patterns of distribution of the flowering plant families are very diverse. This diversity may be partly because they are much better at dispersal across ocean barriers than are the mammals, since successful dispersal may require only a single air-borne seed instead of a breeding pair of mammals or a pregnant female. Most families of flowering plant are therefore much more widely distributed than most families of mammal. For example, almost everywhere in the world, four flowering plant families are among the six most numerous: the Asteraceae (daisies, sunflowers, etc.), Gramineae (grasses), Fabaceae (peas, clover, vetches, etc.) and Cyperaceae (sedges). Only 56 (18%) of the families of flowering plant are endemic to one particular region (and most of these are relatively unimportant families restricted to only a few genera), while 51 (57%) of the families of terrestrial mammal are similarly endemic. The more widespread nature of angiosperm distribution is also shown by the fact that, of the 302 families of flowering plant (excluding the four marine families), 86 are found worldwide, and 28 others are found in each region except the Boreal and Antarctic regions. It is therefore not surprising that it is difficult to characterize the different tropical regions.

However, an even more fundamental problem lies in the comparative uncertainty of flowering plant systematics. Because the fossil record of plants is less complete and less well understood than that of mammals, assessments of the relationships of groups above the level of the genus have been based on subjective judgements of the relevance of morphological features of the flowers, leaves, stem, pattern of branching, etc. Different workers have come to quite different conclusions as to the contents of different groups and as to their

Fig. 7.4 Relationships between the Liliiflorae floras of Takhtajan's floral regions (indicated by numbers, but the names altered to aid geographical identification). After Conran [10]. 1, Circumboreal Northern Hemisphere; 2, Mediterranean; 3, Middle East & Central Asia; 4, Sahara–Arabian; 5, Namib-Karoo region of southern Africa; 6, Cape of Good Hope; 7, sub-Saharan, East and south-central Africa (mainly grassland, woodland and savanna); 8, West African rainforest; 9, Madagascar; 10, Indian; 11, eastern China; 12, South-East Asian; 13, Malaysian; 14, Papua/New Guinea; 15, Northern & Eastern Australia; 16, South-Western Australia; 17, Central Australia; 18, New Zealand; 19, Rocky Mountains; 20, eastern North America; 21, south-western North America; 22, Central America; 23, eastern Brazil; 24, pampas of Argentina; 25, Amazonia; 26, Guyana Highland of northern South America; 27, Andes of tropical South America; 28, Patagonia. Frame and its numbers are axes from Conran's multi-dimensional scaling analysis.

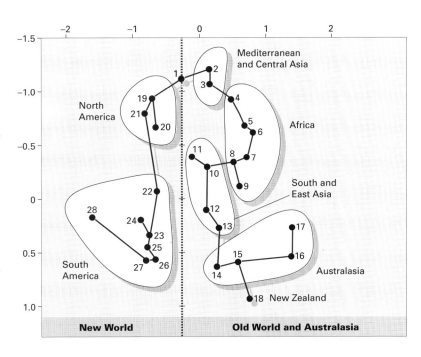

relationships—and each different grouping may, of course, lead to a different pattern of biogeography for the group. The new methods that use molecular information to evaluate relationships are still at an early stage, but they have already suggested the break-up of some traditional family groupings [9]. Some of these had previously included taxa that were widely separated in the Northern and Southern Hemispheres, providing an apparent problem of disjunct distributions. The new methods suggest that the disjunct groups are not, after all, closely related, and so removed the problem.

These problems have also been approached by new methods of analysis. For example, the superorder Liliiflorae, which includes the lily, iris and daffodil (and many other related flowers), is divided into 48–56 families. Although many of these show

interesting biogeographical patterns, these have been seen as isolated, uncoordinated phenomena, rather than as being parts of an overall pattern. However, the Australian botanist John Conran has subdivided the group into 91 smaller taxa that are more narrowly defined [10]. These are more likely to be monophyletic (having a single origin) and therefore to show clear biogeographical patterns resulting from their gradual dispersal. He has then analysed the distribution of these taxa among the 37 floral regions recognized by Takhtajan [2], to show the pattern of relationships between the Liliiflorae of each region. Conran's methods show that it is, after all, possible to identify patterns of distribution that are closely related to the current pattern of continents and climates (Fig. 7.4). The Liliiflorae of South America and North America are linked to one another, while the latter is linked via the high northern latitudes to southern Europe, thence via Africa and South-East Asia to Australia. At a more detailed level, the link between Africa and South-East Asia is seen to lie in similarities between the tropical forests of West Africa and of India, while the link between South-East Asia and Australia is via Papua/New Guinea. However, in addition to these major linkages, there are also a few taxa that show links between southernmost South America and Australia, like that of the tree *Nothofagus* (see p. 144), and which may therefore be a trace of an earlier pattern of distribution across Gondwana before its final break-up. There is no trace of any direct link between South America and Africa, and thus no evidence of any dispersal between these two areas.

Mammalian vs. flowering plant geography: comparisons and contrasts

As might be expected, the systems of faunal and floral regions are very similar to one another (compare Figs 7.1 and 7.3). This is because the continental movements and climatic changes of the Cenozoic have produced patterns of continents and climates that have had similar effects on the distributions of the two groups. However, there are differences in the degree of closeness of relationship between the mammals of the different regions and between the flowering plants of the different regions. As has already been noted, the relationships between the zoogeographic regions are seen more clearly after the deletion of the 11 families of 'wandering' mammal. Similarly, those of the flowering plant regions are also shown more clearly if the families with a worldwide distribution are first excluded. If the resulting patterns are now compared (Fig. 7.5), some very interesting differences can be seen.

The first point is that there is a much greater similarity between the flowering plant floras of South America, Africa, the Oriental region and Australia than between the mammalian faunas of these regions; only in the case of the African/Oriental comparison are the plant and animal figures at a

similar level (see below). Three factors seem to have caused these differences.

First, the families of flowering plants evolved and dispersed earlier than the families of mammals. Recent palaeobotanical techniques have made it possible to retrieve and identify complete and partial flowers from middle Cretaceous sediments [11]. These show that several different extant families had appeared by the middle Cretaceous, about 120 million years ago, at least a dozen by 95–75 million years ago. Therefore the angiosperms commenced their dispersal across the world much earlier than the mammals, and thus had a much greater chance of reaching the different continents before they had drifted far apart. In contrast, the diversification and dispersal of modern mammals only began in the earliest Cenozoic, 66–55 million years ago, by which time the continents had drifted further apart and were more difficult to reach. However, those mammals that did succeed as colonists were able, in the isolation of each continent, to diverge into a number of unique, endemic groups that show little similarity to those in other con-tinents—edentates, New World monkeys and caviomorph rodents in South America; elephants, elephant-shrews, conies and aardvarks in Africa; marsupials in Australia.

Secondly, there has been more extinction and replacement during the history of mammals than during that of flowering plants. For example, in addition to the approximately 100 living families of mammal, over 300 other families evolved and became extinct during the Cenozoic—some 70% of the families of mammal died out completely. Some of these were previously widespread families, which were replaced in the now-separate continents by new, endemic families. In other cases, the family became extinct only in some areas, so that it now had a disjunct distribution, as in the camel–llama group (see p. 155). Another example of the influence of extinction is seen if one compares the similarities between the mammal faunas of North and South America before and after the Pleistocene extinctions (see pp. 154–5). All these phenomena reduced the levels of similarity between the faunal regions. In the

Mammal families

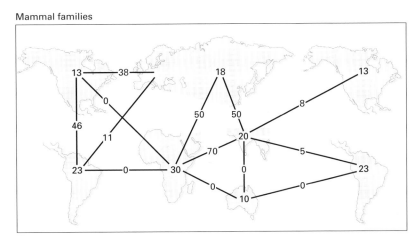

Flowering plant families

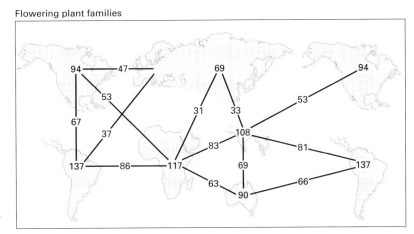

Fig. 7.5 A comparison of the faunal and floral similarities between the different regions. Figures within each region show the number of families found there. (Families found only in Madagascar have been omitted.) Figures linking the regions show the coefficients of biotic similarity, 100 C/N, of the biota of these regions: C being the number of families common to the two regions being compared, while N is the number of families in whichever of the two regions has the smaller fauna. Above, mammal families (excluding the 'Wandering' families). Below, flowering plant families (excluding those with a world-wide distribution); data taken from Heywood [45].

flowering plants, in contrast, there has been much less extinction—in fact, there is as yet no record of the extinction of a major group of angiosperm. Furthermore, the plant taxa are much longer-lived than are the mammalian taxa; for example, the distribution of the southern beech tree, *Nothofagus*, shows that it evolved in the Late Cretaceous, at least 70 million years ago, whilst the average longevity of mammalian genera is only eight million years.

Finally, of course, it would be unwise to assume that all the floral similarities were merely the result of early dispersal across negligible barriers, rather than a later colonization across wider gaps. The extent of the spread of flowering plants across the Pacific (over 200 different immigrant flowering

plants have reached the most isolated island group, Hawaii) shows clearly that they can cross even quite wide stretches of ocean, especially where intermediate island stepping-stones were available.

For all these reasons, it is not surprising that the floras of the different continents show more similarities to one another than do their mammalian faunas. However, there is one exception: the almost identical levels of similarity for the two groups when the African and Oriental regions are compared. The floral similarity here is not surprising, for it is at the same general level as the similarities between the other tropical regions—South America vs. the Ethiopian region, and South America vs. the Indomalesian region. It is therefore the faunal similarity between the African and Oriental regions that is unexpectedly high. This is probably because of the faunal exchange that took place between Africa and Eurasia after the two continents became connected in the Miocene (see p. 146), and before deserts spread through the Middle East.

The greater similarity, noted above, between the floras of the tropical regions also explains another difference. This is the fact that there is more similarity between the floras of North America and Eurasia, which are linked in a single Boreal floral region, than there is between the mammal faunas of these two regions. As we shall see below, both regions lost nearly all of their subtropical biota during the Ice Ages. After the last Ice Age had ended, animals and plants both started to spread northwards to recolonize the newly warmed lands of the Northern Hemisphere. However, there was already much more similarity between the plants of South America and Africa than between their mammals. Their dispersal northwards therefore produced a corresponding similarity between the floras of North America and Europe. No such similarity resulted from the northward spread of the very different mammals of South America and of Africa.

Finally, the high similarity between the floras of North America and Africa is probably because a large number of South American tropical families that are also found in Africa have spread into the almost subtropical south-eastern region of North America.

Floral patterns in the Southern Hemisphere

Another difference between the systems of animal and plant biogeography lies in the southernmost Southern Hemisphere, where plant geographers recognize an 'Antarctic' temperate floral region.

This flora contains descendants of the Southern Gondwana flora already mentioned (p. 128). Its forests consisted of mainly podocarp gymnosperms and mainly evergreen angiosperms (the southern beech tree, *Nothofagus*, being particularly characteristic), plus various herbaceous and shrubby angiosperms. This flora seems to have survived, reasonably intact, in southernmost South America,

in western Tasmania, and perhaps also in New Zealand (however, see also p. 152). Elements of it are also found in scattered locations further north. For example, *Nothofagus* and podocarps are found in the island of New Caledonia and in the mountains of south-eastern Australia and New Guinea (though they only appeared in the latter after its mountains rose in the Middle Miocene). A few flowering plant families (the Cunoniaceae, Gunneraceae, Petiveriaceae, Philesiaceae, Proteaceae and Restionaceae) that are found in this 'Antarctic' flora are also found in various combinations of the Southern Hemisphere continents and islands —southernmost South America, southern Africa, Madagascar, Australia, New Guinea, New Caledonia and New Zealand. As discussed earlier, such patterns may have resulted from spread across a pre-breakup Gondwana, from dispersal across ocean gaps between post-breakup Gondwana fragments, or by extinctions within a formerly more continuous area of presence. The real cause, in each case, will only become clear from the fossil record, and it would be unwise to assume that there is any single, common, cause for all these patterns.

As we shall see below, the flora of the Australian region is mainly derived from the original 'Antarctic' flora, although it has been much altered in adapting to the steadily increasing aridity of that continent. The elements of the old Antarctic flora thus now have merely a relict distribution around its descendant, the highly modified Australian flora.

The Cenozoic historical biogeography of each of the main regions, plus the large islands of New Zealand and Madagascar, will now be considered in turn.

The Old World tropics: Africa, India and South-East Asia

Three original land masses contributed to this area, and all were originally part of Gondwana (see Figs 6.3 and 6.4). South-East Asia may have been the first to unite with Asia, some time in the Triassic, while India collided with southern Asia in the Early Eocene, 52–45 million years ago, its continuing northwards movement throwing up the Himalayan

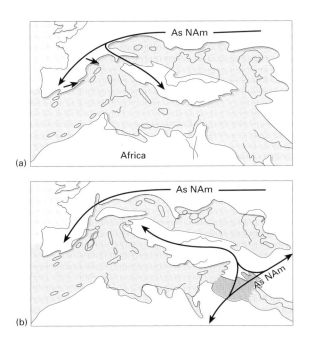

Fig. 7.6 Reconstructions of the Mediterranean area: (a) in the Late Oligocene/Early Miocene boundary; (b) in the middle of the Early Miocene; (c) the beginning of the Late Miocene; (d) near the end of the Late Miocene. Light tint, sea; dark tint, evaporitic deposits laid down as the seas dried up. Arrows show directions of mammal dispersals; As, Asia; NAm, North America. Some present-day geographical outlines have been added to aid recognition and location. After Steininger *et al.* [46], with permission.

Mountains and raising the Tibetan Plateau. Africa was therefore the last of these three to become united with the northern continents, joining them during the Late Cenozoic. But Africa had never lain far from southern Eurasia, and probably acted as a giant stepping-stone for the dispersal of the tropical flowering plants from their origin in the South American/African tropics, to the tropics of southern Asia. Many elements of the tropical biota doubtless became widespread throughout the region and, before the Late Cenozoic cooling of the Northern Hemisphere, would also have ranged northwards into higher latitudes of Eurasia. However, that cooling, together with the spread of seas and deserts in the Middle East, led to a new division of the Old World tropics into a western division, made up of Africa alone, and an eastern 'Oriental' section made up of India and South-East Asia.

It is not yet possible to trace in full the separate contributions of Africa, India and South-East Asia to the final Old World tropical biota. As always, the fossil record of mammals is easier to interpret, and can then be used as a guide for the reconstruction of the histories of the angiosperm floras. However, there was probably one significant difference between these histories. Because flowering plants had evolved and diversified in the middle Cretaceous, and because they have good powers of dispersal, many families probably became widespread throughout Africa and Eurasia. Most of the early types of placental mammal, on the other hand, evolved in the Northern Hemisphere at the beginning of the Cenozoic, and therefore spread southwards into Africa and India as and when that became possible, in the Palaeocene and Eocene.

The best-known and most distinctive of the mammal faunas is that of Africa. The shallow seas that separated Africa from Eurasia (Fig. 7.6a) in the early Cenozoic were not the only barriers to biotic exchange, for northern Africa lay in the northern mid-latitude arid belt. Nevertheless, some placentals managed to enter Africa from the north. An invasion by early insectivorans may have taken place as early as the Cretaceous–Cenozoic boundary, and there was also a later dispersal, involving early primates and creodonts, which may have occurred near the Palaeocene–Eocene boundary [12]. A more complete African mammal record only begins in the Late Eocene to Early Oligocene faunas of northern Africa [13], which contain two slightly different faunal groups. One group consists of the elephants, hyracoids (conies), aquatic sirenians (sea-cows), and the extinct embrithopods; the elephant-shrews, aardvarks and Cape golden mole all seem to be related to this group. They were all the descendants of an ungulate stock that had entered Africa some time earlier, in the Palaeocene or Early Eocene, and had since diversified into these unique mammals, which are all endemic to Africa. The other group of early African placentals apparently entered that continent later. This group included artiodactyls, creodonts, insectivorans, rodents, and the earliest members of the anthropoid primate line (that later evolved into apes and human beings). Marsupials, which had reached the European part of Euramerica during the early Cenozoic, probably via Greenland (see p. 156), are also known to have been present in northern Africa by the Early Oligocene.

The Late Cenozoic geographical and biogeographical relationship between Africa and Eurasia is quite complicated. The first collision between the two continents took place about 19 million years ago, in the Early Miocene, and was between the Arabian and Turkish regions of the continents (Fig. 7.6b). The closure of the seaway between them interrupted the circum-equatorial world oceanic circulation, and this in turn may have been the cause of the climatic deterioration that took place in central Europe at that time. Carnivorans, suids (pigs), bovids (cattle, antelope, etc.) and cricetid rodents passed across the new land bridge from Asia to Africa, while elephants, creodonts and primates passed in the opposite direction.

There was a reopening of the seaway in the Middle Miocene, 16 million years ago, which led to a warmer, moister climate through central Europe. But this marine link was not wide enough, or not constant enough, to prevent the dispersal to Eurasia of new types of primate, elephant and suid that had evolved in Africa. The seaway was finally broken near the beginning of the Late Miocene, 12 million years ago, by rising mountains in Arabia, Turkey and the Middle East, when the early horse *Hipparion*, which had evolved in North America, appears in both Eurasia and Africa, while rhinoceros, hyenas and sabre-toothed cats dispersed from Africa to Eurasia (Fig. 7.6c).

A final, dramatic event was the closure of the western connection between the Mediterranean and the Atlantic near the end of the Late Miocene, six million years ago. It was caused by a worldwide fall in sea levels (caused by an increase in the polar ice caps), as well as by the rise of mountains in both Spain and north-west Africa. Much, perhaps all, of the Mediterranean Sea then dried up completely, leaving an immense plain 3000 m below sea level, covered with a thick deposit of rock salt (Fig. 7.6d). However, this was short-lived, for it seems that the Mediterranean was refilled only about 0.4 million years later.

The drier climates that resulted from the lower sea levels at the end of the Miocene led to the expansion of sclerophyll evergreen woodlands in the Mediterranean region, and may also have led to the evolution of the typical Saharan flora further south. This drier European environment received many Asian mammals that inhabited steppe and savanna conditions, while the African hippopotamuses spread to its river systems, and African rhinoceroses lived on its plains. The fauna of northern Africa was therefore more similar to that of the rest of Africa during the Pliocene and Pleistocene. It was only in the Late Pleistocene that the extending Saharan Desert finally isolated the North African fauna and changed its climate, so that the fauna lost most of its African character.

Although India, because of its geographical origin in southern Gondwana, might have contributed a distinctive element to the tropical biota of Eurasia when it joined that continent, there is no evidence for that. This is probably because India separated from Gondwana in the Early Cretaceous, before placentals had evolved and when the flowering plants had only just appeared, and so may at first have lacked these groups altogether. Some flowering plants dispersed to India, via Madagascar, as those land masses drifted northwards through the Indian Ocean. These, together with India's original non-angiosperm plants, and new endemic forms that had evolved during its long northward drift, were all released into Asia when India met that continent in the Early Eocene [14]. However, after that collision, the flora of India was soon dominated by the varied angiosperms that had, by then, evolved in southern Asia. The Indian flora today is dominated by flowering plants found also in South-East Asia, and it contains no endemic families.

Almost certainly, India originally had early types of Mesozoic mammal, of a more primitive evolutionary grade than the marsupials and placentals, as these primitive types are widespread in the Mesozoic world. Unfortunately, the mammal fauna of India in the Cretaceous and Palaeocene, before it collided with Asia, is unknown. Such primitive Indian mammals would soon have become extinct in the face of the competition from the Asian placentals that would have entered India after the collision, in a wave of extinction rather like that which followed the connection between the Americas in the Pliocene (see p. 154).

The results of the Miocene interchange of mammals between Africa and tropical Eurasia can still be seen today, for a number of groups are found exclusively in these two areas—but, in nearly every case, they contain different genera. For example, the African rhinoceros, elephant and porcupine all belong to genera different from those found in the Oriental region. Similarly, the lemurs of Madagascar and the chimpanzees and gorillas of Africa are not found in the Oriental region, where these groups are represented by the lorises and by the orang-utan and gibbon. The scaly anteater (*Manis*) is an exception, for the same genus is found in both areas.

However, some of the exclusive similarities between the African and the tropical Eurasian mammal faunas that we see today conceal a far more complex earlier history [15]. For example, elephants originated in Africa in the Eocene, migrated into Eurasia in the Early Miocene, and migrated back later in the Miocene. The ancestors of the mammoths and modern Asian elephant probably evolved in Africa four to five million years ago, and migrated back into Eurasia two million years later. Although they spread through much of the world, the mammoths became extinct everywhere during the Ice Ages.

These differences might have developed in any case, because of the distance between Africa and tropical Eurasia. But the two faunas also became isolated from one another by the development of the Red Sea in the Pliocene, and by the extension of deserts in the Middle East. The two faunas also became more different from one another because that of East Africa had to adapt to the uplift and consequent increased dryness of that region in the Late Miocene, which led to it being covered by woodland and bushland with a ground cover of herbs and grass. (The replacement of this environment by the dry grasslands called savanna is probably a result of human activities—overgrazing, cultivation and clearance by fire [16].) The huge herds of browsing and grazing ungulates that now live in that area, such as the many types of antelope (impala, gazelle, gnu, and others), giraffes, buffalo, zebra and wart-hogs, are now thought of as the 'typical' fauna of Africa. But in reality these are late-comers to the African scene; their ancestors are not known in Africa until the Middle Miocene. Our own genus, *Homo*, also seems to have evolved in this environment (see p. 203). Together with the general drying of the world climate at that time, the drying of East Africa reduced the eastward extent of the African rainforest, which now became restricted to West Africa and the Congo Basin. This decrease in its area is probably the reason why the African tropical flora is much less diverse than that of South America and South-East Asia.

A similar increase in aridity, related to the uplift

of the Himalayan Mountains to new heights, affected the northern parts of the Indian subcontinent about three million years ago, and this led to an increase in the numbers of grazing mammals such as horses, antelope and camels, and of elephants [17].

No such climatic changes affected the tropical forests of South-East Asia, which contain a great diversity of primitive families of flowering plant. This fact led the Russian botanist Armen Takhtajan to suggest [18] that the area was the original home in which flowering plants had evolved. However, it is now clear that the diversity is the result of a comparatively recent fusion of two separate angiosperm floras. One of these may have been the original tropical angiosperm flora of southern Asia, while the other is of Gondwanan origin, having dispersed into South-East Asia from India or Australia. But, as always, the nature of the fossil record of angiosperms (see above) makes it difficult to document clearly the historical biogeography of these floras. This same difficulty lies in the way of solving another problem of plant geography in the Old World tropics. The Miocene exchange of mammals between Africa and the Oriental region was doubtless accompanied by a similar exchange of flowering plants. But the inadequate fossil record makes it at present impossible to distinguish which of the angiosperm families that are found in both areas are merely part of a common Late Cretaceous or Early Cenozoic inheritance, and which may have dispersed between the two areas in the Miocene.

At the margins of Africa, two areas contain biota that merit special consideration: the fauna of Madagascar, and the flora of the Cape region.

Madagascar

Much of this large (640 000 km²) island was heavily forested until recently (see Fig. 1.5). Its mammal fauna is notable for two features: most of the families are endemic to the island, and most of them seem to be ancient, primitive offshoots from their respective orders. The primates comprise three endemic families of lemur; the Insectivora comprise an endemic family, the tenrecs; the rodents include an endemic subfamily of cricetine (as well as the rats and mice introduced by humans), and the carnivorans comprise a number of genera of viverrid (civet) that may all belong to a single endemic family. There is also a species of aardvark. The only other land mammals that have reached the islands naturally are a pygmy hippopotamus that became extinct during the Pleistocene, and a river-hog.

These features of the mammal fauna are paralleled in other groups of animals and plants, and suggest that the island obtained its biota at an early stage in the Cenozoic. The lack of more advanced types of mammal suggests that the island was later isolated from Africa before these had evolved, or before they had entered Africa.

The geophysical evidence is, at first sight, inconsistent with this, for it shows that Madagascar separated from Africa in the Middle Jurassic. At first, it moved northwards, still connected to India, but it separated from India in the Early Cretaceous and moved south-westwards to reach its present position. Madagascar was therefore isolated far too early to obtain its mammalian fauna by a direct overland connection to Africa. The puzzle was solved by the discovery that there was a land link between Madagascar and the mainland some time between the Middle Eocene and the Early Miocene [19]. The times of divergence of the Madagascar mammal lineages fits well with this, as does the fact that the groups of mammal that are absent from Madagascar evolved, or entered Africa, later than the Early Miocene. The fact that only a limited selection of the Early Cenozoic mainland groups succeeded in crossing, suggests that the link was a chain of islands rather than a continuous corridor of land.

The Cape flora

It has been customary for plant biogeographers to recognize the flora of the Cape region of southern Africa as a separate floral Kingdom, thus placing it at the same level of importance as the floras of each of the major continents of the world, or of the whole of the temperate Northern Hemisphere (cf.

Fig. 7.3). This was because the Cape region shows a high level of endemicity. It contains six endemic angiosperm families, together with an extremely high rate of generic endemism (19.5%) and a great richness at species level—it contains over 8500 flowering plant species. The vegetation, known as 'fynbos', consists of fine-leaved, bushy, sclerophyll plants, and is dominated by members of the Restionaceae (in particular) and also of the Ericaceae and Proteaceae.

However, recent research has shown that the Cape region is merely one of several regions that have a Mediterranean-type climate, i.e. warm, dry summers and cool, wet winters. The other regions are California, coastal Chile, south-western Australia and the Mediterranean Basin itself [20]. These five regions occupy less than 5% of the earth's surface, yet contain around 48 250 species of flowering plant (almost 20% of the world total), as well as exceptionally high numbers of rare and locally endemic plants The Cape flora is therefore not unique. It is merely one of several floras that have resulted from similar ecological/evolutionary histories. Before the beginning of global cooling and drying in the Pliocene, all five regions were covered by subtropical forest, but now have a mixture of floras that include some relicts of the former forest, plus sclerophyllous shrublands and woodlands, with drought- and fire-adapted lineages predominating. It follows that there is no good reason for recognizing the Cape flora as a floral Kingdom, but merely (like the other regions with a Mediterranean-type climate) as an interesting province within a larger floral Kingdom.

Australia

The characteristics and biogeographical affinities of the biota of Australia are the most unusual and interesting of any in the world, and their explanation necessitates a very rewarding understanding of the interplay between continental movement, climatic change and biotic dispersal [21–24].

When flowering plants and mammals first evolved, radiated and dispersed, in the Cretaceous,

Antarctica was still joined to both South America and Australia (see Fig. 7.5). Warm climates extended to near the ice-free poles, so that the seas surrounding Antarctica were warm, and forests clothed at least its coastal regions as late as the Eocene. As already noted above, the Southern Gondwana flora that evolved in this whole region was adapted to a humid environment and included many podocarp conifers and also ferns. The flowering plants of this flora probably included *Nothofagus* and the families that are still mainly confined to the fragments of Gondwana (see above). In Australia, most of these flowering plants are found today in rainforests—either in temperate cool rainforests or in warm, humid, seasonal rainforests. This suggests that this was also the original environment of the early Southern Gondwana flora of Australia.

The most unusual and characteristic flora of Australia today is that known as the sclerophyll flora. The plants of this flora grow slowly, readily cease growth altogether, and have a small total leaf area, made up of small, broad, evergreen, leathery leaves. This flora is adapted to the unusual soils of Australia, which are highly weathered and low in nutrient minerals: those of the semi-arid zone have about half the levels of nitrates and phosphates of equivalent soils elsewhere. This is because the mountains of Australia are mainly old (around 200 million years) and lie only along the east coast. As a result there has been little erosion to add new sediments and minerals to the vast flat expanses of the continent. The plants of the sclerophyll flora seem to have evolved from the rainforest flora, for all the larger families with sclerophyll types are also found in the rainforests, and about 45% of the sclerophyll genera are endemic to Australia. The most spectacularly successful sclerophyll forms are the gum tree genus *Eucalyptus* (Myrtaceae), which includes about 500 species, and the family Proteaceae.

Although both the rainforest and the sclerophyll floras once covered much greater areas of Australia, they are today found only in isolated, scattered areas. This is because of Australia's plate tectonic history. Although it first split from Antarctica about 96 million years ago, the two continents at

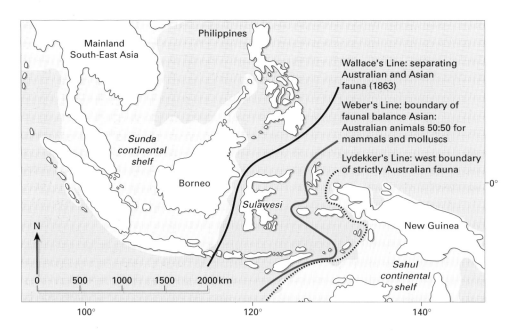

Fig. 7.7 Map of South-East Asia and Australasia. The continental shelves are shown lightly shaded. Three of the 'Lines' of faunal division are shown and explained. From Moss & Wilson [27].

first moved apart quite slowly, and remained parts of a single weather system, so that heat was transferred southwards from Australia to Antarctica. Australia only started to move rapidly northwards 46 million years ago, in the Eocene. By the Early Oligocene, around 34 million years ago, the two continents had separated sufficiently for a deep-water circumpolar current of cooler water and westerly winds to become established. Isolated from the warmth of Australia, Antarctica cooled, and ice sheets started to form. There was also less evaporation from the now-cooler seas around Australia, so that there was less rainfall on the continent, and a consequent increase in its arid, desert areas. This process continued into the Miocene, by which time the aridity of central Australia led to the appearance of open grasslands and savanna containing the varied genus *Acacia* (Fabaceae), which entered the continent from the west in the Early Cenozoic. Australia's continued northward

drift took it, in Late Miocene times, into the 30°S high-pressure zone of low rainfall (see p. 85), further increasing its aridity. All these factors have therefore combined to make Australia the driest of all continents; two-thirds of it have an annual rainfall of less than 500 mm, and one-third has less than 250 mm. Here, then, is the reason for the restriction of its rainforest and sclerophyll floras to isolated, scattered areas.

The mammals of Australia, too, had to adapt to these climatic and vegetational changes. The early history of the Australian marsupials is unknown, for their earliest fossils are of Middle Miocene age, when the grasslands were starting to appear. That early marsupial fauna contained more browsers than the modern fauna, which is mainly composed of grazers (especially the diverse kangaroos and wallabies) that feed on the great expanses of grassland. In the isolation of Australia, marsupials have radiated into a great variety of forms, occupying the niches that placentals have filled everywhere else. Marsupial equivalents of rats, mice, squirrels, jerboas, moles, badgers, ant-eaters, rabbits, cats, wolves and bears all look very like their placental counterparts.

The sclerophyll vegetation of Australia is low in nutrients and high in toxic biochemicals that deter the herbivore. The effects of this in depressing the population density of herbivores are shown by the brush opossum, *Trichosurus vulpecula*, whose density in the very different vegetation of New Zealand is five to six times greater than its density in its native Australia [25]. The consequences of the low nutrients for the carnivores of Australia have already been noted (see p. 139).

As Australia moved northwards, its northern edge came close to a great oceanic trench (see p. 113), where old ocean crust material was sinking downwards. This lighter material therefore came to underlie the northern edge of the Australian continent, causing it to rise, from the Middle Miocene onwards, and form the mountains of New Guinea. These mountains provided a high, cool environment that is the wettest area on earth today, although it is close to the driest continent. It was colonized by the Australian rainforest flora, including *Nothofagus*, though the surrounding lowlands of New Guinea were colonized by a mixture of Asian and Australian plants [14]. Asian placental mammals, too, started to spread eastwards but, apart from the aerial bats, only the rats spread naturally as far as Australia, where their 50 species now form 50% of the Australian land mammal fauna. Human beings probably arrived about 40 000 years ago, and the domestic dog (the ancestor of the dingo) about 3500 years ago.

Wallacea

The series of islands between the mainlands of Asia and Australia contains a transition between the Asian flowering plants and placental mammals, and the Australian flowering plants and marsupial mammals. To plant biogeographers, the whole area is the Malaysian province of the Indomalesian floral region, which extends eastwards to include New Guinea and as far as Fiji. Zoologists, on the other hand, found that New Guinea contained marsupials, but very few placentals, and a predominantly Australian bird fauna. The zoogeographer Alfred

Russell Wallace had, in the nineteenth century, suggested a line of faunal demarcation, later called Wallace's Line, which separated the predominantly Asian bird fauna from the more eastern, predominantly Australian bird fauna. This line, which runs close to the Asian continental shelf, has in the past therefore been recognized as the boundary between the Oriental and the Australian zoogeographic regions. However, the area between the Asian and the Australian continental shelves in fact contains relatively few mammals of any kind or origin. Later zoogeographers proposed six different variants of Wallace's Line [26]. The debate on where to draw a line has until recently diverted attention from the real interest of the area, which is the extent to which animals or plants have been able to enter or cross this pattern of islands from either direction. It is therefore best to draw the boundary of the Oriental and Australian faunal regions at the continental shelves, as is done in the other faunal regions, and to exclude the intervening area, which has sometimes been referred to as 'Wallacea' (see Fig. 7.7).

The biogeographical problems of Wallacea are not as simple as might be expected from a simple inspection of the modern map. One example is provided by the island of Sulawesi, which has a varied mammal fauna and complex geological origin [27]. The northern and western parts were part of Borneo until they rifted away in the Eocene; they may have been a major source of Asian plants for the islands to the east, and for New Guinea [14]. In contrast, the eastern parts of Sulawesi were originally fragments of Australasia, and only joined the rest of the island in the Early Miocene. The Makassar Strait between Sulawesi and Borneo is 104 km (65 miles) wide today. The island has a varied mammal fauna, including bats, rats, shrews, tarsiers, monkeys, porcupines, squirrels, civets, pigs, deer and a fossil pygmy elephant. Many of these are endemic, and so have not been introduced by humans, so they probably crossed the Makassar Strait during the Pleistocene, when lower sea levels would have reduced its width to only 40 km (25 miles). Although some of the types of mammal

found in Sulawesi are found in islands further to
the east, most were probably taken by humans, and
only the bats and rats appear to have made these
additional travels unaided. Sulawesi is also the
only island in Wallacea where there is a natural
overlap between Asian and Australian mammals,
for it contains two species of the marsupial phalangers. (Phalangers are the only marsupials that
have dispersed into Wallacea, being found on several other islands also.)

The island of New Guinea presents an example
of a different problem. Its northern half is made up
of a mosaic of areas that were once island arcs of the
Western Pacific, but that have collided with New
Guinea and become incorporated with it over the
last 40 million years. Some species of insect show
a pattern of scattered, disjunct local endemism in
New Guinea. It has been suggested that the areas
they occupy are the locations of the original islands,
so that the insects have been in these areas before,
during and after the islands were incorporated into
New Guinea [28]. However, the geology (and therefore the soil) of these former islands is different
from that of the surrounding regions. It is therefore also possible that these local conditions have
encouraged speciation, rather than that they represent 'Noah's arks' (see p. 161) embedded in the local
landscape.

New Zealand

The islands of New Zealand have been isolated
ever since they split away from Gondwana in the
Late Cretaceous or Early Palaeocene, 80–60 million
years ago. Their flora is now very impoverished, for
three reasons. First, New Zealand was progressively
submerged by sea during the early Cenozoic; by
the Late Oligocene, there was little sign of land.
Secondly, there was a great deal of volcanic activity
during the Pliocene, with extensive lava flows.
Thirdly, because of its mountainous nature and its
far-south position, New Zealand was extensively
glaciated during the Ice Ages.

New Zealand does possess several genera of plant
or animal (*Nothofagus*, some podocarp conifers, the

reptile *Sphenodon*, the flightless moa and kiwi)
that most biologists believe cannot have reached
the islands by transoceanic dispersal. They therefore also assume that some parts of the islands must
always have remained emergent from the sea, to
provide land on which these organisms might have
survived. However, Mike Pole of the University
of Tasmania has suggested that all of them might
instead have arrived by long-distance dispersal [29].
However this may be, the more warmth-loving of
the New Zealand plants must have arrived by
long-distance dispersal after the end of the Ice Age
glaciations. Their evolution has not been particularly rapid. Although 85% of the New Zealand
angiosperm species are endemic, only 10% of the
300–350 genera are endemic, and there are no
endemic families [30].

New Zealand has no native terrestrial mammals,
and therefore does not form part of the system of
mammalian biogeographical regions.

South America

The mammal fauna of South America today is characterized by having a few marsupials (opossums)
and a diversity of edentate placental mammals
(sloths, armadillos and South American anteaters)
that are hardly known elsewhere. Otherwise, it
does not seem particularly unlike other mammal
faunas, for it includes members of most of the placental orders (see Table 7.1). However, this is largely
because of a wave of extinctions and immigrations
in the Late Cenozoic, most of which were due to
the completion of the Panama Isthmus connecting
South America to North America. But that relationship between the two continents was only the most
recent of three connections between them over the
last 70 million years, which have resulted from
changes in their plate-tectonic relationships [31].

Late Cretaceous/Early Cenozoic

Ever since the appearance of the southern part
of the mid-Atlantic ridge in the Early Cretaceous,
South America had, like North America, been

moving westwards into the Pacific Ocean. A separate small tectonic plate, the Caribbean Plate, formed between them, with a chain of volcanic islands at its eastern and western margins, where old sea floor was being consumed at a subduction zone (cf. Fig. 6.1 and p. 113).

This was the link that allowed some North American dinosaurs to disperse southwards (see p. 125) in the Late Cretaceous, accompanied by some early types of marsupial and placental mammal. By the early Cenozoic, these placentals had evolved in South America into a number of types of unusual herbivorous ungulate, and the edentates had evolved from immigrants from North America. In the earliest Cenozoic, 10 different groups of South American mammal and a giant flightless bird appeared in North America. Their passage across the Caribbean plate may have been aided by low sea levels at that time, which may have changed the chain of volcanic islands into a temporary isthmus between the two continents. Some of these South American groups even reached Europe, for the remains of an ant-eater and of a flightless bird are found in the highly unusual fossils of Messel in Germany. In these deposits, laid down in an Early Eocene lake, even the soft parts and stomach contents of the animals have been perfectly preserved [32].

Later in the Cenozoic

This early fauna was later joined by the ancestors of two other South American groups, the New World monkeys and the caviomorph rodents (the latter include the guinea-pig and the capybara). These first appear there in the Late Oligocene but, because there is a gap in the fossil record of South American mammals in the earlier Oligocene, it is possible that they arrived earlier. They may, like the earlier mammals, have arrived from the north over the chains of islands at either end of the Caribbean plate, which was moving eastwards, pushed from the west by the expanding Pacific tectonic plate. But, because the closest living relatives of both these two groups live in Africa, it has also

been suggested that they might instead have come from that continent, perhaps by way of a chain of volcanic islands along the mid-Atlantic ridge that lay mid-way between north-east Brazil and West Africa in a then-narrower South Atlantic.

This situation has provided an interesting biogeographical conundrum that has been vigorously debated for many years [33].

Some workers consider that the African and South American rodents and monkeys evolved independently from earlier, widely spread, more primitive ancestors, and that they entered South America from North America. The problem with this solution is that the rich North American Early Cenozoic mammal faunas contain no possible ancestor for the South American monkeys [34], and no generally acceptable ancestors for the South American caviomorph rodents [35]. As a result, most recent work has favoured the trans-South Atlantic route.

The South American continent in which all these mammals lived consisted of both tropical and subtropical forest, and of open savanna with occasional trees and shrubs. The savanna, in particular, was the environment that permitted the evolution of a great variety of South American herbivorous ungulates.

The Late Cenozoic/Pleistocene

Ever since it started to separate from Africa in the middle Cretaceous, South America had been moving westwards towards an oceanic trench that lay in the eastern part of the South Pacific. Along this trench, old Eastern Pacific sea floor was being drawn back into the earth (see Fig. 6.1, p. 113). When South America eventually came to lie along this trench in the Miocene, this tectonic movement caused the volcanic and earthquake activity that created the Andean mountain chain. As we shall see, that had a major effect upon conditions in the continent.

From the Middle Eocene to the Middle Miocene, South America also moved roughly 300 km closer to North America. By the Middle Miocene onwards, the water level along the westward margin of the Caribbean plate became steadily shallower. Volcanic islands formed and enlarged to create an increasingly complete link between the two continents (Fig. 7.8). The first biogeographical evidence of this link came in the Late Miocene and Early Pliocene, when two families from each continent crossed this island chain to reach the other continent. The Panama Isthmus became a complete land bridge for the first time between 3.5 and 3.1 million years ago, in the Middle Pliocene, but it may have been temporarily broken again in the Late Pliocene when sea levels rose. The final closure was about two million years ago.

The Pliocene witnessed another geological event that had a major impact upon the biota of South America. This was the great Pliocene uplift of the Andes, that doubled their height from 2000 to 4000 m. These lofty mountains now almost completely interrupted the winds that had previously brought moisture-laden air eastwards from the Pacific to the high-latitude (30°S) south-eastern parts of the continent. In these areas, the savannas were replaced by treeless pampas, cool steppes, semi-deserts and deserts. As a result, the extensive intermingling of the faunas of North and South America that took place after the final completion of the Panama land bridge also took place at a time of profound ecological change in the South American continent. The fascinating interplay of geology and biogeography in what they have called the Great American Interchange has been analysed in rewarding detail, particularly by the American palaeontologists Larry Marshall and David Webb [36–38].

Before the Interchange, each continent had 26 families of land mammal, and about 16 families from each dispersed to the other continent. Of the North American mammals, 29 genera dispersed southwards in the Late Pliocene/Early Pleistocene, about 2.5 million years ago. They included shrews, rodents, felids, canids, bears, mastodont elephants, tapirs, horses, peccaries, llamas and deer. At the same time, genera of ground sloth, armadillo (including the giant *Glyptodon*), porcupine and caviomorph rodent dispersed from South America northwards.

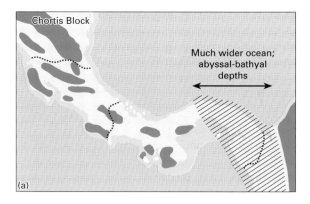
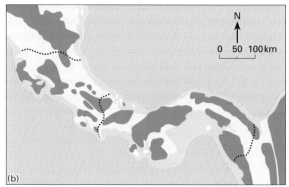
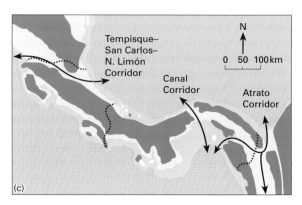

Fig. 7.8 The formation of the Panama Isthmus. Dry land is shown by oblique parallel lines, shallow marine sediments by dots, and deep ocean sediments by horizontal parallel lines. (a) Middle Miocene, 16–15 million years ago. (b) Late Miocene, seven to six million years ago. (c) Late Pliocene, approximately three million years ago. From Coates & Abando [45], with permission of The University of Chicago Press.

These were followed northwards, about 1.5 million years ago, by ant-eaters and opossums.

The fauna that was exchanged between the two continents was one that was adapted to a savanna, open-country environment, suggesting that this was the environment of the connecting Panama Isthmus when the exchange took place. That inference is supported by the fact that several types of birds and xerophilous shrubs that are typical of savannas are now found both north and south of the Isthmus. They, too, presumably passed through the Isthmus when it was still a savanna, before rainforest developed there in the Late Pleistocene and led to their extinction within the Isthmus itself.

Although similar numbers of families emigrated northwards and southwards, the North American emigrants were far more successful than were those from South America. In North America only 29 (21%) of the living land mammal genera are descended from South American stocks, while 85 (50%) of those in South America are derived from North American stocks. Although several other families (such as canids, horses, llamas and peccaries) contributed to the success of North American families in South America, easily the most successful were the cricetid rodents, which diversified into 45 genera in South America.

It has been suggested that the greater success of the North American mammals might have been due to ecological changes, or to the fact that they were the survivors of many millions of years of competition between mammal groups within the whole of the Northern Hemisphere, whereas the South American mammals had been protected in their isolated continent. However, recent statistical analysis shows that size was the only important factor: the types that became extinct were the larger ones. By chance, the South American mammals were larger than the North American ones, so more of them became extinct [39].

A final phase in the transformation of the South American land mammal fauna came during the Pleistocene. The climatic change that in the Northern Hemisphere caused the Ice Ages and the associated biotic changes (see p. 157 and Chapter 9) also caused a number of extinctions in South America. These included the last of the giant ground sloths and giant armadillo-like glyptodonts, as well as the horses and mastodont elephants that had immigrated from North America. However, the tapirs and llamas that had originally dispersed from North America became extinct there but survived in South America. As a result, these two groups now show a disjunct, relict distribution in which their surviving relatives (tapirs and camels) are found in Asia. Perhaps it is not surprising that few of the South American forms that had dispersed to North America when it was warm were able to survive there in the colder environment: only the opossum, armadillo and porcupine.

The final result of the ecological and biogeographical changes is thus a South American mammal fauna that shows little trace of its original inheritance from the Early Cenozoic. The characteristic flowering plant families today include the xerophytic Cactaceae, the Bromeliaceae (the latter include the pineapple, and are often xerophytic), the Tropaeolaceae (including the garden nasturtium) and the Caricaceae (pawpaw tree).

The Northern Hemisphere: Holarctic mammals and Boreal plants

In contrast to the complexity of the geological and geographical history of the Southern Hemisphere, that of the Northern Hemisphere has been far more uniform. Although shallow epicontinental seas, and the developing North Atlantic, have from time to time subdivided the land areas of North America and Eurasia into different patterns, the two continents have never been far apart, so that dispersal between them has usually been fairly easy. Furthermore, their faunas and floras have in the recent past all suffered from the severe climatic effects of the Ice Ages. As a result, they are the two great regions of temperate- and cold-adapted biota. Although the two continents are usually distinguished as being separate 'Nearctic' and 'Palaearctic' zoogeographic regions, they are sometimes considered as a single

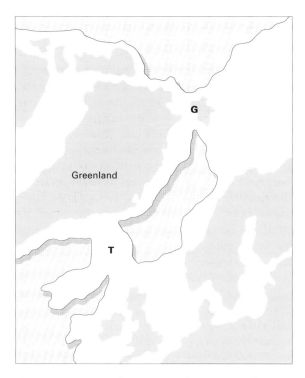

Fig. 7.9 Connections between North America and Europe in the Eocene. G, de Geer route; T, Thulean route (see text). After Tarling [48].

'Holarctic' region, like the single 'Boreal' region of plant geographers.

Although Greenland may have been separated from North America by a sea channel for a short time at the end of the Cretaceous, there was otherwise a continuous connection from North America to Europe via Greenland until the end of the Eocene. Sea levels were then lower, so that the European continental shelf was dry land, and all such present-day islands as the Faroes, Orkneys, Britain and Spitzbergen were merely a part of a more extensive European continent. Greenland was connected to this by two different routes [40] (Fig. 7.9). The Thulean route was along a now-submerged ridge of land that linked eastern Greenland to Europe via Iceland until the end of the Eocene. The de Geer route ran from northern Greenland to the most extreme north-westerly corner of the European continent. This connection

continued until Greenland finally separated from Europe at the end of the Eocene.

Much of the early evolution of mammals appears to have taken place in this Euramerican continent, in which there is a rich series of Late Cretaceous and Early Cenozoic deposits containing fossil mammals. The primates, rodents, bats, artiodactyls, perissodactyls, carnivorans and modern insectivorans are all known first from that area [41]. Although the European Cenozoic fauna is not known until the Late Palaeocene, it was then very similar to that of North America; nearly all the European families are also known in North America, and some genera are found in both continents. On the other hand, a number of North American Late Palaeocene families are unknown in Europe, and the climate of the northern connection between the two continents appears to have acted as a filter. During this time, Asia was isolated from both North America and Europe. Marsupials similar to the living South American opossum are not known in Asia, but are known to have reached Europe from North America by the Early Eocene, and survived in both those continents until the Miocene.

A different pattern of relationship between the northern continents began in the Early Oligocene. The North Atlantic now separated Europe from North America, and such new European groups as palaeothere horses could not cross to North America. On the other hand, the Obik Sea that had lain between Europe and Asia now dried up, perhaps because more water was now becoming locked up in the spreading ice sheets of the Antarctic [42]; this allowed some mammal groups to enter Europe from Asia and led to the faunal change in Europe known as 'la Grande Coupure' [43]. From now on, the high-latitude Bering link was the only route between Siberia and Alaska.

The climatic deterioration that began in the Early Oligocene therefore not only steadily reduced the area occupied by megathermal plants and altered the biota of the two northern continents, but also reduced the variety of plants that could disperse through the Bering link (although the North Pacific was bordered by continuous broad-leaved deciduous forests until at least the Middle Miocene). It led to the expansion of the Northern Hemisphere gymnosperm family Pinaceae—pine, fir, spruce and larch. It was at one time thought that the climatic cooling caused a wholesale southward movement, through the whole of the Northern Hemisphere, of an 'Arcto-Tertiary flora' that had evolved in the Arctic during the Cretaceous and had survived until today, little changed, in south-eastern North America and east-central Asia. However, this concept has not been supported by later knowledge of the floral history of the Alaskan region. Instead, the Northern Hemisphere angiosperm floras appear to have adapted to the Late Cenozoic climatic change in three ways: by the adaptation of some genera to changed, cooler climates; by the restriction of the range of some genera and their replacement by other already-existing genera that preferred a cooler climate; and by the evolution of new genera that preferred those cooler climates.

Much of the Late Cenozoic microthermal vegetation of the Northern Hemisphere appears to have evolved *in situ* from ancestors within the same area. Since there was little exchange of plants between North America and Eurasia during this time, these two floras steadily diverged. Although it has also been suggested that they shared a common 'Madro-Tertiary' dry flora, exchanged via a low- to mid-latitude dry corridor, this concept also now seems erroneous [44]. Instead, the ancestors of the plants found today in the dry region of south-western North America and the Mediterranean also appear to have lived in forests with moderate or high rainfall, and they have, in each region, adapted independently to a drier environment.

The crucial nature of the climate of the Bering region can also be seen in its influence upon faunal exchange between North America and Eurasia. When the climate became cool, as in the Early Oligocene, few mammals crossed. When it improved again a little later in the Oligocene, a number of Asian mammals dispersed to North America. Some of these had evolved within Eurasia, while others had dispersed to that continent from Africa. The final climatic deterioration in the Bering region

began in Miocene times, and may have been related to the increase in Antarctic glaciation at that time. From then on, most of the mammals that dispersed were large forms and, even more significantly, types that are tolerant of cooler temperatures—such warmth-loving forms as apes and giraffes could not reach North America. This climatically based exclusion became progressively more restrictive, until in the Pleistocene only such hardy forms as the mammoth, bison, mountain sheep, mountain goat, musk ox and human beings were able to cross. The final break between Siberia and Alaska took place 13 000–14 000 years ago.

Despite the long history of intermittent connection between the Nearctic and Palaearctic regions, each has certain groups of animals that have never existed in the other, while other groups did reach both regions, but later became extinct in one of them. Prong-horn antelopes, pocket gophers and pocket mice, and sewellels (the last three groups are all rodents) are unknown in the Palaearctic region, whereas hedgehogs, wild pigs and murid rodents (typical mice and rats) are absent from the Nearctic region. The domestic pig has been introduced to North America by human beings, as have mice and rats at various times. The horse became extinct in the Americas during the Pleistocene, but had crossed the Bering connection to Eurasia. Horses were therefore unknown to the American Indians until they were introduced by the Spanish conquistadores in the sixteenth century.

Though the Ice Ages did not commence until the end of the Pliocene, the steadily cooling climates had already exerted a great influence upon the floras of the northern continents. For example, there was a considerable change in the European flora during the Pliocene. Only 10% of the Early Pliocene flora of Europe still survives there, although over 60% of its Late Pliocene flora survives. The intervening three million years had therefore seen a drastic modernization of the flora of Europe, as it started to adapt to climatic changes that now became greatly exaggerated as the Pleistocene Ice Ages commenced. These Ice Ages stripped North America and Eurasia of virtually all tropical and subtropical animals and plants. This happened so recently that the faunas and floras have as yet had no time to develop any new, characteristic groups. Since they also have no old relict groups, such as the marsupials, it is the poverty and the hardiness of their faunas that distinguishes them from those of other regions. Many groups of animals are absent altogether and, of the groups that are present, only the more hardy members have been able to survive. Even these become progressively fewer towards the colder, Arctic latitudes. In North America there is, in addition, a similar thinning-out of the fauna in the higher, colder zones of the Rocky Mountains. This is a general feature of the fauna and flora of high mountains, as described in Chapter 2.

The Palaearctic fauna was almost completely isolated from the warmer lands to the south by the Himalayas and by the deserts of North Africa and southern Asia, and has therefore received hardly any infiltrators to add variety. The situation was, eventually, rather different in the Western Hemisphere. During the Early Cenozoic there had been hardly any exchange of animals between North and South America, presumably because there was still a wide ocean gap between the two continents. After the Panama Isthmus was completed at the end of the Cenozoic (see Fig. 7.8), many North American mammals dispersed to South America. However, North America was successfully colonized by only three types of South American mammal (opossums, armadillos and anteaters), along with a number of birds such as hummingbirds, mockingbirds and New World vultures.

For plants, for some reason, the situation was reversed. Instead of surviving in the Panama Isthmus, few of the North American megathermal plants survived the Late Cretaceous climatic cooling. The bulk of the lowland vegetation of Central America is, instead, of South American origin, perhaps because that great tropical region had produced an enormous variety of tropical plants. The mesothermal plants of North America were more successful at colonizing South America, presumably using the cooler mountainous spine of Cen-

tral America as their route from the Rockies to the Andes.

Summary

1 Biogeographers recognize similar patterns of distribution for living mammals and flowering plants, the major units in these being the individual continents. However, this pattern is the result of comparatively recent tectonic and climatic events. The Ice Ages greatly reduced the variety of animals and plants in the northern continents, while barriers of sea, mountain and desert prevented them from returning northwards when the climate improved. As a result, the tropical and subtropical biota of Africa, India, South-East Asia and South America are very different from the impoverished biota of Eurasia and North America.

2 Australia separated early from the rest of Gondwana and was isolated for a very long time. As a result, it has very few native placental mammals, but a great variety of marsupials. Because of the steadily increasing aridity of Australia, its original 'Antarctic' flora, descended from the Cretaceous Southern Gondwanan cool-temperate flora, was progressively transformed into a sclerophyll flora, which was in turn replaced by grassland and savanna. The 'Antarctic' flora therefore now has a relict distribution, from Patagonia to New Zealand and the mountains of New Guinea.

3 South America was also isolated until only a few million years ago, when the Panama Isthmus was completed. Most of its older fauna of unusual herbivorous placental mammals and carnivorous marsupials then became extinct due to competition from immigrant North American mammals. The rich tropical flora of South America was more successful, and colonized the lowlands of Central America.

4 The floras of North America and Eurasia each adapted to the climatic changes of the Late Cenozoic and to the Ice Ages of the Pleistocene, both by evolutionary change and by changes in their patterns of distribution.

Further reading

Goldblatt P. *Biological Relationships between Africa and South America*. New Haven: Yale University Press, 1993.

Hall R, Holloway JD. *Biogeography and Geological Evolution of SE Asia*. Leiden: Backhuys, 1988.

Jackson JCB, Budd AF, Coates AG. *Evolution and Environment in Tropical America*. Chicago: University of Chicago Press, 1988.

References

1 Good R. *The Geography of the Flowering Plants*, 4th edn. London: Longman, 1974.

2 Takhtajan A. *Floristic Regions of the World*. Berkeley: University of California Press, 1986.

3 Darlington P. *Zoogeography: the Geographical Distribution of Animals*. New York: John Wiley & Sons, 1957.

4 Tiffney BH. An estimate of the Early Tertiary paleoclimate of the southern Arctic. In: Boulter MC, Fisher HC, eds. *Cenozoic Plants and Climates of the Arctic*. NATO Advanced Science Institutes Series I, 27, 1994: 267–95.

5 Graham A. Development of affinities between Mexican/Central American and northern South American lowland and lower montane vegetation during the Tertiary. In: Churchill SP, Balslev H, Forero E, Luteyn JL (eds). *Biodiversity and Conservation of Neotropical Montane Forests*. New York: New York Botanical Garden, 1995: 11–22.

6 Pickford M, Morales J. Biostratigraphy and palaeobiogeography of East Africa and the Iberian peninsula. *Palaeogeogr Palaeoclimatol Palaeoecol* 1994; 112: 297–322.

7 Eisenberg JF. *The Mammalian Radiations. An Analysis of Trends in Evolution, Adaptation and Behavior*. Chicago: University of Chicago Press, 1981.

8 Flannery T. The mystery of the Meganesian meat-eaters. *Aust Nat Hist* 1991; 23: 722–9.

9 Soltis DE, Soltis PS, Nickrent DL *et al*. Angiosperm phylogeny inferred from 185 ribosomal DNA sequences. *Ann Missouri Bot Garden* 1997; 84: 1–49.

10 Conran JG. Family distributions in the Liliiflorae and their biogeographical implications. *J Biogeogr* 1995; 22: 1023–34.

11 Crane PR, Herendeen PS. Cretaceous floras containing angiosperm flowers and fruits from eastern North America. *Palaeoecol Palaeoclimatol Palaeoecol* 1996; 90: 319–37.

12 Gheerbrant E. On the early biogeographical history of the African placentals. *Hist Biol* 1990; 4: 107–16.

13 Coryndon SC, Savage RJG. The origin and affinities of African mammal faunas. *Spec Paper Palaeontol* 1973; 12: 121–35.

14 Morley RJ. Palynological evidence for Tertiary plant dispersals in the SE Asian region in relation to plate tectonics and climate. In: Hall R, Holloway JD, eds. *Biogeography and Geological Evolution of SE Asia*. Leiden: Backhuys, 1998: 211–34.

15 Kalb JE. Fossil elephantids, Awash paleolake basins, and the Afar triple junction, Ethiopia. *Palaeogeogr Palaeoclimatol Palaeoecol* 1995; 114: 357–68.

16 Andrews P, Van Couvering JAH. Palaeoenvironments in the East African Miocene. *Contrib Primatol* 1975; 5: 62–103.

17 Singh G. History of aridland vegetation and global climate. *Biol Rev* 1988; 63: 156–90.

18 Takhtajan A. *Flowering Plants, Origin and Dispersal*. London: Oliver & Boyd, 1969.

19 McCall RA. Implications of recent geological investigations of the Mozambique Channel for the mammalian colonization of Madagascar. *Phil Trans R Soc London B* 1997; 264: 663–5.

20 Cowling R, Rundel PW, Lamont BB, Arroyo MK, Arianoutsou M. Plant diversity in Mediterranean-climate regions. *Trends Ecol Evol* 1996; 11: 362–6.

21 Keast A. *Ecological Biogeography of Australia*, Vols I–III. The Hague: Junk, 1981.

22 Gressitt JL (ed.) *Biogeography and Ecology of New Guinea*, Vols I, II. The Hague: Junk, 1982.

23 Whitmore TC (ed.) *Biogeographical Evolution of the Malay Archipelago*. Oxford Monographs in Biogeography no. 4, 1987.

24 Beadle NCW. Origins of the Australian flora. In: Keast A. (ed.) *Ecological Biogeography of Australia*. Vols I, II. The Hague: Junk, 1981: 407–26.

25 Tyndale-Biscoe CH. Ecology of small marsupials. In: Stoddart DM, ed. *Ecology of Small Mammals*. London: Chapman & Hall, 1979: 342–79.

26 Simpson, GG. Too many lines; the limits of the Oriental and Australian zoogeographic regions. *Proc Am Phil Soc* 1977; 121: 107–20.

27 Moss SJ, Wilson MEJ. Biogeographic implications of the Tertiary palaeogeographic evolution of Sulawesi and Borneo. In: Hall R, Holloway JD, eds. *Biogeography and Geological Evolution of SE Asia*. Leiden: Backhuys, 1998:133–63.

28 Polhemus DA. Island arcs, and their influence on Indo-Pacific biogeography. In: Keast A, Miller SE, eds. *The Origin and Evolution of Pacific Island Biotas, New Guinea to Eastern Polynesia: Patterns and*

Processes. Amsterdam: SPB Academic Publishing, 1996: 51–66.

29 Pole MS. The New Zealand flora—entirely long-distance dispersal? *J Biogeogr* 1994; 21: 625–55.

30 Godley EJ. Flora and vegetation. In: Kuschel C, ed. *Biogeography and Ecology in New Zealand.* Amsterdam: Junk, 1975: 177–229.

31 Marshall LG, Sempere T. Evolution of the Neotropical Cenozoic land mammal fauna in its geochronologic, stratigraphic, and tectonic context. In: Goldblatt P, ed. *Biological Relationships Between Africa and South America.* New Haven: Yale University Press, 1993: 329–92.

32 Storch G. 'Grube Messel' and African–South American faunal connections. In: George W, Lavocat R, eds. *The Africa–South America Connection.* Oxford Monographs in Biogeography no. 7, 1993: 76–86.

33 George W, Lavocat R (eds) *The Africa–South America Connection.* Oxford Monographs in Biogeography no. 7, 1993.

34 Aiello LC. The origin of the New World monkeys. *In:* George W, Lavocat R, eds. *The Africa–South America Connection.* Oxford Monographs in Biogeography no. 7, 1993: 100–18.

35 George W. The strange rodents of Africa and South America. In: George W, Lavocat R, eds. *The Africa–South America Connection.* Oxford Monographs in Biogeography no. 7, 1993: 119–41

36 Marshall LG. The Great American Interchange—an invasion-induced crisis for South American mammals. In: Nitecki MH, ed. *Third Spring Systematic Symposium: Crises in Ecological and Evolutionary Time.* New York and London: Academic Press, 1981: 133–229.

37 Marshall LG, Webb SD, Sepkoski JJ, Raup DM. Mammalian evolution and the Great American Interchange. *Science* 1982; 215: 1351–7.

38 Webb SD. Late Cenozoic mammal dispersals between the Americas. In: Stehli FG, Webb SD, eds. *The Great American Biotic Interchange.* New York: Plenum, 1985: 357–86.

39 Lessa EP, Fariña RA. Reassessment of extinction patterns among the Late Pleistocene mammals of South America. *Palaeontology* 1996; 39: 651–9.

40 McKenna MC. Cenozoic paleogeography of North Atlantic land bridges. In: Bott MHP *et al.*, eds. *Structure and Development of the Greenland–Scotland Ridge.* New York: Plenum, 1983: 351–99.

41 Cox CB. Vertebrate palaeodistributional patterns and continental drift. *J Biogeogr* 1974; 1: 75–94.

42 Zachos JC, Breza JR, Wise SW. Early Oligocene ice-sheet expansion in Antarctica. *Geology* 1992; 20: 569–77.

43 Prothero DR. *The Eocene-Oligocene Transition. Paradise Lost.* New York: Columbia University Press, 1994.

44 Wolfe JA. Some aspects of plant geography of the Northern Hemisphere during the Late Cretaceous and Tertiary. *Ann Missouri Bot Garden* 1975; 62: 264–9.

45 Heywood VG. *Flowering Plants of the World.* Oxford: Oxford University Press, 1978.

46 Steininger FF, Rabeder G, Rogl F. Land mammal distribution in the Mediterranean Neogene: a consequence of geokinematic and climatic events. In: Stanley DJ, Wezel FC, eds. *Geological Evolution of the Mediterranean Basin.* New York: Springer, 1985: 559–71.

47 Coates AG, Obando J. The geologic evolution of the Central American Isthmus. In: Jackson JBC, Budd AF, Coates AG, eds. *Evolution and Environment in Tropical America.* Chicago: University of Chicago Press, 1996: 21–56.

48 Tarling DH. Land bridges and plate tectonics. *Mémoir Spéciale Géobios* 1982; 6: 361–74.

CHAPTER 8: *Interpreting the past*

Although it is obvious that there is a continuous spectrum of biogeographical problems from the most recent to the most ancient, most workers distinguish between two different approaches— ecological biogeography and historical biogeography. Of course, there is an overlap between these two [1]; for example, the biogeographical problems arising from the Ice Ages might be considered either as ecological, if they concern the Pleistocene ancestors of living groups, or as historical if they instead concern organisms that are less well understood because they have left no living descendants.

Because it is concerned with the more distant past, historical biogeography has been greatly affected by the 'revolution in the earth sciences' that resulted from the acceptance of the theory of plate tectonics. Previously, it had been assumed that the gaps in the distribution of organisms represented inhospitable areas that had existed before the organisms had evolved. They had therefore been obliged to disperse actively across such barriers, which had thereafter isolated them as separate units, within each of which independent genetic change had led to divergence into distinct species—and perhaps ultimately into distinct genera etc. as these differences became more numerous, comprehensive and fundamental.

But plate-tectonic theory suggested a greater role for geographical changes in subdividing and isolating populations of organisms. Whole continents could split, each fragment carrying away its cargo of both living organisms and buried fossils—acting, as the American palaeontologist Malcolm McKenna has so vividly described them [2], both as Noah's arks carrying living organisms, and as Viking funeral ships carrying the dead (fossils). In this new approach, known as 'vicariance biogeography', the emphasis is on new barriers, subdividing a previously continuous range of distribution, rather than on species

dispersing across a pre–existing barrier as in the older 'dispersal' approach.

As sometimes happens in science, the supporters of these two approaches became antagonistic, and the argument became polarized. Much biogeographical literature from the 1960s contains distasteful and unnecessary attacks on other biogeographers extending as far back as Charles Darwin. However, it is obvious that both phenomena not only exist, but also can take place simultaneously. For example, the completion of the Panama land bridge allowed dispersal of some northern mammals southwards into South America, at the same time as it provided a vicariance event in dividing a previously single population of marine organisms into separate Caribbean and East Pacific populations.

Because the event in question usually took place during past geological time, the only direct evidence is historical—either geological or palaeontological. If it is possible to demonstrate a clear correlation between a geological or climatic event and the subsequent divergence of biota on the appropriate geographical units, that is strong evidence for vicariance. For example, as Malcolm McKenna has shown [3], the mammal fauna of North America and that of Europe were very similar until the Middle Eocene. Geological and geophysical research suggests that the North Atlantic finally separated these two areas at just that time. But this evidence is nevertheless circumstantial. The absence of data supporting one explanation does not prove that it did not take place, and the same is true of the alternative explanation. Except for comparatively recent events, the evidence in many cases is not adequate to allow us to make a firm decision between dispersal and vicariance explanations.

On the other hand, although this may be true of the individual case viewed in isolation, there is a fundamental difference between the likely results of the two processes when a number of examples of disjunct distributions are considered together. A vicariant event, caused by tectonic changes or by the spread of shallow seaways, will simultaneously affect a considerable variety of organisms, whose subsequent evolutionary history is likely to be altered as a result of the ensuing isolation. Any such parallelism in apparent biogeographical history is therefore good prima-facie evidence for vicariance. Dispersal, on the other hand, is by its very nature a rare and comparatively isolated event, which will only very occasionally affect more than one type of organism—as, for example, when a violent cyclone or tidal wave carries a number of organisms from one area to another. Dispersal is therefore likely to appear only as the occasional aberrant event in the biogeographical record.

The realization of this difference has to a large extent calmed the argument between the proponents of dispersalism and of vicarianism. Even though several of the current techniques of analysis of historical biogeography originally arose from a primarily vicariance approach, the data are examined without any preconceived ideas, and judgement as to the original cause of the disjunction is made subsequently, on the basis outlined above.

The current vicariance theories can be divided into two groups, one of which relies primarily on 'cladistic' analysis of the characteristics of the organisms involved, while the other instead commences with analysis of the patterns of distribution. The cladistic method originated as a new approach to taxonomy, propounded by the German worker Willi Hennig in 1950. This treated the process of evolutionary change as a series of branching events, or 'dichotomies', at each of which a single group divides into two daughter groups. At each dichotomy, one or more of the characteristics of the group changes from a primitive or 'plesiomorphic' state into a derived or 'apomorphic' state. The evolutionary history of the group can then be portrayed as a branching 'cladogram'. Thus, in Fig. 8.1, characters a–g evolved after the divergence between group 1 and groups 2–6. They are therefore derived

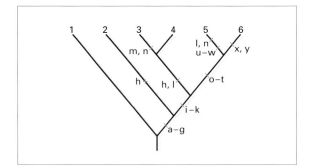

Fig. 8.1 Cladogram of the relationships between six groups, using characteristics a–y. (See text for explanation.)

or apomorphic relative to the characters of group 1 (in which these characters have remained primitive), but primitive or plesiomorphic for groups 2–6. Other new, apomorphic characters then evolved at different points within the evolutionary history of groups 2–6, and can therefore be used to analyse their patterns of relationship.

As far as possible, it is assumed that each apomorphic evolutionary event only occurred once in the history of each group of related taxa (a concept known as economy of hypothesis, or 'parsimony'), and the taxa are arranged on the cladogram in such a way as to minimize the number of parallelisms. For example in Fig. 8.1 it is most parsimonious to believe that character h has evolved twice, because that involves the assumption of only that single additional evolutionary event. The alternative is to transfer the origin of group 2 to near the base of groups 3/4, with the consequent need to assume that characters i–k had been lost in group 2—an assumption of three additional evolutionary events, instead of only one.

Phylogenetic biogeography

The potential of this method as a tool for biogeographical analysis was first realized by the Swedish entomologist Lars Brundin, who in 1966 analysed the distribution of chironomid midges belonging to three subfamilies found in the Southern

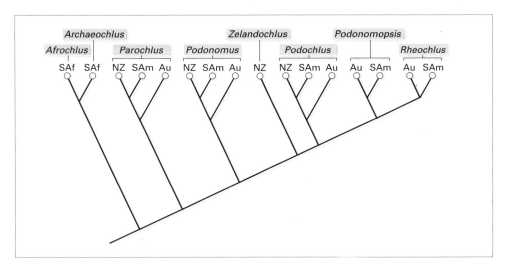

Fig. 8.2 Simplified taxon-area cladogram of some of the Gondwana genera of podonominine chironomid midges studied by Brundin [5]. The names in italics are those of the genera involved, while the circles represent individual species. The initials indicate the continent in which each species is found: Au, Australia; NZ, New Zealand; SAf, South Africa; SAm, South America. The African genera appear to have diverged first. In each of the other genera, the divergence of the New Zealand species preceded the divergence between the South American and the Australian species.

Hemisphere. He first produced a cladogram of the evolutionary relationships of the species. In place of the name of each species he then instead inserted the name of the continent on which it is found, transforming the phyletic cladogram into a taxon-area cladogram (Fig. 8.2). The result was a consistent pattern, in which the African species appeared to have diverged first, followed in turn by those of New Zealand, South America and Australia [4]. This sequence, based upon the evolutionary relationships of the midges, was independently supported by geophysical data on the sequence of breakup of the Gondwana supercontinent. (India and Antarctica do not appear in this analysis, because these subfamilies of midge are not found in those continents.) The divergences between the midges of the different continents can therefore be explained as the result of vicariance, the ocean barriers between the continents having appeared after the midges had colonized them. This in turn had useful implications as to the apparent geological ages of the different groups of midge, because the dates of separation of the continents were known from the geophysical data. (Divergence between species within each continent could, of course, have been the result of either vicariance or dispersal.) Like any method based upon an evolutionary, phyletic cladogram, Brundin's method depends entirely upon the accuracy of that cladogram, which in turn depends on the taxonomist's judgement in assessing which of each pair of divergent characters is primitive and which is derived. Where there is doubt on this, there arises the possibility of alternative phyletic cladograms and of corresponding variations in the taxon-area cladogram.

Cladistic biogeography

Some biologists have attempted to avoid the above limitations of phylogenetic cladistics by discarding any assumptions as to the evolutionary relationships of the groups under analysis. This approach is variously known as 'transformed cladism' or 'pattern cladism', and it forms the basis for cladistic

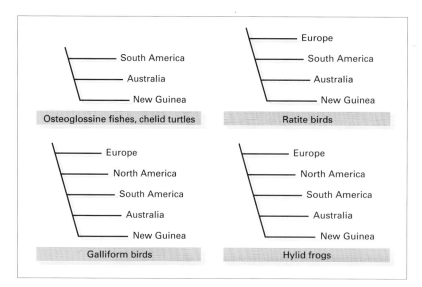

Fig. 8.3 Area cladograms of several vertebrate groups. After Patterson [5].

biogeography, which itself includes two different approaches.

First, if the phyletic and deduced biogeographical histories of several groups that occupied the area in question over the same period of time are compared, significant points of similarity or difference may emerge. The British palaeontologist Colin Patterson [5], for example, has shown interesting parallels between the taxon–area cladograms of several groups, suggesting that these all reflect the same sequence of separation of the areas concerned (Fig. 8.3). Secondly, one can avoid the subjectiveness of the taxonomist's judgement of evolutionary change by using as many different characteristics as possible, in the reasonable expectation that any anomalies in individual characteristics will become obvious and can be evaluated against the general pattern.

In applying the resulting information to biogeographical problems, data from as many groups as possible are used, again in the hope that individual anomalies will become statistically unimportant. This method, and the variety of methods that can be used to evaluate and integrate the data, are well described by Humphries *et al.* [6]. These authors also give an interesting and satisfying example of the application of this method to the biogeographical history of 25 species of the eucalypt tree

Monocalyptus along the coastal region of southern Australia, correlating this with the climatic and geological history of the area (Fig. 8.4). It is hypothesized that the genus was distributed continuously through this area in the Early Cenozoic (a), and that climatic change caused the progressive reduction in its range until there are now 25 different species distributed in the separate areas A–H (b).

Panbiogeography

The alternative approach to historical biogeography commences with the analysis and comparison of patterns of distribution of organisms rather than with their relationships (cladistic or otherwise). It originated with the work of the Venezuelan botanist Leon Croizat in the middle of the 1950s. Like many biologists, Croizat was puzzled by the widely disjunct distributions of many taxa, especially in the Pacific and Indian Oceans. At that time, before the emergence and eventual acceptance of the theory of plate tectonics, with its implications of the movement of continents or of continental fragments, most biologists believed that these patterns could only be explained by assuming that organisms had, by chance, been able to cross the gaps in their present patterns of distribution. Croizat, however, amassed

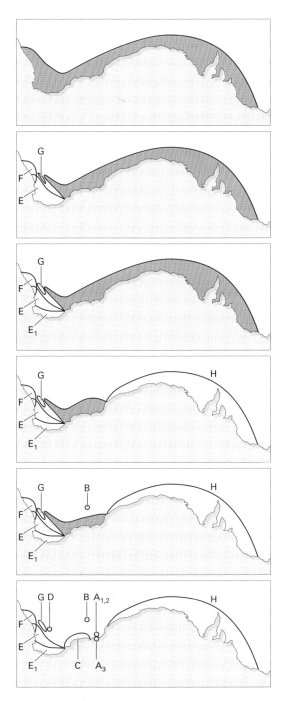

a vast array of distributional data, representing each biogeographical pattern as a line, or 'track', connecting its known areas of distribution. He found that, in many cases, the tracks of many taxa, belonging to a wide variety of organisms, could be combined to form a 'generalized' track. Some of these generalized tracks also converged on particular areas that Croizat called 'major nodes', and which he interpreted as regions where more than one biogeographical system had converged [7]. Croizat found that his generalized tracks did not conform to what might have been expected if these organisms had evolved in a limited area and had dispersed from there over the modern pattern of geography, as other biologists then believed. Croizat felt that it would be surprising if any single taxon had managed by chance to cross the intervening gaps, and incredible that a considerable variety, with different ecologies and methods of distribution, should have been able to do so. He therefore rejected both the concept of origin in a limited area, and that of dispersal. Instead, he believed that the organisms had always occupied the areas where we now see them, together with the intervening areas, and that they had colonized all these areas by slow spread over continuous land. (Unfortunately, and confusingly, Croizat described this process as 'dispersal', but in his usage it refers to the gradual extension of the range of an organism, not involving the crossing of barriers. Other biogeographers use it specifically to describe the extension of the range of an organism across a barrier.)

Croizat believed that any barriers, such as mountains or oceans, that exist today within the pattern of distribution of the taxa had appeared after that pattern had come into existence, so that these taxa had never had to cross them by chance dispersal. He therefore assumed that the biota of such isolated island chains as the Hawaiian Islands had spread there at a time when land had extended from North

Fig. 8.4 The hypothetical sequence of changes in the distribution pattern of the eucalypt tree *Monocalyptus* in southern Australia. The living species are found in areas A–H (bottom figure), which resulted from the progressive subdivision of an original single continuous area (top figure). This original area and its remnants are coloured dark blue. After Humphries *et al.* [6].

America to the islands. Finally, he believed that disjunct patterns of distribution of plants along the Pacific margins of North and South America had arisen because they had originally been present on a group of Pacific islands that had moved eastwards to fuse with those continents. To this extent, Croizat's theorizing anticipated the way in which plate tectonics would provide a geological contribution to the spread of organisms. However, even after plate tectonics became well documented and widely accepted, Croizat for a long time refused to accept that theory, preferring other geological explanations, and he never integrated it into his methodology.

Today, most biogeographers agree that plate tectonics have played a major role in producing the disjunct distributions that Croizat so assiduously catalogued. Similarly, most of his 'nodes' are areas such as the Panama Isthmus, South-East Asia or New Guinea, where plate-tectonic movements have brought together two areas of land of quite different biogeographical histories, with a consequent interaction and exchange of biota. However, Croizat's technique of assembling generalized tracks has been found useful, both in building up a series of tracks of different ages to explain the historical biogeography of an area (see below, 'Phyletic tracks'), and in analysing the pattern of spread of the species within a taxon [8]. Croizat's ideas also served a useful function in provoking the debate that led to the emergence of the whole vicariance school of historical biogeography. For Croizat was correct, and ahead of his time, in believing that, in many cases, barriers had emerged within an existing area of distribution of a taxon, so that dispersal across a pre-existing barrier was not the only explanation for disjunct patterns of distribution.

Croizat also concentrated primarily upon the distribution patterns of living organisms, and paid little attention to the implications of the fossil record or of changes in patterns of geography or climate. His 'generalized tracks', directly linking the present-day areas within which the taxa are distributed, therefore often cross ocean basins. For example, a generalized track runs directly across

the North Pacific between northern North America and Asia, rather than through the intervening land regions of Siberia, Alaska and northern Canada. This is because most forms of life are absent from these high northern latitudes—although that is only a comparatively recent phenomenon, having been caused by the recent Ice Ages. This aspect of Croizat's approach has more recently been emphasized by a group of mainly New Zealand biogeographers such as R. Craw [9], who specifically require that the track should be a 'minimum spanning graph or tree' in which the different localities are connected by the shortest possible series of straight lines. These tracks cross the Indian Ocean or Pacific Ocean basins and are therefore regarded as examples of a set of distribution patterns related to those ocean basins. These sets are referred to as 'ocean baselines' (Fig. 8.5), and the panbiogeographers of this group believe that this system is more useful and important than the conventional system of continental zoogeographical and plant-geographical regions. However, the number of organisms that may have been involved in trans-oceanic patterns of distribution is trivial compared with the number that have evolved and diversified upon the main continents, or that have dispersed from continent to continent when tectonic movements provided new land pathways between North and South America or between Europe, Asia and Africa. It is this, mainly stable, biota that has been used to characterize the classic biogeographical regions, and it is far more practical to use this wide-ranging and diverse data as the basis for biogeography, rather than the occasional, hypothetical trans-oceanic event. This panbiogeographical method also considers the area where a taxon is most diverse in numbers, genotypes or morphology as the area from which the track for that particular taxon radiated —a dangerous assumption (see p. 172). For example, because six species of the bird *Aegotheles* are found in New Guinea, but only one in Australia, Craw [9] concludes that the related fossil genus *Megaegotheles* of New Zealand was the result of a dispersal there from New Guinea, rather than from Australia. Finally, this school of panbiogeography

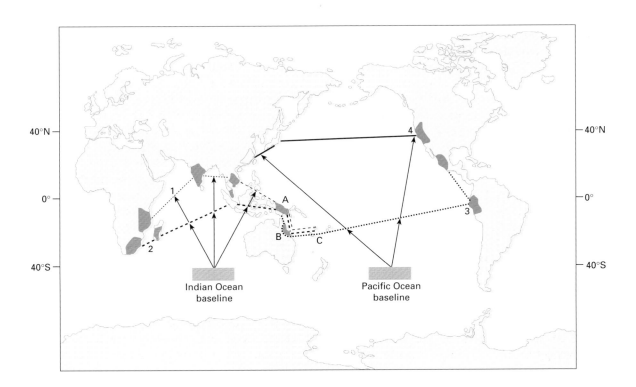

Fig. 8.5 Craw's panbiogeographic method. Tracks 1 and 2 cross the Indian Ocean, and represent examples of an Indian Ocean baseline, while tracks 3 and 4 similarly represent examples of a Pacific Ocean baseline. After Craw [9].

also ignores the increasingly detailed understanding of the ways in which the earth's geography has changed over geological time. Their refusal to take this area of research into account makes it very difficult to test their theories.

The whole history and development of the pan-biogeographical school has recently been reviewed [10].

Phyletic tracks and patterns

In contrast to the difficulties that arise from the approach of the New Zealand school of panbiogeographers, patterns of distribution that are based on monophyletic groups (i.e. in which all the members

are descended from a single ancestral taxon) and that take past geological and climatic history into account can, in contrast, provide useful insights—although a persistent problem for any taxonomist or cladistic biogeographer is that comparatively few groups have yet been subjected to the detailed analysis that is required to establish monophyly. The American zoologist Donn Rosen has used this technique to identify a series of generalized tracks involving the Caribbean region, which imply the relationships of its terrestrial biota to North America, South America and West Africa, and of its marine fauna to the east Pacific and west Atlantic [11]. He then attempts to integrate these data with contemporary theories of the geological evolution of the Caribbean and of the Panama Isthmus to provide a vicariance model for the historical biogeography of the region over the last 150 million years (Fig. 8.6).

Monophyletic taxa are also used by the British biogreographer Jeremy Holloway in analysing the

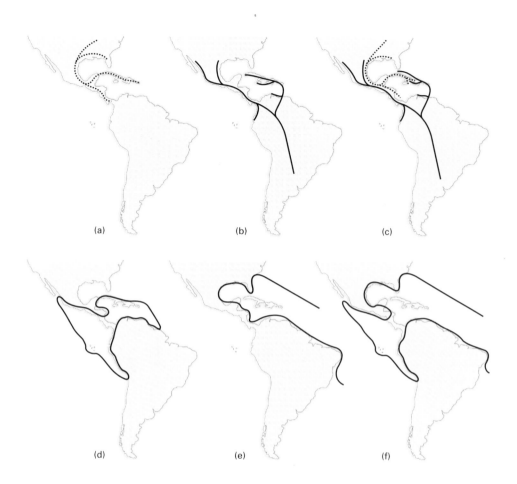

Fig. 8.6 Summary of the generalized tracks representing different elements in the historical biogeography of the Caribbean biota. (a) North American–Caribbean track. (b) South American–Caribbean track. (c) Overlapping of (a) and (b), to enclose the Caribbean Sea. (d) Eastern Pacific–Caribbean track. (e) Western Atlantic–eastern Atlantic track. (f) Eastern Pacific–eastern Atlantic track. From Rosen [11].

patterns of distribution shown by lepidopterans (butterflies and moths) and cicadas in the East Indies [12]. The taxa are linked according to the degree of closeness of their patterns of distribution, giving weight to both presence vs. absence in the different areas, and to the patterns of species abundance. These relationships can be displayed both by a branching dendrogram and by a two-dimensional diagram. Analysis of the butterfly taxa, which are good at dispersal, shows a basic dichotomy between the taxa with a more western, Oriental, pattern and those with a distribution centred on New Guinea and its surrounding islands (Fig. 8.7). Other clear geographical groupings are those that are geographically widespread, those that are widespread but close to the New Guinea grouping, and taxa that are particularly species-rich in the Philippines and Sulawesi. Finally, the analysis also identifies taxa that are either geographically sympatric (related species being found in the same area) or geographically allopatric (related species being found in different areas). Holloway suggests that the allopatric patterns are the result of vicariant speciation, caused by the division of previous land areas by rising

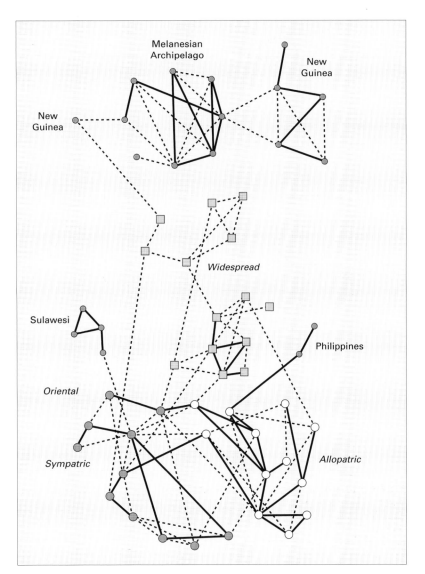

Fig. 8.7 Linkage diagram of the similarities in patterns of distribution between lepidopteran taxa in the Indo-Australian tropics. Degree of similarity is indicated by solid lines (60–100% similarity) *not* by relative closeness of the symbols. Taxa that are widespread are indicated by squares (the upper group is particularly close to the New Guinea region), those that are sympatric (see text) are indicated by filled circles, and those that are allopatric (see text) are indicated by open circles. Other taxa are particularly characteristic of Sulawesi, the Philippines, New Guinea or the Melanesian archipelago of the islands in the Southern Pacific. After Holloway [12].

Pleistocene sea levels. The similar analysis of patterns in cicadas, which have poorer powers of dispersal, showed some intriguing additional features. For example, there are some signs of separate relationships between, on the one hand, the northern part of Sulawesi with the Philippines (see Fig. 7.7, p. 150) and, on the other hand, the south-western part of Sulawesi with the more eastern islands. These relationships are particularly interesting in view of

geological evidence that Sulawesi is geologically composite, the present island being made up of some sections split off from Borneo and of others that arrived from a more easterly direction.

Endemicity and history

A completely different approach that, like panbiogeography, is not based upon phylogenetic analysis,

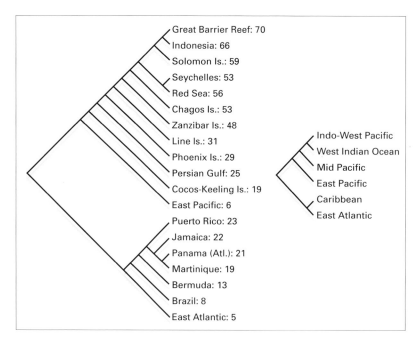

Fig. 8.8 Left, locality cladogram of living reef corals. Numbers after the locality names indicate the number of taxa that this locality shares with at least one other locality. Right, simplified cladogram, showing the general area relationships that emerge from the detailed cladogram. After Rosen [13].

takes as its starting point the pattern of endemicity of a group of related taxa. The British palaeontologist Brian Rosen [13] used the method of parsimony (see above) to set up a cladistic tree of the similarities between different localities, two localities being more similar if they share a greater number of taxa, and less similar if they share fewer taxa (Fig. 8.8). A closer similarity between two localities may result either from greater ecological similarity or from a more recent biotic linkage. Choice between these two possibilities as the preferred explanation depends upon the extent to which the groups concerned share a limited range of ecological requirements (suggesting an ecological explanation), or range widely across a variety of environments (suggesting a historical explanation). This method is known as parsimony analysis of endemicity.

The method has been used by Alan Myers of University College, Cork, to analyse the biogeographical affinities of the Hawaiian amphipods (a type of marine crustacean), some of which live there in the reef coral while others live in marine algae [14]. He found an interesting change of pattern over time, in which the biogeographical relationship of the Hawaiian *species* is closest to those of the south-west Pacific, in which there are many islands that could act as a source for the Hawaiian taxa. But the affinities of the Hawaiian *genera* instead lay with the eastern Pacific and the Caribbean. On average, different genera are likely to have diverged from one another earlier than their different species. It would therefore appear that the affinities of the Hawaiian genera reflect an earlier period of time, before the Panama Isthmus had closed during the Pliocene, and when there was free connection between the East Pacific and the Caribbean.

Brian Rosen's method has also been used to great effect in a study of the biogeography of the flowering plants that belong to the group known as the Liliiflorae (see p. 141).

Pleistocene problems

The interpretation of the relationship between modern biogeography and that of the Pleistocene

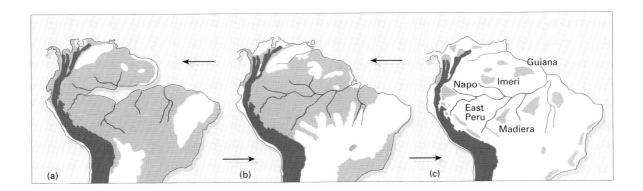

Fig. 8.9 Maps of northern South America, the shaded areas representing the lowland forests during: (a) the glacial minima; (b) today; and (c) glacial maxima. Coastlines have been adjusted to reflect the changes in ocean levels. Dark areas represent land above 1000 m. The pattern of Amazon Basin drainage is left unaltered in all three maps, to facilitate comparison. After Lynch [19].

provides unique problems. The amount of data on Pleistocene distributions is quite considerable, and much greater than that on more distant geological periods. The organisms themselves are usually closely similar to those of today, so that their ecological preferences and modes of dispersal are well understood. Furthermore, there was little significant large-scale geographical change during and after the Pleistocene, so that it can be assumed that the present-day connections between continents were in existence—even if their potential for individual groups was different, owing to climatic alterations. But those climatic alterations were severe, affected most of the world's biota, and caused large-scale alterations in their patterns of distribution over a comparatively short space of time. As a result, the scale of the problems of interpretation that arise is more similar to those of historical palaeobiogeography, although our knowledge of the organisms themselves is more like that experienced in ecological biogeographical problems.

The identification of regions within which a number of taxa are endemic (see above) also provides a possible method of reconstructing the biogeography of areas in the recent past, during the

Pleistocene glaciations. Organisms whose ecological preferences were adversely affected by these climatic changes are likely to have found their once continuous patterns of distribution changed. They would, instead, be restricted to smaller and probably scattered areas that still preserved the climatic conditions that these organisms required. The phenomenon has long been recognized in the Northern Hemisphere, where such organisms are known as glacial relicts (see p. 45).

More recently, this scenario has been proposed as a potential explanation for the unparalleled biotic diversity of the tropical rainforests of the Amazon Basin. In 1969 the American zoologist Jurgen Haffer suggested that the biogeography of many Amazon forest birds showed two features [15]. First, there are a few (about six) areas that each contain clusters of endemic species with rather restricted and similar ranges; these centres of endemism together contain about 150 species of bird, which make up 25% of the forest bird fauna of the Amazon Basin. Secondly, Haffer also found evidence, between these centres, of zones where the related species of the centres of endemism hybridized. He hypothesized that, during the glacial periods of the Northern Hemisphere, the centres of endemicity had been islands of persistently high rainfall (and therefore of rainforest), surrounded by areas of grassland (Fig. 8.9). Many bird species had been able to survive in these refugia, in each of which they had evolved towards becoming separate species. However, they had not yet become fully separate species before the climatic improvement had allowed the forest and its fauna to spread again over the intervening grasslands. As a result, when the related forms met again, they were still able to hybridize. Similar patterns of endemism and hybridization in Amazon birds have been found by other workers on Amazon frogs, lizards, butterflies and flowering plants.

Although he agrees with Haffer that climatic change may have caused these patterns, the British ecologist Paul Colinvaux [16] has suggested that its nature was very different. He points out that the pollen record shows that closed forest remained the dominant environment throughout the Amazon Basin during the Pleistocene, and therefore does not support Haffer's suggestion of extensive areas of grassland. However, ice-core data from Greenland and Antarctica show that the concentrations of carbon dioxide in glacial periods were only two-thirds those of today, and there is also evidence that the tropical lowlands were then cooled by about 6°C. Colinvaux suggests that these two effects are likely to have caused invasion of the tropical lowlands of Brazil by species from higher elevations. This effect would have been most marked in elevated areas, corresponding to Haffer's refugia, where it would have been exaggerated by the effects of altitude, and would have been the cause of vicariant speciation in these cool, carbon dioxide-starved areas.

Centres of dispersal

At one time some biogeographers believed that the area in which a group was represented by the largest number of species was also likely to be the area from which the group dispersed. This hypothesis, however, assumes that new species will appear at a constant rate, whatever the environmental conditions, and that the presence of a large number of species in a particular area therefore indicates that the group has existed for a long time in that area. In fact the rate of speciation depends upon ecological opportunity as well as upon the appearance of new features by genetic change. A group may first appear in an area where opportunities

for its particular way of life are limited, and few species will evolve. Later, it may gain access to an area in which the opportunities are far greater, and will there undergo rapid speciation. For example, although cichlid fishes are more varied in the African Great Lakes than anywhere else (see p. 65), no biogeographer today would argue that these lakes were the centres from which these fish dispersed over Africa and South America.

Some biogeographers have used the term 'dispersal centre' in a different way. By plotting the breeding ranges of many South American species, belonging to different groups, Paul Müller found that these overlap in one or a few smaller areas [17]. He suggests that these were centres to which these species were confined during the Ice Ages and from which they later dispersed as the climate improved. However, Colinvaux's criticisms of Haffer's theory (above) largely apply to Müller's theory also.

Fossils and historical biogeography

In the case of living organisms, comparisons between the biota of different localities can be based upon firm data as to the definitions of the localities, the presence or absence of the taxa concerned, and upon the ecology and dispersal potential of the taxa. The biota itself often includes a wide range of organisms, so diminishing the impact on the data of the occasional aberrant dispersal phenomenon or incorrect taxonomic assignment. The organisms themselves also provide a wealth of data that can be used to construct cladistic phylogenies or patterns of endemicity. The resulting biogeographical data can then be used to suggest past patterns of geographical relationship, or to provide support for particular geographical patterns that have been suggested on geological grounds. The existence of a fossil relative of one of these groups may then provide further support for such a theory of historical biogeography, or indicate that the theory requires modification. But these fossils alone will not have been used in the initial formulation of the theory. This is because the density of documentation of the pattern of fossil distribution is usually not sufficiently detailed to provide significant data on the past geographical relationships of localities within a single modern continent. At the other extreme, at the larger scale of intercontinental relationships, there have been few changes in these geographical relationships over the last 50 million years, during which the dominant terrestrial groups have achieved their current patterns of distribution, so that there have been few incompatibilities between geography and biogeography. The Great American Interchange provides an interesting exception to this, for the total dissimilarity between the mammal faunas of North and South America over most of the 65 million years of the Cenozoic Era had long convinced biogeographers that the Panama Isthmus connection between the two continents was of comparatively recent origin (see pp. 152–5).

Palaeobiogeography

The interpretation of the historical biogeography, or 'palaeobiogeography', of the distant ancestors of living groups or of ancient, totally extinct groups is more difficult, because the data are more limited. It is often spread over a great period of geological time, during which the earth's geography was radically different from that of today, and involves groups whose ecology, powers of dispersal and phylogenetic relationships are imperfectly known. These past patterns, such as the Permo-Carboniferous patterns of plant geography (see p. 117) were historically important in provoking the realization that continents might have changed their positions and interrelationships in the past. More recently, palaeontological studies have similarly led to the realization that some areas down the western flank of North America are fragments of originally more westerly placed, isolated packages of Pacific sea floor or islands, that later collided with North America. These areas are known as 'displaced terranes' [18]. More frequently, however, analysis of the historical biogeography of fossil groups depends on a theory of historical palaeogeography to provide the units in time and space between which biotic comparisons may

provide support for the geologists' theories. A combination of knowledge of plate-tectonic movements and of the shallow epicontinental seas that form additional barriers to the dispersal of terrestrial organisms can identify land areas within which useful analyses of palaeobiogeography can be made. For example, the existence of two separate Late Cretaceous Northern Hemisphere land masses, 'Asiamerica' and 'Euramerica', is supported both by geological evidence and by evidence from the distribution of contemporary dinosaurs, early mammals and plant spores, while the earliest, Permo-Carboniferous land vertebrates seem to have been limited to an earlier version of the Euramerican continent (see pp. 121–8).

Summary

1 There are several different approaches to the study of past patterns of biological distribution. Cladistic analysis of the characteristics of a group of related organisms, comparison of the relationship between the patterns of evolution and of distribution of unrelated groups, comparison of the apparent pathways of dispersal of unrelated groups, and comparison of the patterns of endemicity of different areas, are all used in attempting to identify how the patterns that we see today arose.
2 Since one of the aims of these techniques is to reveal anomalies that may indicate unusual events in the past and, since all of these techniques have their own limitations, it is often useful to use more than one technique (as far as possible). Concordance of results will then strengthen the resulting interpretation, while differences will highlight areas that require further consideration.
3 Approaches to the interpretation of past patterns of distribution have been greatly affected by acceptance of the theory of plate tectonics. This has led to the realization that, although in some cases species had to disperse across existing barriers, in other cases these barriers arose within their area of distribution, leading to speciation by a process known as 'vicariance'.
4 At a more global level, plate tectonics theory has

also been useful in identifying past patterns of geography, and the times at which these changed. Analysis of the biota of the appropriate palaeocontinental areas, over the periods of time during which these remained constant in their geographical content and relationships, can then show where biological groups appear to have originated, and also when they extended their distribution to other areas.

Further reading

Myers AA, Giller PS (eds) *Analytical Biogeography*. London: Chapman & Hall, 1988.

References

1 Rosen BR. Biogeographical patterns: a perceptual overview. In: Myers AA, Giller PS, eds. *Analytical Biogeography*. London: Chapman & Hall, 1988: 23–55.
2 McKenna MC. Sweepstakes, filters, corridors, Noah's arks and beached Viking funeral ships in paleobiogeography. In: Tarling DH, Runcorn SK, eds. *Implications of Continental Drift to the Earth Sciences*, Vol. I. London: Academic Press, 1973: 295–308.
3 McKenna MC. Fossil mammals and Early Eocene North Atlantic land continuity. *Ann Missouri Bot Garden* 1975; 62: 335–53.
4 Brundin LZ. Phylogenetic biogeography. In: Myers AA, Giller PS, eds. *Analytical Biogeography*. London: Chapman & Hall, 1988: 343–69.
5 Patterson C. Methods of paleobiogeography. In: Nelson G, Rosen DE, eds. *Vicariance Biogeography: a Critique*. New York: Columbia University Press, 1981: 446–89.
6 Humphries CJ, Ladiges PY, Roos M, Zandee M. Cladistic biogeography. In: Myers AA, Giller PS, eds. *Analytical Biogeography*. London: Chapman & Hall, 1988: 371–404.
7 Croizat L, Nelson G, Rosen DE. Centers of origin and related concepts. *Syst Zool* 1974; 23: 265–87.
8 Morrone JJ, Lopretto EC. Distributional patterns of freshwater Decapoda (Crustacea; Malacostraca) in southern South America; a panbiogeographic approach. *J Biogeogr* 1994; 21: 97–109.
9 Craw R. Panbiogeography: method and synthesis in biogeography. In: Myers AA, Giller PS, eds. *Analytical Biogeography*. London: Chapman & Hall, 1988: 405–35.
10 Cox CB. From generalized tracks to ocean basins— how useful is Panbiogeography? *J Biogeogr* 1998; 25: 813–28.
11 Rosen DE. A vicariance model of Caribbean biogeography. *Syst Zool* 1975; 24: 431–64.
12 Holloway JD. Geological signal and dispersal noise in two contrasting insect groups in the Indo-Australian tropics: R-modes analysis of pattern in Lepidoptera and cicadas. In: Hall R, Holloway JD, eds. *Biogeography and Geological Evolution of South East Asia*. Leiden: Backhuys, 1998: 291–314.
13 Rosen BR. From fossils to earth history: applied historical biogeography. In: Myers AA, Giller PS, eds. *Analytical Biogeography*. London: Chapman & Hall, 1988: 437–81.
14 Myers AA. How did Hawaii accumulate its biota? A test from the Amphipoda. *Global Ecol Biogeogr Lett* 1990; 1: 24–9.
15 Haffer J. Speciation of Amazonian forest birds. *Science* 1969; 165: 131–7.
16 Colinvaux P. A new vicariance model for Amazon endemics. *Global Ecol Biogeogr Lett* 1998; 7: 95–6.
17 Müller P. *The Dispersal Centres of Terrestrial Vertebrates in the Neotropical Region*. The Hague: Junk, 1973.
18 Cox CB. New geological theories and old biogeographical problems. *J Biogeogr* 1990; 17: 117–30.
19 Lynch JD. Refugia. In: Myers AA, Giller PS, eds. *Analytical Biogeography*. London: Chapman & Hall, 1988: 311–42.

CHAPTER 9: *Ice and change*

Many landform features in the temperate areas of the world show that major, geologically rapid changes in climate have taken place since the Pliocene. The general cooling of world climate that started early in the Tertiary continued into the Quaternary; the boundary between the two is placed at about two million years ago, but difficulties in definition as well as in dating techniques and geological correlation leave this date open to some doubt. The definition of the boundary comes from Italian marine sediments, where the appearance of fossils of cold-water organisms (certain foraminifera and molluscs) suggests a fairly sudden cooling of the climate which has been dated at 1.8 million years. Similar evidence of cooling has been found in sediments from the Netherlands, and this is believed to mark the end of the final stage of the Pliocene (locally termed the Reuverian) and the first stage of the Pleistocene (the Pretiglian) [1]. Evidence from sediment cores in the North Atlantic indicates that debris was being carried into deep water by ice rafts as long ago as 2.5 million years. Evidence from Norway suggests that there were Scandinavian glaciers extending down to sea level as long as 5.5 million years ago [2], therefore the point in time when we mark the opening of the Quaternary is inevitably disputed.

At various stages during the Pleistocene, ice covered Canada and parts of the United States, northern Europe and Asia. In addition, independent centres of glaciation were formed in low-latitude mountains, such as the Alps, Himalayas, Andes, the East African mountains and in New Zealand. A number of present-day geological features show the effects of such glaciations; one of the most conspicuous of these is the glacial drift deposit, boulder clay or till covering large areas and sometimes extending to great depths. This is usually a clay material containing quantities of rounded and scarred boulders and pebbles, and geologists consider it to be the detritus deposited during the melting and retreat of a glacier. The most important feature of this till, and the one by which it may be distinguished from other geological deposits, is that its constituents are completely mixed—the finest clay and small pebbles are found together with large boulders. Often the rocks found in such deposits originated many hundreds of miles away, and were carried there by the slow-moving glaciers. Fossils are rare, but occasional sandy pockets have been found that contain mollusc shells of an Arctic type. Some enclosed bands of peat or freshwater sediments within these tills provide evidence of the warmer intervals. They often show that there were phases of locally increased plant productivity, and they may contain fossils indicative of warmer climates.

Many of the valleys of hilly, glaciated areas have distinctive, smoothly rounded profiles, because they were scoured into that shape by the abrasive pressure of the moving ice. In places, the ice movement has left deep scratches upon the rocks over which it has passed, and tributary valleys may end abruptly, high up a main valley side, because the ice has removed the lower ends of the tributary valleys. Such landscape features provide the geomorphologist with evidence of past glaciation.

Immediately outside the areas of glaciation were regions which experienced periglacial conditions (Fig. 9.1). These were very cold and their soils were constantly disturbed by the action of frost. When water freezes in the soil it expands, raising the surface of the ground into a series of domes and ridges. Stones within the soil lose heat rapidly when the temperature falls, and the water freezing around them has the effect of forcing them to the surface,

Fig. 9.1 Contorted birch tree on the tundra in northern Lapland, marking the northern limit of tree growth.

where they often become arranged in stone stripes and polygons. Similar patterns are produced by ice wedges that form in ground subjected to very low temperatures. Sometimes these patterns, which are so evident in present-day areas of periglacial climate, can be found in parts of the world that are now much warmer. For example, they have been discovered in eastern parts of Britain as a result of air photographic survey. Such 'fossil' periglacial features show that, as the glaciers expanded, so the periglacial zones were pushed before them towards the Equator.

Climatic wiggles

The Pleistocene Epoch, however, has not been one long cold spell. The careful examination of tills and the orientation of stones embedded within them soon showed that several advances of ice have taken place during the Pleistocene, often moving in different directions. Occasional layers of organic material were sometimes discovered trapped between tills and other deposits, and these have provided fossil evidence of warm periods alternating with the cold. Where sequences of deposits are reasonably complete and undisturbed, as in the eastern part of England (East Anglia) and parts of the Netherlands, it has been possible to construct schemes to describe these alternations of warm

and cold episodes, to name them and to determine their relationships in time. But in many parts of the world this has not proved at all easy, and the correlation of events between different areas has often been speculative and unsatisfactory, mainly because of the difficulty experienced in obtaining secure dates for the deposits. At one time, for example, geologists considered that there were four episodes of ice advance in Europe, defined mainly by sequences of tills in the Alps. The four glaciations were named Günz, Mindel, Riss and Würm. This is now regarded as a simplification of the true situation, and it seems likely that there have been far more climatic fluctuations in the Pleistocene than this simple model suggests [3].

Because of the difficulties experienced in climatic reconstruction using land-based (terrestrial) evidence, attention has turned to the seas, where marine sediments provide a more complete and uninterrupted sequence. The retrieval of long, deep ocean cores of sediment has provided an opportunity to follow the rise and fall of various members of plankton communities in the past, particularly those, like the foraminifera, which, although tiny, have robust outer cases that survive the long process of sedimentation to the ocean floor and there accumulate as fossil assemblages. Some members of the foraminifera, like some species of *Globigerina* and *Globorotalia*, are sensitive to ocean temperature, so their relative abundance in the fossil record provides evidence of past climates.

An even more powerful tool for reconstructing long-term climatic changes has been the use of oxygen isotopes retained in the fossil material of the sediments. 'Normal' oxygen (^{16}O) is far more abundant than the heavier form of oxygen (^{18}O). For example, the heavy form comprises about 0.2% of the oxygen incorporated into the structure of water (H_2O). Water evaporates from the sea, but those molecules containing ^{18}O condense from a vapour form rather more readily than their lighter counterparts, so this heavy form tends to return rapidly to the oceans. Water containing ^{16}O, on the other hand, remains in the atmosphere as vapour for longer and is more likely to fall eventually over the ice caps and to become incorporated into these as ice. Under cold conditions, the volume of global ice increases and this (since it is formed largely from precipitation) tends to lock up more of the ^{16}O, leaving the oceans richer in ^{18}O. So the ratio of $^{18}O : {}^{16}O$ left in the oceans increases during periods of cold. This ratio is then reflected in the skeletons of foraminifera and other planktonic organisms and is deposited in the ocean beds. Analysis of the oxygen isotope ratios in ocean sediments thus provides a long and continuous record of changing water temperatures going back millions of years [3]. It has even been possible to use these methods for the analysis of oxygen isotope ratios in inland areas, as in the gradual deposition of calcite in the Devil's Hole fault in Nevada [4].

As a result of such oxygen isotope studies of a series of cores in the Caribbean and Atlantic Oceans, Cesare Emiliani of the University of Miami, Florida, was the first researcher to be able to construct palaeotemperature curves for the ocean surface waters [5]. A summary curve for the past 700 000 years is shown in Fig. 9.2 and it is quite obvious from this diagram that the climatic changes in this latter part of the Pleistocene have been numerous and complex. Figure 9.2 also shows a long sequence derived from the analysis of a deep-sea core from the equatorial Pacific Ocean by Nick Shackleton of the University of Cambridge, England. This analysis covers the past two million years, and the upper part can be correlated with the Emiliani curve. Such correlation ('wiggle matching') is assisted by the magnetic reversals that have occurred during the course of the Quaternary and which provide a basic time framework. A magnetic reversal is a situation where the polarity of the earth's magnetic field is switched, leaving a record in the rocks because of those particles incorporated in the sediments that had become magnetized by the earth's field. These reversals have been occurring throughout the earth's history, and five have taken place in the last million years, the latest one about 700 000 years ago. They provide very useful datum horizons for the correlation of sediments.

Examination of the long core in Fig. 9.2(a) shows that the fluctuations in temperature become stronger as one advances through the Quaternary. Gentle wanderings around a mean value gradually become more pronounced as greater extremes of temperature are experienced. Sharper changes also become more apparent, with the temperature likely to change both radically and rapidly. One can also see that the course of change, although showing a general wave-like form, is not simple or regular, but has many minor wiggle patterns imposed upon it. This latter point implies that we should not expect the terrestrial record in currently temperate lands such as North America or central Europe to show a simple alternation of temperate conditions with arctic ones, but a much more varied pattern in which cold and warmth alternate in varying degrees and in which intermediate conditions are often found.

Interglacials and interstadials

The fluctuations in global temperature represented in the ocean cores are reflected in the terrestrial geological sequence by glacial and interglacial deposits. A warm episode (usually represented by an organic, peaty deposit), sandwiched between two glacial events (often represented by tills) and which achieved sufficient warmth for a sufficient duration for temperate vegetation to establish itself, is termed an interglacial. The sequence of events demonstrated in the fossil material of such an interglacial shows a progressive change from high arctic conditions (virtually no life), through subarctic (tundra vegetation) to boreal (birch and pine forest) to temperate (deciduous forest) and then back through boreal to arctic conditions once more. If the warm event is of only short duration, or if the temperatures attained are not sufficiently high, then the vegetation changes may only reach a boreal stage of development. In this case it is termed an interstadial. We are currently living in the most recent interglacial (termed by geologists the Holocene).

Interglacials are often times of increased biological productivity (except in the more arid parts of the world) and they are often represented in temperate

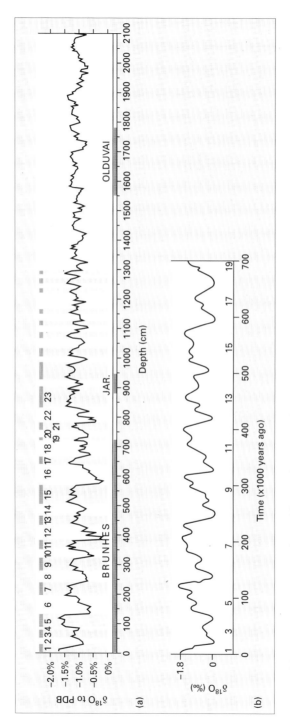

Fig. 9.2 (a) Oxygen isotope curve derived from sediments in the Pacific Ocean, which record over two million years of climatic history. Lower (more negative) $^{18}O : ^{16}O$ ratios indicate higher temperatures. Brunhes, Jar (Jaramillo) and Olduvai dark bars indicate periods of reversal of the earth's magnetic field. (b) A more detailed record from another Pacific core, covering only the last 700 000 years. After Emiliani [5]. Data from N.J. Shackleton.

Fig. 9.3 Organic sediments from an interglacial (dark band), resting on gravel of a former raised beach and covered by deposits laid down under periglacial conditions. A cliff exposure at West Angle, Pembrokeshire, Wales [7].

geological sequences as bands of organic material (Fig. 9.3). This material usually contains the fossil remains of the plants, animals and microbes that existed at or near the site during its formation, and it is this evidence, often stratified in a time sequence, that allows us to reconstruct past conditions and habitats. One of the most valuable sources of fossil evidence for this purpose have been the pollen grains from plants that are preserved in the sediment and that reflect the vegetation of the area at that time. Pollen grains are sculptured in such a way that they are often recognizable with

a considerable degree of precision; they are also produced in large numbers (especially those of wind-pollinated plants) and are widely dispersed; finally, they are preserved very effectively in water-logged sediments, such as peats and lake deposits. Therefore the analysis of pollen grain assemblages can provide much information about vegetation and hence about climate and other environmental factors [6].

Fossil pollen data are usually presented in the form of a pollen diagram, and Fig. 9.4 shows a pollen diagram belonging to the last (Ipswichian) interglacial in Britain. The vertical axis is the depth of deposit, which is inversely related to time, so the diagram should be read from the bottom up. The sequence begins with boreal trees, birch and pine, which are then replaced by deciduous trees such as elm, oak, alder, maple and hazel. Later in the sequence these trees go into decline, to be replaced by hornbeam, and then pine and birch once more. The details of such a sequence will obviously vary from one locality to another (Fig. 9.5 shows the pollen diagram from the same interglacial further east in Europe, in Poland, where lime, spruce and fir play a more important role) and varies quite considerably between different interglacials, but there is a consistent pattern to all such sequences, for they all pass through a predictable series of developmental stages. Often these are shown on such diagrams by dividing them into four zones (usually labelled in Roman numerals I–IV), pretemperate, early temperate, late-temperate and post-temperate, respectively. Pollen diagrams from our present interglacial suggest that we are well advanced into the late-temperate stage.

The last interglacial is the most easy to identify stratigraphically because it occurred in the very recent past. Placing precise dates upon it is difficult because the radiocarbon method becomes less accurate as one proceeds back in time, and dating methods relying on argon isotopes are dependent on the presence of volcanic material [8]. Nevertheless, it is believed that the last interglacial began about 130 000 years ago (although a recently studied and well-dated site from Nevada suggests

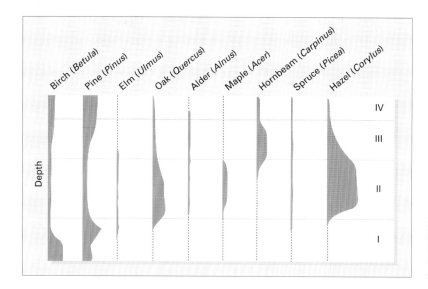

Fig. 9.4 Pollen diagram from sediments of the last (Ipswichian) interglacial in Britain. Only tree taxa are shown. The depth axis is related to age.

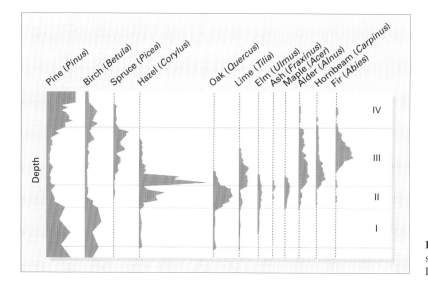

Fig. 9.5 Pollen diagram from sediments of the last (Eemian = Ipswichian) interglacial in Poland.

commencement at 147 000 years ago, and this has caused considerable controversy amongst geologists and palaeoclimatologists [9]) and ended some 115 000 years ago with the commencement of the last major glaciation. This means that we can identify the terrestrial record of the last interglacial with oxygen isotope stage 5 (see Fig. 9.2b) in the marine sediments. Earlier interglacials are much more problematic; they are usually given local names when they are first described, and attempts are later made to correlate them, often on the basis of the fossils they contain. Some general correlations are shown in Fig. 9.6 and a more detailed attempt to correlate the glacial and interglacial sequence of the Netherlands and East Anglia (Britain) is shown in Fig. 9.7. As can be seen from the second of these schemes, the problem of correlation is made even more difficult by gaps in the

Temp.	North America	Alps	North-west Europe	Britain	Approx. date
C	Wisconsin	Würm	Weichsel	Devensian	← 10 000
					← 70 000
W	Sangamon	Riss/Würm	Eamian	Ipswichian	
C	Illinoian	Riss	?Saale	?Wolstonian	← 250 000
W	Yasmouth	Mindel/Riss	?Holstein	?Hoxnian	
					500 000
C	Kansan	Mindel	?Elster	?Anglian	
W	Aftonian	Günz/Mindel	?Cromerian complex	?Cromerian	
C	Nebraskan	Günz	?Menapian		← 1 × 10⁶
W		Donau/Günz	?Waalian		
C		Donau	?Eburonian		
W		Biber/Donau	?Tiglian		
C		Biber	?Pretiglian		← 2.4 × 10⁶
W			Reuverian		

Fig. 9.6 Conventional correlations assumed between local glacial and interglacial stages in the Pleistocene. Local complexities render such correlations tentative, particularly in the earlier stages. See Fig. 9.7 for a suggested correlation of European stages in the early part of the Pleistocene. C, cold; W, warm.

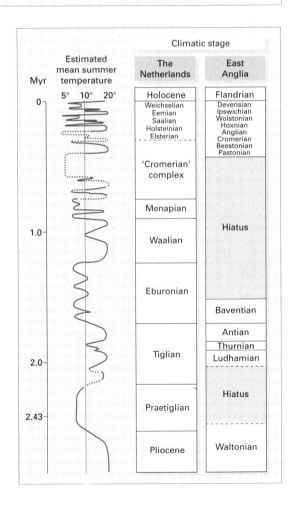

Fig. 9.7 Correlation of Pleistocene stages in Britain and the Netherlands as proposed by Zagwijn. A suggested temperature curve is also shown. After West [1].

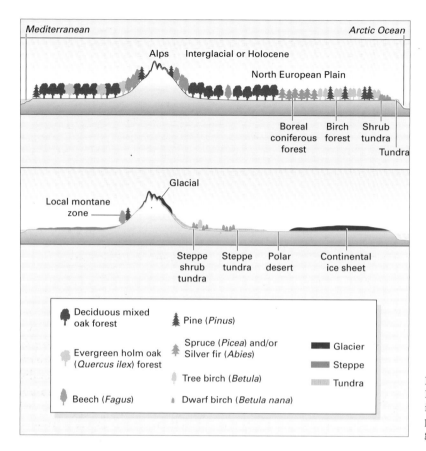

Fig. 9.8 The vegetation belts of Europe during glacial and interglacial time. A very different pattern of biomes is apparent during glacial times.

record in some localities. Any attempt at a global synthesis of terrestrial data is premature without a system for supplying absolute dates. The estimated dates on Fig. 9.6 are based on magnetic reversals.

Biological changes in the Pleistocene

With the expansion of the ice sheets in high latitudes, the global pattern of vegetation was considerably disturbed. Many areas now occupied by temperate deciduous forests were either glaciated or bore tundra vegetation. For example, most of the north European plain probably had no deciduous oak forest during the glacial advances. The situation in Europe was made more complex by the additional centres of glaciation in the Alps and the Pyrenees. These would have resulted in the isolation and often the ultimate local extinction of species and, indeed, of whole communities of warmth-demanding plants during the glacial peaks. Figure 9.8 shows the broad vegetation types that occupied Europe during interglacial and glacial times. During the interglacials, tundra species would have become restricted in distribution due to their inability to cope with such climatic problems as high summer temperatures or drought, and their failure to compete with more robust, productive species. High-altitude sites and disturbed areas would have served as refugia within which groups of such species may have survived in isolated localities. Similarly, during glacials particularly favourable sites which were sheltered, south-facing, or oceanic and relatively frost-free may have acted as refugia for warmth-demanding species [10].

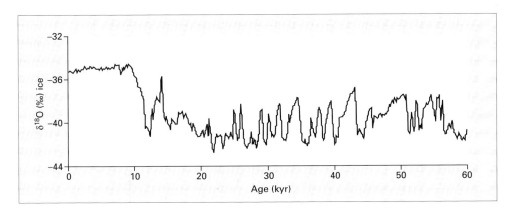

Fig. 9.9 Oxygen isotope curve analysed from an ice core extracted from the Greenland ice cap. (Less negative values of ¹⁸O in ice indicates higher temperature.) It covers a period of 60 000 years, approximately the latter half of the last glacial event. The instability of the temperature, as reflected by the isotope curve, is evident, the cold period being frequently interrupted by short episodes of higher temperature. From Dansgaard *et al.* [12].

Animal species with good dispersal mechanisms were able to shift their populations to cope with the new conditions. Plants were dependent on the movement of their fruits and seeds to spread into new areas and meanwhile old populations died out as conditions became less favourable. The fact that each species faced its own peculiar problems, both in terms of climatic requirements and dispersal capacity, means that species must have become shuffled into new assemblies (or 'communities', see Chapter 4) during times of change. Figure 9.8 expresses vegetation as life forms (with examples), so the picture is one of major biome shifts under different climatic regimes. The movements of mammals does not seem to have been as predictable as that of plants. An extensive study [11] of the ranges of North American mammals during Late Quaternary times has shown that individual species shifted at different times, in different directions and at different rates during the climatic swings. Only during the last few thousand years have the modern groupings of mammals assembled themselves. The idea that species assemblages moved as intact 'communities' must be abandoned in light of these findings.

Extinctions did occur in this process of successive climatic changes and distributional shifts. In Europe many of the warmth-preferring species so abundant in the Tertiary Epoch were lost to the flora. The hemlock (*Tsuga*) and the tulip tree (*Liriodendron*) were lost in this way, but both survived in North America, where the generally north–south orientation of the major mountain chains (the Rockies and the Appalachians) allowed the southward migration of sensitive species during the glacials and their survival in what is now Central America. The east–west orientation of the Alps and Pyrenees in Europe permitted no such easy escape to the south. The wing nut tree (*Pterocarya*) was also extinguished in Europe, but it has survived in Asia, in the Caucasus, and in Iran.

The last glacial

The most recent glacial stage lasted from approximately 115 000 years ago until about 10 000 years ago, so our current experience of a warm earth is really rather unusual as far as recent geological history is concerned. However, even a glacial is not a period of uniform cold, and there have been numerous warmer interruptions to the prevailing cold climate. Many interstadials are recorded within the last glacial. Figure 9.9 shows a detailed oxygen isotope curve of an ice core taken from the Greenland ice cap [12]. The instability of the temperature during the glacial is evident, and it is clear that there were many short-lived warm episodes right the way through the glacial. Mobile animals, such as flying insects, may well have been able to respond to these short-term and rapid fluctuations, but slow-spreading organisms, such as most plants, would not have been able to take advantage of them.

Continuous sediment records that contain fossil pollen and allow us to trace terrestrial vegetation back through the last glacial into the preceding interglacial are relatively uncommon. Many possible sites were actually scoured by glaciers, so have lost their record, but there are some deep lake sites, often associated with old volcanic craters, where a complete record is available. One such profile that takes the vegetation record back about 125 000 years is shown in Fig. 9.10. This is a pollen diagram from Carp Lake, situated on the eastern side of the Cascade Mountains in the Pacific north-west of America [13]. It lies just outside the limit of the great ice sheet (the Laurentide Ice Sheet) that covered much of the northern part of North America during the Wisconsin glacial (see Fig. 10.2, p. 199). It currently lies at the boundary of two biomes, the sagebrush (*Artemisia*) steppe and the the montane pine forest (*Pinus ponderosa*), and its closeness to this boundary (called an ecotone) means that the vegetation will have been particularly sensitive to climatic swings in the past. The basal part of the diagram (Zone 11) shows the open pine and oak (*Quercus*) forest of the last interglacial, with some indications that conditions were then warmer and drier than at present. There follows, for the bulk of the diagram, alternating peaks of *Artemisia* and pine, showing the constant alteration of the vegetation over this sensitive ecotone as the climate warmed and cooled. At one stage (Zone 7, about 85 000–74 000 years ago), there was an episode of such warmth that an open mixed forest including oak established itself. It is likely that this interlude was warmer and wetter during the summers than at present. Cooler conditions then returned until the glacial ended and the current vegetation established itself in the last 10 000 years (Zone 1).

This remarkable record demonstrates that the temperature fluctuations recorded in the ice-core profile of oxygen isotopes (see Fig. 9.9) were reflected in the response of vegetation, and that the instability of the glacial climate caused constant adjustments in the ranges of species and in the boundaries of biomes.

In the tropics the effect of ice was not, of course, experienced directly except on the very high mountains, but the climate was generally colder, and changes in vegetation reflected in pollen diagrams suggest that there were important variations in precipitation. In Queensland, Australia, for example, Peter Kershaw [14] has analysed the sediments of a volcanic crater lake from a rainforest area, and the results are shown in Fig. 9.11. Dating at this site is difficult, but the total span of the diagram is thought to cover about 120 000 years, extending back into the closing stages of the last interglacial —a similar time coverage to that of the Carp Lake diagram from the Cascades. During the last interglacial this Australian site was occupied by many of the rainforest tree genera that currently occupy the area, but at the time when the final glacial began in high latitudes, this forest was replaced by forest of simpler structure in which *Cordyline* played an important part. There was then a brief reversion (dated at about 86 000–79 000 years ago— very close to the warm spell in the Cascades) when rainforest became re-established. But then much more arid conditions set in and the forest became increasingly dominated by gymnosperm trees, such as the monkey puzzle (*Araucaria*). Between about 26 000 and 10 000 years ago the climate became very dry and the former forests took on a sclerophyllous form, being dominated by *Casuarina*, a tree currently associated with hot dry conditions. At the end of the 'glacial', however, there was a very rapid change in vegetation, with the invasion of rainforest trees once again.

The period of maximum aridity at this site, 26 000–10 000 years ago, encompasses the period of maximum extent of the Northern Hemisphere glaciers at about 20 000 years ago. Tropical aridity seems to have been widespread at this time, as is

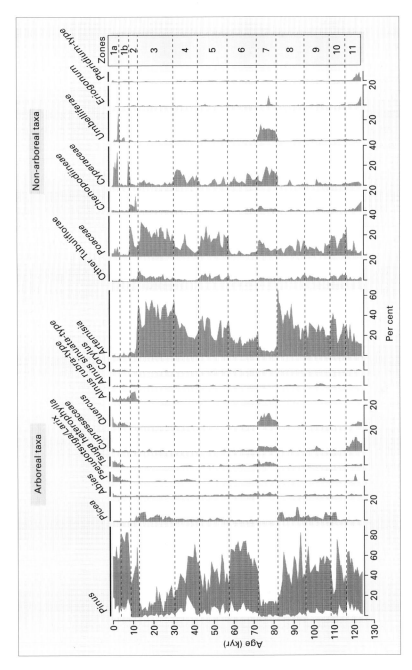

Fig. 9.10 Pollen diagram from Carp Lake, a crater lake in the eastern Cascade Mountains of the north-west United States. The sediments cover approximately 125 000 years of record and display the erratic response of vegetation to climatic instability during the last (Wisconsin) glacial. From Whitlock & Bartlein [13].

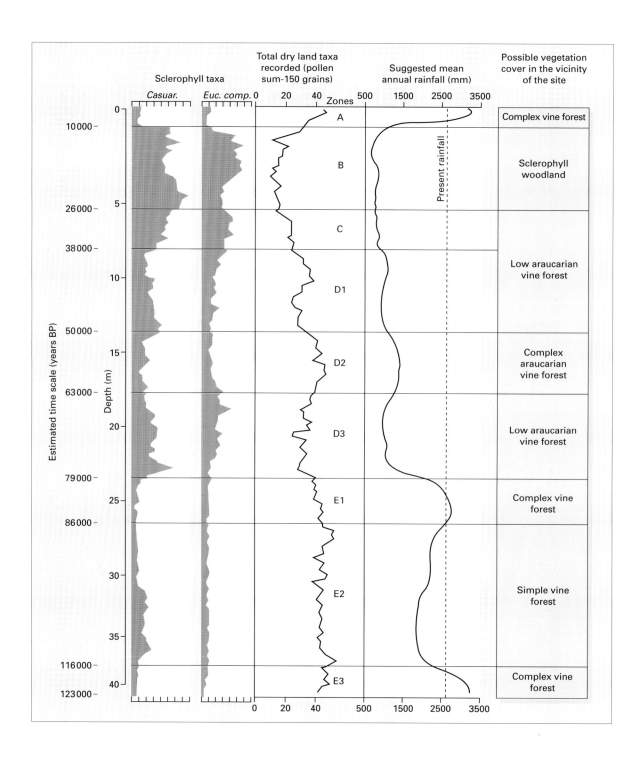

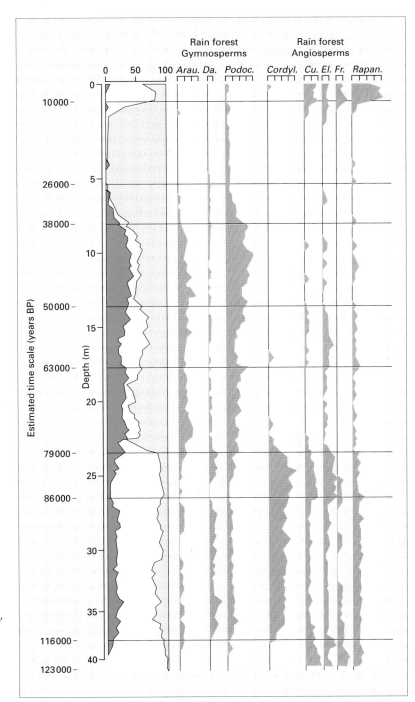

Fig. 9.11 (*Right and facing page.*) Pollen diagram from Lynch's Crater, Queensland, Australia. The column on the left shows the percentage of rainforest gymnosperms (grey); rainforest angiosperms (white); sclerophyll taxa (blue). The frequencies of pollen of all taxa are shown as percentages of the dry land plant pollen total; each division represents 10% of the pollen sum. Abbreviations for taxa: Arau., *Araucaria*; Da., *Dacrydium*; Podoc., *Podocarpus*; Cordyl., *Cordyline* comp.; Cu., Cunoniaceae; El., *Elaeocarpus*; Fr., *Freycinettia*; Rapan., *Rapanea*; Casuar., *Casuarina*; Euc., *Eucalyptus*. After Kershaw [14].

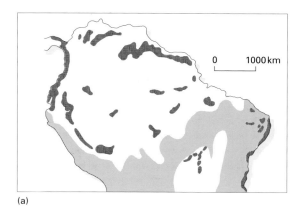

(a)

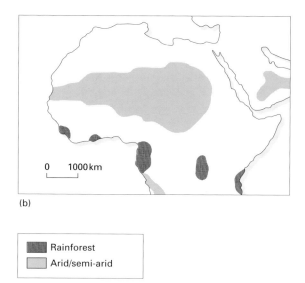

(b)

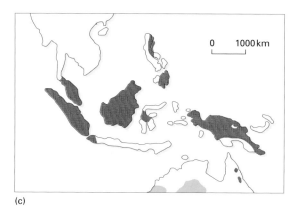

(c)

Rainforest
Arid/semi-arid

Fig. 9.12 Possible areas of humid rainforest (black) and arid/semi-arid areas (hatched) 20 000 years ago, when the glaciation of the higher latitudes was at its maximum. From Tallis [16].

confirmed by evidence from tropical Africa, India and South America. Much of the humid tropical forest of the Zaire basin was probably replaced by dry grassland and savanna during the glacial, although some forest may well have survived in riverine and lakeside situations. The temperature records from tropical locations, such as East Africa, follow closely the trends observed in other parts of the globe. Pollen analyses from lake sediments along the Uganda/Zaire border have shown that rainforest was replaced by dry scrub and grassland to the east of the Western Rift Valley during the height of the last glacial episode. Any refugial fragments must have been situated further west, in the lowlands of the Congo Basin.

Figure 9.12 shows a proposed reconstruction of the approximate areas of rainforest that existed at the time of maximum glacial extent in the high

latitudes (about 20 000 years ago). From this it can be seen that rainforest was much fragmented as a result of the glacial drought in tropical latitudes [16]. Many areas currently occupied by rainforest were reduced to savanna woodlands at this time, and this is an important point to bear in mind when considering the high species diversity of the rainforests (see Chapter 2). Most rainforests have not enjoyed a long and uninterrupted history but have been disrupted by global climatic changes. Even some of the regions (like the Uganda/Zaire borders) which are now hotspots of biodiversity bore a very different vegetation during the last glacial. But whereas the temperate forests were forced to occupy new areas at lower latitudes, the tropical forests had nowhere to which they could retreat, so became fragmented and dismembered. Some ecologists believe that this fragmentation may even have

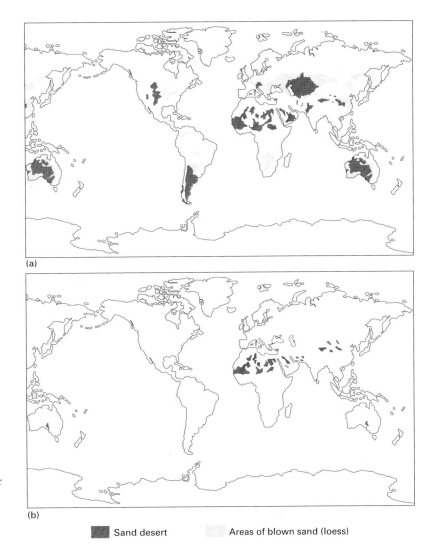

Fig. 9.13 (a) Proposed distribution of sand deserts at the height of the last glacial compared with (b) their modern distribution. Mobile sand (loess) was a feature of many areas during the glacial episode. From Wells [17].

Sand desert   Areas of blown sand (loess)

added to their diversity by permitting the isolation of populations and the development of new species.

Figure 9.12 also shows possible areas of drought at the time of maximum glacial extent, and the global pattern of dry regions is shown in greater detail in Fig. 9.13, where it is compared to modern desert distribution [17]. The global aridity of the last stages of the last glacial is very apparent. Blown sand (loess) from these deserts is found fossil in many parts of the eastern United States and in central Europe. Lake-level studies have similarly shown that the period between 15 000 and 20 000 years ago was particularly dry. For example, the lake levels from tropical Africa were at a low point during this time period, as shown by the data summarized in Fig. 9.14.

It would be misleading, however, to give the impression that drought prevailed throughout the earth at this time. Just as at the present day some areas are wetter than others, so it was in glacial times, and the western part of the United States enjoyed a time of wet climate and high lake levels

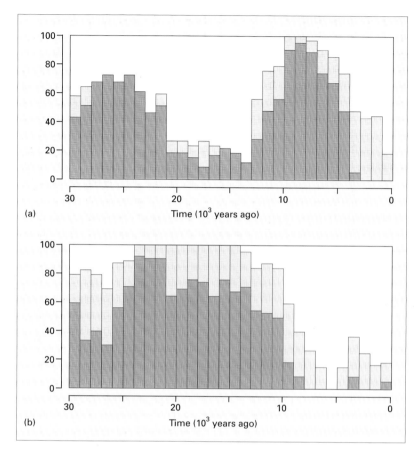

(a)

Time (10³ years ago)

(b)

Time (10³ years ago)

Fig. 9.14 Lake levels over the past 30 000 years in (a) tropical Africa and (b) the western United States. The bars represent the proportion of lakes studied with high (dark shading), intermediate (light shading) or low (white) lake levels. From Tallis [16].

between 10 000 and 25 000 years ago. Such wet periods are called 'pluvials', and many parts of the world experienced pluvials at various times in Quaternary history. In Africa, for example, 55 000 and 90 000 years ago seem to have been times of pluvial climate.

Evidence for the existence of pluvial lakes in western North America is provided not only by the geological deposits but also by the present-day distribution of certain freshwater animals. In western Nevada there are many large lake basins that are now nearly dry, but in the remaining waterholes there live species of the desert pupfish (*Cyprinodon*) (Fig. 9.15). Over 20 populations of the pupfish are known in an area of about 3000 square miles [18]. The isolated populations have gradually evolved

into what are considered four different species, each adapted to its own specific environment, rather like the Hawaiian honey-creepers (see Chapter 13). In many respects the wet sites in which these fish live can be regarded as evolutionary 'islands' separated from each other by unfavourable terrain. The species have probably been isolated from one another since the last pluvial at the beginning of the present interglacial, whereas populations within each species may still be in partial contact during periods of flooding. The pluvial periods of the Pleistocene provided the conditions necessary for a wider and more continuous distribution of these aquatic animals. A similar Old World example is provided by the waterbug genus *Corixa*, which is now widespread in Europe, but in Africa exists in only a few scattered

Males

Females

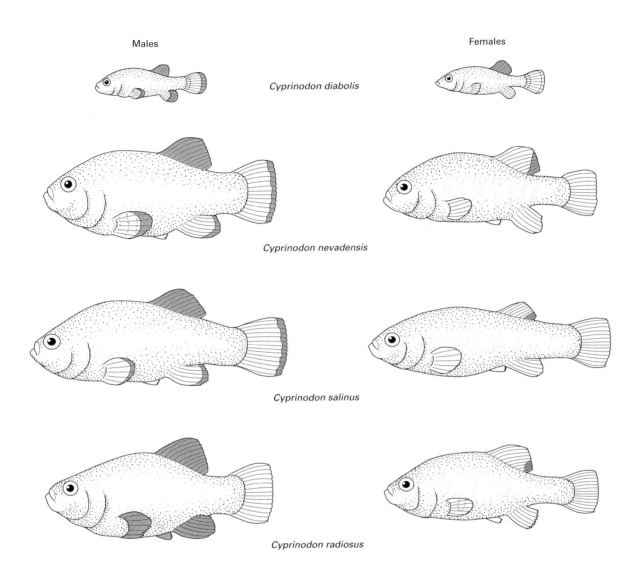

Cyprinodon diabolis

Cyprinodon nevadensis

Cyprinodon salinus

Cyprinodon radiosus

Fig. 9.15 The four species of desert pupfish that have evolved in the streams and thermal springs of the Death Valley region. Males are bright iridescent blue; females are greenish. After Brown [18].

localities extending south to the Rift Valley in Kenya. Presumably it was much more abundant when what is now the Sahara was dotted with lakes, and has since become restricted in distribution as the climate has grown drier.

Although pluvial conditions prevailed in western North America during the glacial maximum, this was not generally the case, as has been described. Glacial conditions in the high latitudes were associated with colder and drier conditions over much of the earth, as can be seen from the modelled reconstruction of glacial conditions in Fig. 9.16 [19]. Of particular note in this diagram is the greater reduction in temperature apparent in the high latitudes of the northern hemisphere, and also the increased wind stress during the glacial in the mid-latitudes of the north. Fossil evidence from various

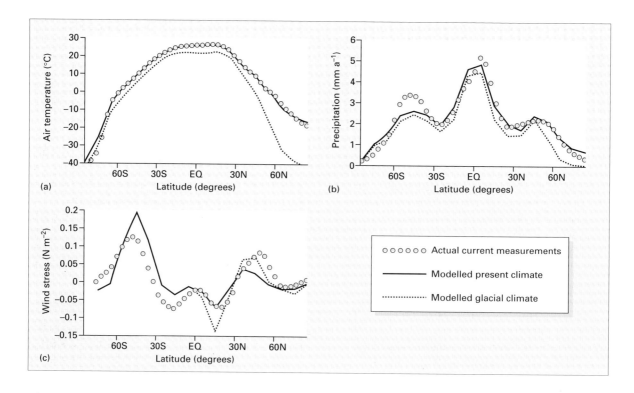

Fig. 9.16 Models of glacial and current climatic conditions (annual means) at different latitudes compared with actual modern measurements. Note the colder, drier, windier conditions during the glacial is especially pronounced in the Northern Hemisphere. From Ganopolski *et al.* [19].

parts of the world dating from the last glacial confirm that the glacial stages represented times of severe disruption to the entire biosphere. This fact, coupled with the assurance that we are still locked into the oscillating climatic system that has operated over the past two million years, makes it imperative that we should understand the mechanisms that have generated the glacial/interglacial cycle.

Causes of glaciation

Ice Ages are relatively rare events in the earth's 4.5 billion years of history. Although the polar regions receive less energy from the sun than the equatorial regions, they have been supplied with warmth by a free circulation of ocean currents through most of the world's history. Only occasionally do land masses pass over the poles, or form obstructions to the movement of waters into the high latitudes. The movement of Antarctica into its position over the South Pole led to the development of a Southern Polar ice cap perhaps as long ago as 15 million years. Rearrangements of the Northern Hemisphere land masses have subsequently led to the isolation of the Arctic Ocean so that it received relatively little influence from warm-water currents, leading to its becoming frozen some three to five million years ago. The presence of two such polar ice caps increased the albedo, or reflectivity, of the earth, for whereas the earth as a whole reflects about 40% of the energy that falls on it, the ice caps reflect about 80%. The formation of two such ice caps therefore significantly reduced the amount of energy retained by the earth. The scene was set for the development of an 'Ice Age'.

It is still necessary, however, to explain why the Ice Age has not been a period of uniform cold, but has consisted of a sequence of alternating warm and cold episodes. A proposal to explain this pattern was put forward in the 1930s by the Yugoslav physicist Milutin Milankovich, and has become widely accepted by climatologists. Milankovich constructed a model based on the fact that the earth's orbit around the sun is elliptical and that the shape of the ellipse changes in space in a regular fashion, from more circular to strongly elliptical. When the orbit is fairly circular there will be a more regular input of energy to the earth through the year, whereas when it is strongly elliptical the contrast between winter and summer energy supply will be much more pronounced. It takes about 100 000 years to complete a cycle of this change in orbital shape.

A second source of variation is produced by the tilt of the earth's axis relative to the sun, which again affects the impact of seasonal changes, with a cycle duration of about 40 000 years. The third consideration is a wobble of the earth's axis around its basic tilt angle, which shows a cycle of about 21 000 years. All of these cycles affect the strength of solar intensity that is received by the earth, and the pattern of climatic change should, according to Milankovich, be a predictable consequence of summing the effects of these three cycles.

Geophysicists working on the chemistry of ocean-floor sediments have expended much effort in seeking evidence for cyclic periodicity in past ocean temperatures and in checking the cycles found against those proposed by Milankovich. In 1976 Jim Hays, John Imbrie and Nick Shackleton [20] were able to confirm that all three levels of Milankovich cycle, 100 000, 40 000 and 21 000 years, could be detected in the sediments.

The Milankovich pattern has now been found in sediments dating back eight million years, but one important change has been detected in their effects. Whereas eight million years ago the effects of the 100 000-year cycle were weak, in the last two million years it is this cycle that has been very strong and that has dominated the glacial/interglacial

sequence. There must be additional factors that have amplified this particular cycle in recent times.

One possible explanation for the current exaggeration of the effects of the 100 000-year cycle is that ice masses themselves are responsible. The ice grows slowly and decays relatively quickly, which can itself modify the global climate. Computer models have been constructed which take account of this effect, and they produce a better fit to observed data than the Milankovich cycles on their own.

Detailed testing of the Milankovitch theory is, of course, dependent on well-dated records, and at present the dating of climatic fluctuations in the Pleistocene is crude. The difference in the dating of the commencement of the last interglacial, for example, varies over 17 000 years depending on the materials used. Until dating is more firmly established, full confirmation of the correlation between orbital cycles and climate cannot be achieved.

A further complicating factor in the elucidation of the causes of climatic change is that a shift in temperature can induce changes in the global carbon cycle that have feedback effects. The importance of this process has been revealed by direct measurements of the past atmosphere by the analysis of gas bubbles trapped in the Greenland ice cap [21]. Figure 9.17 shows the results of the analysis of carbon dioxide (top curve) and methane (bottom curve) in these gas bubbles, and the curves are compared with the proposed temperature curve (middle) for the last 160 000 years. It is very evident that both methane and carbon dioxide were elevated during the last interglacial to levels very similar to those of the present interglacial, and that they were depressed during the glacial episode. Even the minor fluctuations in these gases correspond quite closely to the proposed variations in temperature during the last glacial.

It is not surprising that an increase in warmth should be accompanied by increases in the atmospheric load of these two gases, for both are associated with the activities of organisms. Carbon dioxide is produced by respiratory activity of plants, animals and microbes, and one might well anticipate elevated levels during interglacials. The ocean also

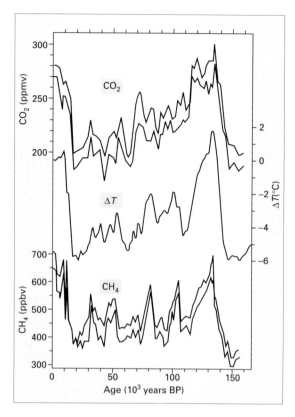

Fig. 9.17 Analyses of gas bubbles in an ice core taken from the Antarctic ice cap. These are regarded as fossil samples of the ambient atmosphere of the past. The concentrations of carbon dioxide (upper) and methane (lower) are shown together with a reconstruction of temperature change derived from oxygen isotope studies of the same area (middle). From Lorius *et al.* [21].

this process is that warm periods generate methane and carbon dioxide, and the presence of these gases in the atmosphere further elevates global temperature, so that there is a positive feedback and the temperature fluctuation is further amplified. Some climatologists believe that the rising levels of methane and carbon dioxide slightly precede the development of a warm episode and actually drive the climatic variations.

A possible consequence of these fluctuations in atmospheric carbon dioxide is that the anatomy of some plants may have altered in response to the changes. Since plants take up the gas through the pores (stomata) in their leaves, they may have needed fewer stomata under conditions of high carbon dioxide concentration. Ian Woodward of Sheffield University, England, has shown that plants have changed their stomatal density in parallel with carbon dioxide changes in the last century [22], and Jenny McElwain [23] has followed this up by looking at stomatal densities in more ancient plant materials. She has found a close link with supposed carbon dioxide levels in the distant past, so the technique may be sufficiently robust to use for long-term atmospheric reconstructions.

Wallace S. Broecker of Columbia University has emphasized the importance of the circulation of the earth's ocean currents, coupled with those of the atmosphere, to explain the rapidity with which some of the climatic shifts in the Pleistocene have come about [24]. His theory does not contradict the ideas of Milankovitch, but supplements them. Much of the warmth of the North Atlantic is brought into the region by highly saline waters heading northwards at intermediate depths of about 800 m. This current effectively redistributes the tropical warmth into the high latitudes, as seen in Chapter 4 (Fig. 4.10, p. 88). Broecker estimates that this input of energy into the North Atlantic is equivalent to 30% of the annual solar energy input to the area. Fossil evidence, however, indicates that this oceanic 'conveyor belt' became switched off during the glacial episodes, and the reduction in the heat transfer to high latitudes led to the development of the extended ice sheets. If this were so, it would

acts as an important reservoir for carbon, and raised carbon dioxide levels in the atmosphere may be related to a different behaviour of the oceanic mixing processes in glacials and interglacials. Methane is generated by partial decomposition of organic matter in wetlands and is also produced by ruminant animals and by termites in significant quantities, so it too might be expected to increase in warm periods. But both gases are strongly absorptive of infra-red radiation, so can contribute strongly to a 'greenhouse effect' in which they insulate the earth and enhance its energy retention. The outcome of explain the severe drop in temperature in the high latitudes of the northern hemisphere projected in Fig. 9.16a.

One further suggestion is that massive volcanic eruptions may precede and initiate the process of glaciation [25]. There is certainly some evidence for a correlation between glacial advances and periods of volcanic activity during the last 42 000 years in New Zealand, Japan and South America. Volcanic eruptions produce large quantities of dust, which are thrown high into the atmosphere. This has the effect of reducing the amount of solar energy arriving at the earth's surface, and dust particles also serve as nuclei on which condensation of water droplets occurs, thus increasing precipitation. Both of these consequences would favour glacial development. Attempts to correlate volcanic ash content with evidence of climatic changes in ocean sediments, however, have not met with much success, except to show an increase in the general frequency of volcanic activity during the last two million years.

Summary

1 The overall downward trend in global temperature observed in the Cenozoic finally resulted in the Quaternary Ice Age of the last two million years.
2 Within the Quaternary, glacials have alternated with interglacials in a cyclic pattern, but generally the cold phases have dominated.
3 Biomes have altered their patterns of distribution in response to changes, and communities of plants and animals have broken up and reconstituted in new assemblages. Some species have become extinct in the process.
4 Many areas that are now regarded as biodiversity hotspots were greatly altered during cold episodes, including the rainforests of equatorial regions, which are likely to have been reduced in area and fragmented.
5 Many factors have contributed to the development of the glacial/interglacial cycle, primarily external, astronomic factors, but also including internal changes, such as ocean/atmospheric circulation

patterns, feedbacks with living ecosystems, and perhaps volcanism.

Further reading

Delcourt HR, Delcourt PA. *Quaternary Ecology: A Palaeoecological Perspective*. London: Chapman & Hall, 1991.
Lowe JJ, Walker MC. *Reconstructing Quaternary Environments*. London: Longman, 1996.
Moore PD, Chaloner B, Stott P. *Global Environmental Change*. Oxford: Blackwell Science, 1996.
Tallis JH. *Plant Community History*. London: Chapman & Hall, 1991.

References

1 West RG. *Pleistocene Geology and Biology*, 2nd edn. London: Longman, 1977.
2 Jansen E, Sjoholm J. Reconstruction of glaciation over the past 6 Myr from ice-borne deposits in the Norwegian Sea. *Nature* 1991; 349: 600–3.
3 Moore PD, Chaloner B, Stott P. *Global Environmental Change*. Oxford: Blackwell Science, 1996.
4 Winograd IJ, Szabo BJ, Coplen TB, Riggs AC. A 250000-year climatic record from Great Basin vein calcite: implications for Milankovitch theory. *Science* 1988; 242: 1275–80.
5 Emiliani C. Quaternary paleotemperatures and the duration of high temperature intervals. *Science* 1972; 178: 398–401.
6 Moore PD, Webb JA, Collinson ME. *Pollen Analysis*, 2nd edn. Oxford: Blackwell Scientific Publications, 1991.
7 Stevenson AC, Moore PD. Pollen analysis of an interglacial deposit at West Angle, Dyfed, Wales. *New Phytologist* 1982; 90: 327–37.
8 Lowe JJ, Walker MJC. *Reconstructing Quaternary Environments*, 2nd edn. London: Longman, 1996.
9 Kerr RA. Second clock supports orbital pacing of the ice ages. *Science* 1997; 276: 680–1.
10 Delcourt HR, Delcourt PA. *Quaternary Ecology: A Paleoecological Perspective*. London: Chapman & Hall, 1991.
11 FAUNMAP Working Group. Spatial response of mammals to Late Quaternary environmental fluctuations. *Science* 1996; 272: 1601–6.
12 Dansgaard W, Johnsen SJ, Clausen HB, Dahl-Jensen D, Gundestrup NS, Hammer CU, Hvidberg CS, Steffensen JP, Sveinbjornsdottir AE, Jouzel J, Bond G. Evidence for general instability of past

climates from a 250-kyr ice-core record. *Nature* 1993; 364: 218–20.

13 Whitlock C, Bartlein PJ. Vegetation and climate change in northwest America during the past 125 kyr. *Nature* 1997; 388: 57–61.

14 Kershaw AP. A long continuous pollen sequence from north-eastern Australia. *Nature* 1974; 251: 222–3.

15 Jolly D, Taylor D, Marchant R, Hamilton A, Bonnefille R, Buchet G, Riollet G. Vegetation dynamics in central Africa since 18,000 yr BP. pollen records from the interlacustrine highlands of Burundi, Rwanda and western Uganda. *J Biogeogr* 1997; 24: 495–512.

16 Tallis JH. *Plant Community History*. London: Chapman & Hall, 1991.

17 Wells G. Observing earth's environment from space. In: Friday L, Laskey R, eds. *The Fragile Environment*. Cambridge: Cambridge University Press, 1989.

18 Brown JH. The desert pupfish. *Sci Am* 1971; 225 (11): 104–10.

19 Ganopolski A, Rahmstorf S, Petoukhov V, Claussen M. Stimulation of modern and glacial climates with a coupled global model of intermediate complexity. *Nature* 1998; 391: 351–6.

20 Hays JD, Imbrie J, Shackleton NJ. Variations in the earth's orbit: pacemaker of the ice ages. *Science* 1976; 194: 1121–32.

21 Lorius C, Jouzel J, Raynaud D, Hansen J, Le Treut H. The ice-core record: climate sensitivity and future greenhouse warming. *Nature* 1990; 347: 139–45.

22 Woodward FI. Stomatal numbers are sensitive to increases in CO_2 from pre-industrial levels. *Nature* 1987; 327: 617–18.

23 McElwain JC. Do fossil plants signal palaeoatmospheric CO_2 concentration in the geological past? *Phil Trans R Soc Lond B* 1998; 353: 83–96.

24 Broecker WS, Denton GH. What drives glacial cycles? *Sci Am* 1990; 262 (1): 43–51.

25 Bray JR. Volcanic triggering of glaciation. *Nature* 1976; 260: 414–15.

CHAPTER 10: *The making of today*

Following its maximum extent at about 21 000 years ago, the climate began to warm and the great ice sheets started to recede. Every indication pointed to the commencement of a new interglacial. Our present interglacial has been a very brief episode in the history of the earth, yet it has seen changes over the face of the planet that have taken place at a faster rate than ever previously observed. The last 10 000 years or so differ from the rest of the earth's history in one important respect; they have seen the rise to a position of ecological dominance of a single, influential species, *Homo sapiens*. During this interglacial our species has emerged from being one among a number of socially organized predators into a position from which it can and does modify whole global cycles and climates. It is no exaggeration to state that there is no habitat or ecological process on earth that can now be studied without reference to the impact of this species. It is natural therefore that this short, final part of the earth's history demands a biogeographer's special attention.

The current interglacial: a false start

Before one can undertake a survey of the rise of our species, it is worth pausing to consider the climatic instability that typifies the closing stages of the last glaciation and the opening of the current interglacial, for it tells us something of the rapidity with which climate can change, and may serve as a warning about future changes. This instability was first noted by geologists in Denmark, who found that the sediments of lakes dating from the transition between the glacial and the current interglacial exhibited some unusual features. The inorganic clays, typical of sediments formed in lakes surrounded by arctic tundra vegetation where soils are easily eroded and the vegetation has a low organic productivity, were replaced by increasingly organic sediments as the development of local vegetation stabilized the mineral soils, and the aquatic productivity of the lakes led to an increasing organic content in the sediments. But this process, reflecting the increased warmth of the climate, was evidently interrupted, for the sediments then reverted to heavy clay, often with angular fragments of rock denoting a return to severe climatic conditions. Above this layer, the organic sediments reappeared and the climatic warming became evident once again, this time leading to the development of our present interglacial, but the interruption proved to be a consistent feature of sediments of this age throughout north-west Europe.

The cold episode that caused the deposition of these sediments was severe enough to cause the regrowth of many glaciers, and geomorphologists have shown that the glaciers of Scandinavia and Scotland extended considerably during this episode, while small glaciers began to form on north-facing slopes in more southerly mountains, from Wales to the Pyrenees. The event was called the 'Younger Dryas', because the fossil leaves of the arctic plant the mountain avens (*Dryas octopetala*) were found in abundance in the clay layers during the original Danish studies (Fig. 10.1). Radiocarbon dating of the Younger Dryas from a range of sites indicates that it took place somewhere between 10 800 and 10 000 years ago (measured on the 'radiocarbon time scale' which, at this age, is about 1000 years younger than true, solar time [1]), interrupting a warming process that had begun around 14 000–13 000 years ago. In a number of sites in continental Europe a shorter and less severe interruption to the warming process (called the 'Older Dryas') was also detected and dated to around 12 000–11 800

Fig. 10.1 Mountain avens (*Dryas octopetala*), an arctic–alpine plant, remains of which were found in sediments from Denmark and which has lent its name to the Younger Dryas cold event.

seaboard of North America, is difficult to detect when one moves inland. There is some evidence for a cooling at this time in the northern Pacific, but again it is less strong than in the Atlantic. The evidence therefore suggests that the event was centred upon the North Atlantic, and oceanic information from fossils in the sediments shows that a cold front of polar water did indeed extend southwards as far as Iberia (Spain and Portugal) at this time.

One possible explanation of how this climatic reversion, centred in the North Atlantic, may have come about was put forward by Claes Rooth of the University of Miami and was later supported by the data of Wallace Broecker [3] and his coworkers. They suggest that the change was brought about by an alteration in the pattern of discharge of melt-water from the great Laurentide glacier of North America. As this ice mass, covering the whole of eastern Canada, melted, its waters flowed initially down to the Gulf of Mexico. During the Younger Dryas, however, it is proposed that the meltwater was re-routed via the St Lawrence into the North Atlantic (Fig. 10.2). Not only would this bring large volumes of cold water into the North Atlantic, but the fresh water would also dilute the saline waters on the oceanic conveyor belt (Chapter 4) and could disrupt the global movement of this conveyor. Indeed, modifications to the conveyor could prove an important component in many climatic changes [4]. Such modification would explain why the Younger Dryas was most strongly felt in the regions around the North Atlantic, where the warm influence of the conveyor has its greatest impact. But if meltwater flow and salinity changes were responsible for switching off the conveyor, we would expect the onset of the Younger Dryas to be accompanied by a rapid rise in sea level. The work of R.G. Fairbanks [5] suggests that ice sheets were not melting rapidly at the time of the onset of the Younger Dryas, so the meltwater hypothesis is not supported. Subsequent work on the growth of deep sea corals in the North Atlantic also confirms these findings [6] and the Broeker model looks unlikely. The cold arrived and the oceanic

years ago, but its impact was geographically more confined and its influence was much weaker than that of the Younger Dryas.

Although stratigraphic and fossil evidence for the Younger Dryas cold phase was abundant in north-west Europe, it was more difficult to discern in the sediments of the Alps, especially on the southern side. Analyses of lake sediments from Italy [2] and Bulgaria show that vegetation even in these more easterly regions was affected by the Younger Dryas, and sediments from the Pyrenees show a strong impact. On the other side of the Atlantic, however, its influence, though apparent on the eastern

Fig. 10.2 The extent of the Laurentide ice sheet in the final stages of the last Ice Age (about 12 000 years ago).

circulation changed, but which is cause and which is effect still needs to be clarified.

If the oceanic conveyor was disrupted, however, one might expect its effects to be felt globally, and evidence is mounting that this was indeed the case. Although its effects are stronger in some parts of the world than in others, the Younger Dryas was undoubtedly a time of global cooling. Its occurrence was clearly linked to changes in the circulation of the oceans and related movements of the atmosphere. Not only did this cause climatic cooling, but also droughts in some parts of the world, such as California, where lake levels fell and saline playas became more widespread during Younger Dryas times [7].

Perhaps the most interesting and important aspect of the Younger Dryas is that it demonstrates how quickly the earth's climate can flip from interglacial to glacial mode and back again. The brief warm spell before the Younger Dryas was sufficiently hot to bring beetles with Mediterranean affinities as far north as the middle of England. Yet within a few centuries the area was back in glacial conditions. Perhaps even more remarkable is the change at the end of the Younger Dryas. The work of Willie Dansgaard [8] of the University of Copenhagen, and his colleagues, on the Greenland ice cap have shown that the Younger Dryas came to an abrupt end 10 720 years ago, and that it was followed by a warming of about 7°C in only 50

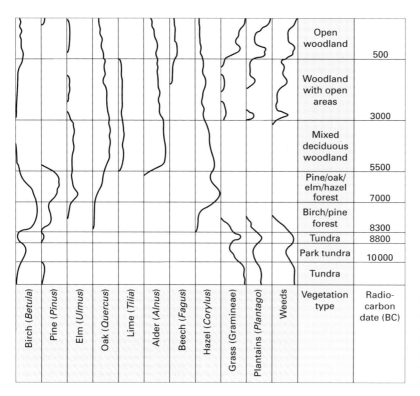

Birch (*Betula*)	Pine (*Pinus*)	Elm (*Ulmus*)	Oak (*Quercus*)	Lime (*Tilia*)	Alder (*Alnus*)	Beech (*Fagus*)	Hazel (*Corylus*)	Grass (*Gramineae*)	Plantains (*Plantago*)	Weeds	Vegetation type	Radio-carbon date (BC)
											Open woodland	500
											Woodland with open areas	3000
											Mixed deciduous woodland	5500
											Pine/oak/elm/hazel forest	7000
											Birch/pine forest	8300
											Tundra	8800
											Park tundra	10000
											Tundra	

Fig. 10.3 Generalized pollen diagram from the southern part of the British Isles, showing the changing proportions of tree pollen and some non-arboreal types during the final stages of the Pleistocene and through the Holocene.

years. It is possible that even this remarkable rate of change may be an underestimate of what took place. Deep-sea coral records from the Atlantic indicate a very high level of climatic instability and the swing from interstadial climate to the cold Younger Dryas may have occurred in as little as 5 years.

One might ask why this particular event, many millenia ago, should attract so much current research attention. But in days when the climate may well have been destabilized by human activity, it is important for us to understand what factors control the climate and whether there are certain thresholds which, when crossed, may lead to profound changes in the climatic balance of the entire world—as in the case of the Younger Dryas event. There is another aspect to the palaeoecolgical work on the closing stages of the last glacial. The detailed study of these times could help us to understand how plants and animals react to very rapid and very considerable changes in climate.

Forests on the move

After this faltering start to the current interglacial, the general rise in temperature (recorded in the oxygen isotope ratios of the ocean sediments and in the ice of the ice caps) was consistently maintained, and the vegetation and animal life of the earth once again had to adjust to a new set of conditions. The pollen grains preserved in the accumulating lake sediments of 10 000 years ago provide detailed records of the arrival and expansion of tree populations as species distribution patterns changed and forests were reconstituted. Figure 10.3 shows a generalized pollen diagram for southern Britain, compiled from a range of sites.

In part, the sequence of trees (birch, pine, elm plus oak, lime and alder) reflects the changing climatic tolerances of species as the climate warmed, but it is also a function of their relative speed of migration and the distances which they had to cover to invade land exposed by the retreating glaciers. Birch, for example, has light, airborne fruits which can travel considerable distances. It is also able to produce fruits when only a few years old, which permits a rapid expansion of its range. Add to this the fact that it is cold-tolerant and may have survived on the mainland of Europe quite close to the British Isles during the last glaciation, and it is not surprising that birch should be the first tree to appear in any abundance in the postglacial pollen record. One must also remember that birch produces large quantities of pollen and that one species, *Betula pubescens*, is likely to have been growing locally around the lake sites in which pollen-bearing sediments are found. Many factors therefore must be considered before a pollen sequence such as that displayed in Fig. 10.3 can be interpreted in terms of changing climate. Some elegant techniques have been employed to try to translate pollen densities in sediments into estimates of the population densities. In this way one can trace the population expansion of trees as they invaded new areas [9].

Overall, it is evident from the data that there was a warming of the climate during the early stages of the interglacial, reaching a maximum between 7000 and 5000 years ago (5000–3000 BC). This warmth maximum is marked by the invasion of Britain by lime (*Tilia*), which extended its range as far as southern Scotland and locally even further north at this time.

Large numbers of pollen diagrams covering the current interglacial are now available from all over the world and many of them are firmly dated by means of radiocarbon. This has made it possible to study the movement of individual species and genera of plants by constructing pollen maps for particular periods in the past. In this way one can follow the spread of trees, for example, and observe the routes along which they have dispersed and the way in which they have reassembled themselves into reconstituted forests. Figure 10.4 shows some examples of this type of work taken from the data analyses of George Jacobson, Tom Webb and Eric Grimm on the recolonization of America by trees after the last glaciation [10].

Just three tree taxa have been selected from their extensive studies to illustrate the patterns of tree movement in the last 18 000 years. Spruce survived the glacial maximum in the mid-west and along a broad front immediately to the south of the Laurentide ice sheet. It followed close on the retreating ice front, eventually settling in Canada. Pollen levels suggest that it has increased in density over the past 6000–8000 years, just as it did in the later stages of many earlier interglacials. Pine is more difficult to interpret because there are several species all with very different climatic requirements, and they cannot be effectively separated on the basis of their pollen grains. Other fossil material, such as their needles, however, indicates that both southern and northern pine types survived the glaciation in the south-east of the United States, mainly on the Atlantic coastal plain. The two groups subsequently separated, one group heading north to invade areas left bare by retreating ice, and the other group becoming firmly established in Florida. The oaks had found refuge from the glaciation in Florida, where they had achieved a marked dominance by 8000 years ago. Subsequently, they have declined in Florida and have moved into their current stronghold to the south of the Great Lakes area.

The technique of mapping tree movements has provided some valuable information about the rapidity with which different species can respond to climatic change, which may well prove useful when we are concerned with predicting future responses to our currently changing climate. The rates of spread in response to climate change, even of large-seeded species such as the oaks, are surprisingly rapid. In Europe, for example, oak spread reached 500 m per year in the early part of the current interglacial. Rates of 300 m per year are common for many tree and shrub species.

(a)

	Ice sheet
	5–20% pollen
	20–40% pollen
	>40% pollen

(b)

	Ice sheet
	20–40% pollen
	40–60% pollen
	>60% pollen

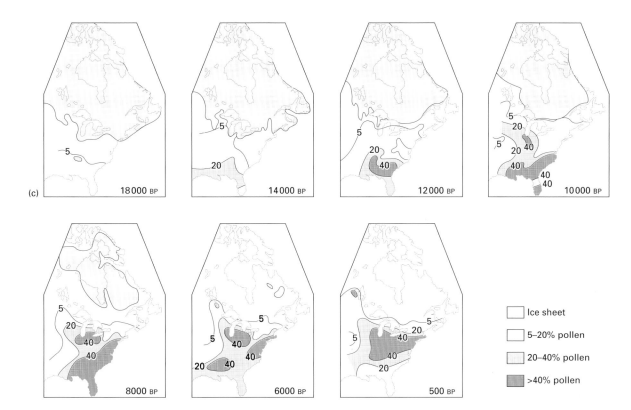

Fig. 10.4 (*Above and facing page.*) Spread of selected trees through North America since the last glacial maximum. Dates are in radiocarbon years Before Present (BP), they have not been correct to solar years. Contours ('isopolls') join sites of equal pollen representation in lake sediments from the appropriate time period. (a) Spruce (*Picea*). (b) Pine (*Pinus*). (c) Oak (*Quercus*). From Jacobson *et al.* [10].

The emergence of humans

It was not only trees and other plants that moved into the higher latitudes from their glacial refuges to take advantage of the new lands being released from their former ice cover. Many animal species expanded into the new regions and among them was our own species. *Homo sapiens* evolved during the late Pleistocene, and the arrival of this species on the scene proved of such dramatic ecological importance that it is necessary to interrupt the narrative of the current interglacial to trace the emergence of our species.

The fossil history of humans is very incomplete, but each year brings new fossil material which helps to fill in the gaps and provides a more detailed picture of how anatomically modern humans emerged. The primates of the New World and the Old World became separated some 40 million years ago, and it is the Old World branch that was ancestral to humans. The hominoids, which include such living primates as the orang-utan, the chimpanzee and gorilla, first began their development in Africa and spread into Asia when the two continents collided at the end of the Miocene, some 17 million years ago (see Chapter 7). The last common ancestor of the true hominids and our nearest living relative, the chimpanzee (with which we share almost 99% of our genetic make-up), lived about 20 million years ago [11].

The hominids themselves are first known in East Africa, in Tanzania and in Ethiopia, about 6 million years ago or more. Among the fossils of this age is

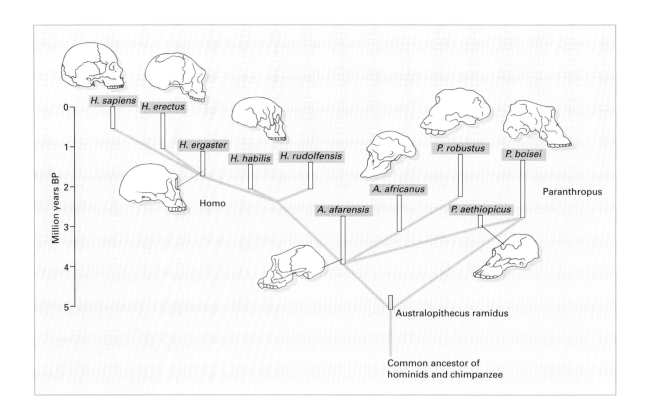

Fig. 10.5 A scheme showing the possible interrelationships between hominids over the past five million years. The diversification of hominids over the past two million years is apparent. Based on the scheme of Bernard Wood [15].

the partial skeleton which has come to be known as 'Lucy', and which has supplied a great deal of anatomical information about the early hominids. These fossils have been assigned the name *Australopithecus afarensis*, and it is believed that they walked upright on their hind legs—a conclusion that is supported by the extraordinary discovery of fossil human footprints in volcanic ash from Tanzania [12]. The habitat in which these organisms lived was open woodland and savanna, far from the dense tropical forests, but very little is known of their precise way of life and the ecological niche which they occupied in the ecosystem.

Walking on two legs, it is probable that they used tools with their free hands, much in the way that a chimpanzee is capable of doing, but there is no evidence that they were able to make extensive modifications to natural tools. In fact, analysis of the bone structure of Lucy and other fossils of *A. afarensis* indicate that they did not have the appropriate hand structure for toolmaking [13]. They probably contented themselves with using the sticks and stones that they found. Whatever its ecological role in the ecosystem, we must regard this primitive hominid as one more species in the complex food web of the ecosystem that it occupied. There is no reason to believe that it was more influential than any other species.

By 2.5 million years ago, evolutionary developments had taken place in the hominid line and there were at least three species of hominids present in East Africa (Fig. 10.5): two of these were species of *Australopithecus*, but the third belonged to our own genus and has been called *Homo habilis*. This species had a larger brain than its relatives, and there is evidence that it was able to modify stone tools [14], thus showing a marked cultural change which was to continue with remarkable consequences along the *Homo* line of development. Surprisingly, following the development of a tool-making culture, no real change in technology occurred for the next million years. Knowledge of the diet and the ecological role of *Homo habilis* is still fragmentary, but it is believed that the meat content of the diet increased.

Some time prior to one million years ago a new hominid appeared on the scene, *Homo erectus*, and there is every reason to believe that this new species was the direct evolutionary product of *Homo habilis*. There are two aspects of *Homo erectus* that are of biogeographical significance: first its cultural developments and secondly its distributional changes. On the cultural side we find much more sophisticated stone tools, including structures which could be termed hand axes, and even more significantly there is evidence that some populations of *Homo erectus* used fire. Clearly this latter development provided the species with the capacity to modify its environment more profoundly than any other cultural development to date. Although fire would have been used as a means of food preparation, its potential as an aid in hunting must surely have been appreciated by this intelligent species, and the consequences must have been felt in terms of the new constraints placed upon the flora and fauna of its surroundings. The distribution of the species is also of interest, for this is the first species of our genus to be found outside Africa, and by one million years ago it had spread into eastern Asia, Java and southern Europe. Some excavations in Java indicate that *Homo erectus* may have survived in South-East Asia as late as the last Ice Age (50 000 years ago), in which case it would have overlapped with our own species in the area [16].

From this widespread species, modern humans have emerged during the last million years. There are still disputes about the precise relationship between modern *Homo sapiens* and the type known as Neanderthal. A possible common ancestor has been described from Spain [17] and the two lines of evolution, based on DNA analyses, clearly developed independently for at least half a million years [18]. It is now apparent that the two human types co-existed during the last glaciation, but only one emerged from the glacial, and that was our own species. Whether this replacement actually involved the deliberate destruction of the Neanderthals, or whether it was a consequence of interbreeding remains uncertain, but the fact remains that only *Homo sapiens* remained when the Holocene began around 10 000 years ago.

The spread of *Homo sapiens* into the New World probably took place during the last glacial episode, when sea levels were sufficiently low to expose a land bridge across what are now the Bering Straits. The exact date of arrival is still disputed, but indirect evidence from the geography of human languages suggests that the invasion must have taken place before the major advance of the last glaciation at about 20 000 years ago. The linguistic research of R.A. Rogers [19] has revealed three distinct groups of native American languages, and these are centred on the three ice-free refugial areas of North America during the height of the last glaciation (Fig. 10.6). It seems likely that human populations were isolated in these three areas during the glacial maximum and subsequently spread to other regions. The extinct language of Beothuk, once spoken in Newfoundland, could belong to another population isolated in that eastern refugium.

An alternative explanation is that there have been three separate invasions of North America from Asia, thus accounting for the three language groups [20]. A complication in all theories of the peopling of the Americas is the existence of early human occupation in southern Chile, far from the proposed Beringian point of entry [21]. This settlement at Monte Verde dates from around 12 500 years ago, so it is possible that an early wave of immigrant peoples from eastern Asia made their way as far as the southern part of South America.

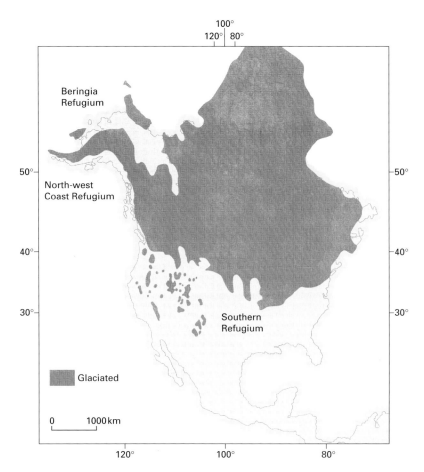

Fig. 10.6 Maximum extent of the last (Wisconsin) glacial in North America, showing the three ice-free areas (refugia) which correspond to the language groups of native Americans. From Rogers *et al.* [19].

Modern humans and the megafaunal extinctions

The spread of the human species during the last glaciation was accompanied by an alteration in the fauna of North America and of Europe, namely the extinction of many species of large mammals —the megafauna. It was long assumed that these extinctions were the result of the climatic changes, but the American anthropologist Paul Martin suggested [22] that humans may have been the culprits. Martin pointed out that most of the animals that became extinct were large herbivorous mammals or flightless birds, weighing over 50 kg body weight —precisely the part of the fauna that humans might have been expected to hunt. He also pointed out that similar extinctions had taken place in other, more southern areas than North America, and suggested that the timing of these extinctions varied, in each case the time corresponding with the evolution, or arrival, of a race of humans with relatively advanced hunting techniques. In Africa, for example, where humans probably evolved, the extinctions of many large herbivores apparently took place before those in the Northern Hemisphere. However, it is difficult to date these precisely, and even more difficult to correlate them with any changes in the hominid cultures of that continent. The same is true of the extinctions that took place in South America.

It is in North America that the record of extinctions has been studied in the greatest detail. Martin suggests that 35 genera of large mammal (55 species) became extinct in North America at the end of the last (Wisconsin) glaciation—over twice as many as had taken place during all the earlier glaciations, and this at a time when the climate was already improving. This certainly seemed to support the idea that some agent other than climate had been responsible, and it seemed reasonable to suspect the hunting activities of humans. But the American anthropologist J.E. Grayson [23] has shown that there was a similar rise in the level of extinctions of North American birds (ranging from blackbirds to eagles) at that same time. Since it is unlikely that early humans were responsible for the extinction of these birds, this observation throws doubt upon the whole hypothesis of the dominant role played by humans in Pleistocene extinctions in general.

On the other hand, the fact that so many North American species become extinct at the same time (11 500–10 500 years ago) with the arrival of hunting peoples, whereas in Europe the extinctions were spread over a longer period, provides a strong body of circumstantial evidence to support Paul Martin's claim that humans are to blame [24]. The debate continues, and the extinction of so many different species of mammal may not have been caused by one single factor, but there is an increasing number of studies in which the time of extinction can be precisely correlated with the arrival or intensification of settlement by human populations. An example is given in Fig. 10.7, showing the dates of the last recorded dung of the Shasta ground sloth from a variety of American sites [25]. Extinction seems to have taken place at many localities at about 11 000 years ago, which is when hunting cultures—Clovis Man—were most active in the region. It is difficult to escape the conclusion that the marks of human activities first became noticeable very early in the present interglacial. It is also clear that the early impact of human cultures was essentially negative with respect to biodiversity. Perhaps the negative impact was even greater than is apparent from the megafaunal extinctions as some

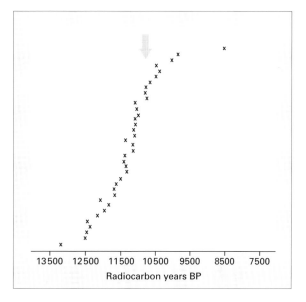

Fig. 10.7 Radiocarbon dates of samples of the last recorded dung from the Shasta ground sloth in the south-west of the United States. The arrow represents the time at which the Clovis hunters were most active in the region. From Lewin [25].

of these large herbivores may have acted as keystone species with many other species dependent upon them. Major habitat alterations could have accompanied the loss of the megafauna and many smaller species of animal, plant and microbe may well have followed them unnoticed into extinction.

Domestication and agriculture

It was while these extinctions were taking place in various parts of the world that populations of humans living in south-west Asia were experimenting with a new technique for enhancing their food supplies. In the fertile region of Palestine and Syria grew a number of annual grasses with edible seeds—the ancestors of our wheat and barley. Like many other plants, these species were advancing north with the improving climate, and they may have proved particularly adept at exploiting the clearings that human groups made for their camps and settlements in the developing woodland. No

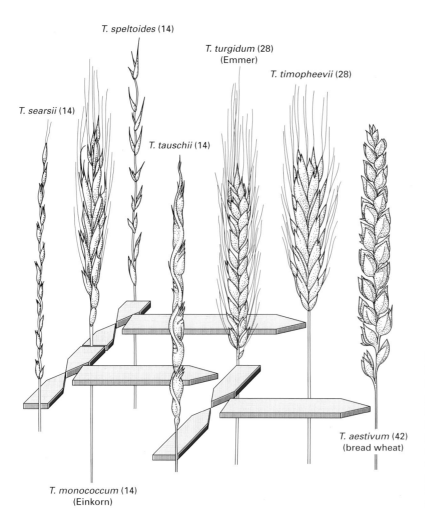

T. speltoides (14)

T. turgidum (28)
(Emmer)

T. timopheevii (28)

T. searsii (14)

T. tauschii (14)

T. aestivum (42)
(bread wheat)

T. monococcum (14)
(Einkorn)

Fig. 10.8 The evolution of modern bread wheat. This reconstruction is tentative, but represents the probable course of crossings among the wild wheat species that led to the early domesticated forms of the genus *Triticum*, and the subsequent crossings of domesticated wheats with wild species and chromosome doubling that led to bread wheat. Figures in parentheses after names represent chromosome numbers.

doubt the inhabitants of such settlements soon discovered the nutritional value of their seeds and learned to propagate and ultimately cultivate them.

Tracing the ancestry of our modern species of cereals is difficult, but Fig. 10.8 represents a possible scheme for the evolution of modern wheat, based on a study of the chromosome numbers of the various wild and cultivated species. The original wild wheats undoubtedly had a total of 14 chromosomes (seven pairs) in each cell. *Triticum monococcum* (einkorn) was the first wild wheat to be extensively used as a crop plant. It probably hybridized with other wild species, but the hybrids

would have proved infertile because the chromosomes could not pair up prior to the formation of gametes. Faulty cell division in one of these hybrids, however, could solve this problem because once the chromosome number had doubled (polyploidy) chromosome pairing could take place and the species would become fertile. Fertile polyploid hybrid species formed in this way included another important crop species, emmer (*T. turgidum*) with 28 chromosomes. But the bread wheat of modern times (*T. aestivum*) has 42 chromosomes and this probably arose as a result of polyploidy following the hybridization of emmer with another wild species, *T. tauschii*. This species comes from Iran and it probably interbred with emmer as a result of the transport of that wheat species to Iran by migrating human populations. Thus, early agricultural peoples began not only to modify their environment but also to manipulate the genetics of their domestic species.

More sophisticated studies have used not simply chromosome number but DNA fingerprinting of wheat seeds. Manfred Heun, from the Agricultural University of Norway, and his colleagues [26] have analysed 338 samples of the most primitive wheat species, einkorn, which still grows wild in the Near East. Their aim was to determine the precise locality where wheat domestication first took place. They made two important assumptions as a basis of their work; first, that the DNA constitution of the species had not altered substantially since domestication took place, and secondly that they could determine how closely populations of the plant were related by reference to marker sections of the DNA sequence. They were able to locate wild einkorn populations in a region of the Karacadag Mountains in south-eastern Turkey and close to the Euphrates River that provide genetic evidence of being the progenitors of domesticated einkorn (Fig. 10.9). The application of molecular techniques in the study of domestication and of biogeographical origins of many plants and animals will undoubtedly become increasingly important in solving some important and some long-unanswered questions.

Other wild plants of the Middle East were also grown and cultivated by humans (Fig. 10.10), among them barley, rye, oat, flax, alfalfa, plum and carrot [27]. Further west, in the Mediterranean basin, yet more native plants were domesticated, including pea, lentil, bean and mangelwurzel. An analysis of all the accumulated data from European archaeological investigations shows how these crop plants gradually became used more extensively across the continent and how new species were adopted in regions where climatic factors limited the use of some early types. Some of these results are shown in Fig. 10.11. *Lens culinare*, *L. orientalis* and *L. nigricans*, the lentils, have proved useful

crop species in the Near East, but climatic factors have prevented their being used extensively in northern Europe. Barley (*Hordeum vulgare*), however, proved much more suitable for cultivation in Europe than it had in the early domestication sites in south-west Asia.

From information such as this, it is possible to trace the spread of the agricultural idea from the Near East across the continent of Europe during the Holocene (Fig. 10.12). It is still disputed whether this process involved the movement of peoples, or the diffusion of a new technique. If agriculture had proved an efficient means of stabilizing food supplies and avoiding some of the chance catastrophes of hunting and gathering, then it could have led to population expansion and the need to move to new areas. Some archaeologists, such as Colin Renfrew from Cambridge, England, have investigated the association between agricultural spread and the assumption of dominance by certain language groups. In Europe and Asia, for example, 144 of its languages belong to the group known as Indo-European, and the reason for this dominance could be its spread with agriculture in the early Holocene [28]. The languages of the early agriculturalists may well have become 'steamroller' tongues as they spread out from the sources of domestication.

The idea of plant domestication seems to have evolved independently in many different parts of the globe and at many different times. In each area, appropriate local species were exploited: in south-west Asia there were millet, soybean, radish, tea, peach, apricot, orange and lemon; central Asia had spinach, onion, garlic, almond, pear and apple; in India and South-East Asia there were rice, sugar cane, cotton and banana. Maize, New World cotton, sisal and red pepper were originally found in Mexico and the rest of Central America, while tomato, potato, common tobacco, peanut and pineapple first grew in South America.

In some cases there may have been independent cultivations of the same or similar species in different parts of the world. Thus, emmer wheat may well have originated quite independently in the Middle East and in Ethiopia.

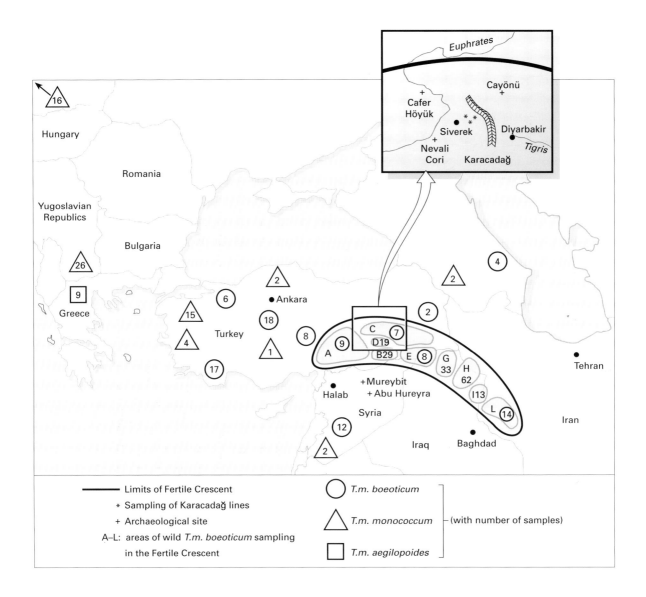

Fig. 10.9 Map of the Near East showing the 'Fertile Crescent' region and, within it, the probable location of the site of domestication of einkorn wheat (the Karacadag Mountains of southern Turkey), based upon studies of molecular genetics. Data from Hean *et al.* [26].

New World agriculture is thought to have begun in Central America with the cultivation of three major crops, maize (*Zea mays*), bean (*Phaseolus vulgaris*) and squash (*Cucurbita pepo*). Some argument has surrounded the time of this independent agricultural development, especially in relation to agricultural origins in other parts of the world. Radiocarbon dates of squash seeds from caves in Oaxaca, Mexico, place their cultivation at about 10 000 years ago, so an early origin for New World agriculture is now well established [30].

One of the most puzzling problems in the study

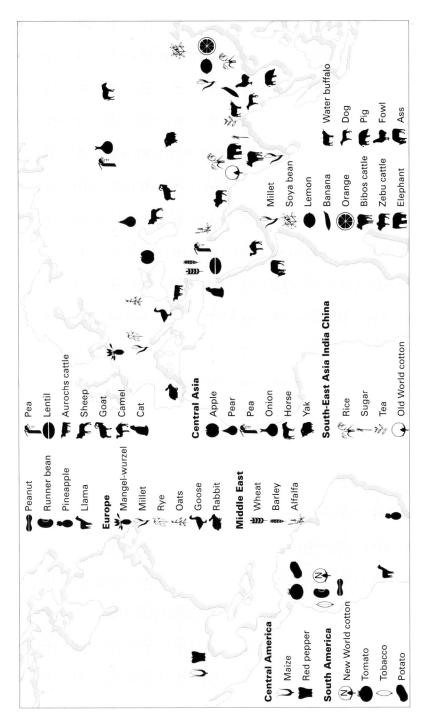

Fig. 10.10 Areas of probable origin of domesticated animals and plants.

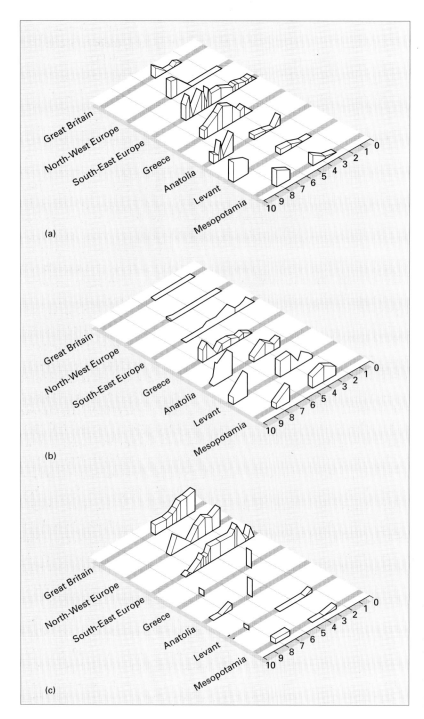

(a)

(b)

(c)

Fig. 10.11 Diagrams illustrating the spread of crop species from the Near East through Europe. The vertical scale represents the percentage of sites in which the species is present and the horizontal scale is in thousands of radiocarbon years before present. Blanks indicate a lack of data. The species represented are: (a) *Triticum monococcum*, einkorn wheat; (b) *Lens culinare, L. orientalis* and *L. nigricans*, lentils; (c) *Hordeum vulgare*, six-row hulled barley. From Hubbard [27].

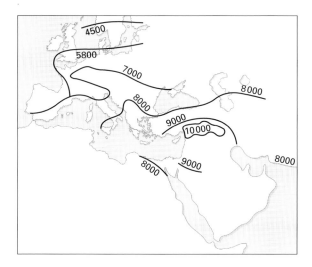

Fig. 10.12 Map of Europe showing the spread of agriculture from the area of the Fertile Crescent. Dates are in years before present. The pattern is greatly simplified here, and there are problems concerning the precise time and direction of agricultural spread in such regions as the Balkans [29].

of plant domestication is the origin of maize. Unlike most of the cereals of Eurasian origin, there is no wild species that can be regarded with certainty as its progenitor. Most research workers into this subject are now agreed that the most likely ancestor is the Mexican annual grass, teosinte. Indeed, both maize and teosinte are now regarded as subspecies of *Zea mays*, but structurally they are very different, and in particular the evolution of the all-important flower and fruit structure is still in dispute. The real problem is that there are no intermediate types between teosinte and maize, and neither are there any archaeological records of wild teosinte grains being collected and used for food. One explanation of the sudden appearance of maize without any intermediate form being recorded has been put forward by the botanist Hugh Iltis [31], who claims that an abnormality arose in teosinte, possibly set off by some environmental factor, in which male inflorescences became female on the lateral branches, thus producing a kind of quantum leap in evolution. There are currently four wild species

of the maize genus know, all found in Mexico, and their genetic constitution and their conservation must be high priorities for research [32].

Very often maize appears quite suddenly in the archaeological record of the diet of early human populations, and it may then become the dominant food resource. This can be traced by an elegant chemical technique in which the ratio of carbon-13 to carbon-12 in fossil human bone structures is analysed. Maize is a C_4 plant (see Chapter 3) and one of the characteristics of these plants is that the ratio of ^{13}C to ^{12}C in the sugars produced by photosynthesis differs from that found in C_3 plants. This difference is retained in animals feeding on the plant, so we can trace the importance of C_4 plants in an animal's diet [33]. In this case it can be used to trace the adoption of maize cultivation during the history of a human population.

The domestication of certain animals may have preceded that of plants. There is some evidence, for instance, that earlier cultures had domesticated the wolf or, in some North African communities, the jackal. It is likely that such animals were of considerable use in driving and tracking game and hunting down the wounded prey, but determining when the dog was domesticated has proved very difficult using conventional archaeological techniques. Bones of dog/wolf associated with human settlements (found as far back as 400 000 years) could, after all, mean that people ate the animal rather than domesticated it. Joint burials, which provide a reasonable indication of domestication, are known that date from the early Holocene in the Near East, but little is known of earlier associations. Molecular studies may again give the best clue. Carles Vila, from the University of California, Los Angeles, and his colleagues [34] have analysed mitochondrial DNA samples from 162 wolves and 140 dogs, representing 67 different breeds. They support the idea that the dog evolved from the wolf, but the differences between the two groups suggest that the evolutionary separation (presumably associated with domestication and isolation of dogs from wolves) took place about 100 000 years ago.

This would place dog domestication at least as far back as the last interglacial.

However, it is likely that many of the other animals that became associated with humans, such as sheep and goats, were domesticated during the Early Neolithic Period soon after the first cultivation of plants. These were initially herded for their meat and hides, but would have also been a source of milk, once tame enough to handle. The first traces of domesticated sheep come from Palestine around 6000 BC. These may have originated from one of the three European and Asiatic sheep, or may have resulted from interbreeding among these species. The Soay sheep that have survived in the Outer Hebrides in Scotland almost certainly originated from the moufflon, either the European *Ovis musimon* or the Asiatic *O. orientalis*. Domestication of these animals may have resulted from the adoption of young animals orphaned as a consequence of hunting activity. In Israel there is a marked shift in diet between 10 000 and 8000 years ago, gazelle and deer being replaced by goat and sheep. Almost certainly this was a consequence of domestication [35].

The aurochs, *Bos primigenius*, was a frequent inhabitant of the mixed deciduous woodland which was spreading north over Europe during the post-glacial period. In many of the sites where remains of these forests have been preserved, such as in buried peats and submerged areas, the bones of this animal have been found. It was probably first domesticated by the Neolithic farmers of Anatolia around 7000 BC. Other animals, such as the reindeer, were gradually brought into a state of semi-domestication as migrant hunters followed the herds through the seasons and became increasingly dependent on them.

The dry lands

At the time of the first agricultural experiments at the opening of the Holocene, the Middle East was experiencing a time of mild, humid climate. Pollen diagrams from Syria show that oak forest was advancing into the dry, steppic vegetation that

had persisted during the glacial maximum. In fact, many of those parts of the world currently occupied by desert or semi-arid scrub suffered a similarly dry climate during the glacial maximum, but the close of the glaciation brought renewed rains to many of these dry areas. Studies of lakes in the vicinity of the Sahara suggest that conditions became more humid in a series of stages, commencing around 14 000 years ago, but there was a short period of aridity during the Younger Dryas. The early part of the Holocene then provided a time of wetness for many currently desert areas extending from Africa through Arabia to India. Lakes existed in the middle of the very arid Rajasthan Desert of north-west India, and the surge of fresh water down the Nile created stratified waters in the eastern Mediterranean, the low-density fresh water lying over the top of the high-density salt water. As a result the lower layers became depleted in oxygen, and black anoxic sediments, called sapropels, were deposited [36].

The wetness permitted the northward extension of savanna and rainforest into formerly dry areas, and Fig. 10.13 summarizes data from many pollen diagrams taken from sites in the southern Sahara [37]. The expansion of the more humid biomes can be seen, followed by their contraction when aridity set in once more about 5000 years ago. The Sahara Desert is quite rich in ancient rock paintings, some dating back over 8000 years, and these older ones depict big game animals now associated with the savanna grasslands, confirming the evidence of the pollen. Between 7500 and 4500 years ago the painters of the pictures were evidently pastoralists who depicted their cattle on rocks in locations where cattle could certainly not graze today. After that time the pictures of cows were replaced by camels and horses as the arid climate became more severe, and the habits and domesticated animals of the local peoples altered accordingly.

Many of the great deserts of the world were evidently initiated by climatic changes in the latter half of our interglacial. The involvement of increased human activity during this time, however, has obviously complicated the picture, and

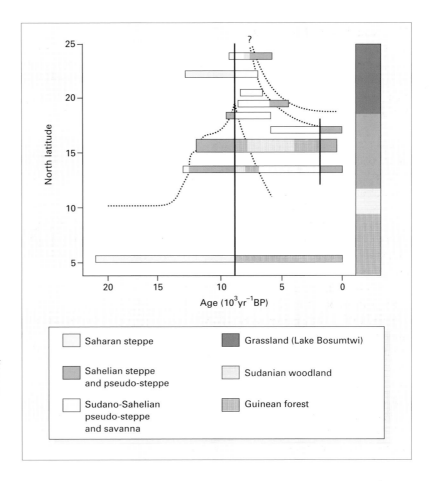

Fig. 10.13 Changes in the vegetation in northern tropical Africa over the past 20 000 years. The present latitudinal zonation of vegetation is shown on the right of the diagram and it can be seen that this zonation was displaced about 5°N during the period 9000–7000 years ago, in the early stages of the current interglacial. From Lezine [37].

some researchers believe that human exploitation, even in prehistoric times, may have contributed to the development of deserts, as in the case of the Rajasthan Desert, where the increasing aridity was accompanied by the cultural development of the Indus Valley civilization. It can be difficult to determine the role of humankind in the development of desert spread under such circumstances.

Changing sea levels

Interglacials are times of changing sea levels, and the current interglacial is no exception. The melting of ice has released considerable quantities of water into the oceans resulting in a (eustatic) rise in sea level relative to the land. This may have amounted to as much as 100 m in places. On the other hand, the loss of ice caps over those land masses which acted as centres of glaciation relieved the earth's crust of a weight burden, resulting in an (isostatic) upwarping of the land surface with respect to sea level. The relative importance of these processes varied from one place to another, depending upon how great a load of ice an area had borne. In western Europe the result was a general rise in the level of the southern North Sea and the English Channel with respect to the local land surface. In this way Britain, which was a peninsula of the European mainland during the glaciation, gradually became an island. Evidence from submerged peat beds in the Netherlands suggests a rapid rise in sea level between about 10 000 and 6000 years ago,

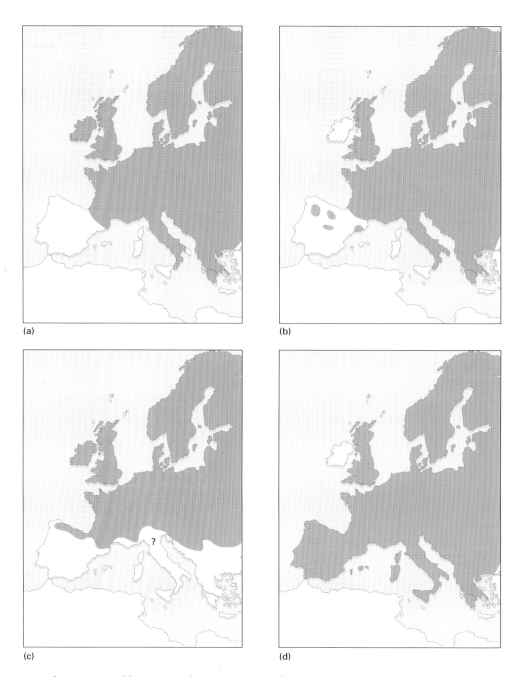

Fig. 10.14 Distribution maps of four mammal species in Europe. (a) Pygmy shrew (*Sorex minutus*) (b) common shrew (*Sorex araneus*) (c) stoat (*Mustella ermines*) and (d) weasel (*Mustela nivalis*). Of these, the pygmy shrew and the stoat reached Ireland, but the common shrew and the weasel failed to do so.

which has subsequently slowed down gradually. By this latter date England's links with continental Europe had been severed. At this time many plants with slow migration rates had still not crossed into Britain and were thus permanently excluded from the British flora. The separation of Ireland from the rest of Britain occurred rather earlier because there are deeper channels between them, and many species native to Britain have not established themselves as far west as Ireland. As a result, plants such as the lime tree (*Tilia cordata*), and herb paris (*Paris quadrifolia*) are not found growing wild in Ireland.

There is, however, one group of plants of great interest to plant geographers that did succeed in reaching Ireland before the rising sea level separated that country from the rest of the British Isles, and this is known as the Lusitanian flora. Lusitania was the name for a province of the Roman Empire consisting of Portugal and part of Spain and, as its name suggests, the Lusitanian flora has affinities with that of the Iberian peninsula. Some of the plants—such as the strawberry tree (*Arbutus unedo*) (see Chapter 3) and giant butterwort (*Pinguicula grandiflora*)—are not found growing wild in mainland Britain. Others, such as the Cornish heath (*Erica vagans*) and the pale butterwort (*Pinguicula lusitanica*), are found in south-western England as well as in Ireland. It therefore seems likely that these plants spread from Spain and Portugal up the Atlantic seaboard of Europe in postglacial times but were subsequently cut off by the rising sea levels. It is not impossible that some or all of the species may have survived the last glaciation in oceanic south-west Ireland, but direct proof of this is lacking.

Some mammals also failed to make the crossing into Ireland, and it is difficult to explain why they failed when closely related species succeeded. For example, the pygmy shrew reached Ireland, yet the common shrew did not (Fig. 10.14a,b). Perhaps this is related to their habitat preferences, for the pygmy shrew is found on moorlands and might have survived better than the common shrew if the conditions of the land bridge were wet, peaty and acidic. The stoat also reached Ireland, but the weasel did

not (Fig. 10.14c,d). Arrival in this case may have been due to sheer chance. A single pregnant female arriving on a raft of floating vegetation could have been sufficient to populate the island. Discoveries in some archaeological sites in Ireland of small mammals like the wood mouse raise the possibility that some plants and animals could have been carried over the water by prehistoric humans. The arrival of some large mammals on isolated islands could also be an outcome of human transport, even in pre-agricultural times. Mesolithic (Middle Stone Age) people in the British Isles, for example, may have been responsible for carrying the red deer (*Cervus elaphus*) to Ireland and to other offshore islands such as Shetland. The red deer was the major prey animal of these people, who did not have truly domesticated animals and plants, and the transport of young animals would have presented few difficulties even in small, primitive boats.

Rising sea levels during the present interglacial were responsible for the severing of land connections in many other parts of the world also. For example, Siberia and Alaska were connected across what are now the Bering Straits, which in places are only 80 km (50 miles) across and 50 m deep. This high-latitude land bridge would have been a suitable dispersal route only for arctic species, but it is believed to have been the route by which humans entered the North American continent (see p. 205).

Time of warmth

The period of maximum warmth during the present interglacial lasted from about 7000 to 5000 years ago. At this time, warmth-demanding species extended farther north than they do at present. For example, the hazel (*Corylus avellana*) was found considerably further north in Sweden and Finland than it is today. This indicates that conditions have become cooler since that time. The remains of tree stumps, buried beneath peat deposits at high altitude on mountains and far north of the tree line in the Canadian Arctic, also bear witness to more favourable conditions in former times. Things are not always what they seem, however, and one has to remember the possible involvement of humans in the clearance of forests and the modification of habitats. Humans may have played an important part, for example, in the forest clearance that led to the formation of many of the so-called 'blanket mires' of western Europe [38]. By clearing the trees they created a new set of hydrological conditions and the saturated soils began to accumulate peat. Continued burning and grazing by prehistoric pastoralists ensured that the forests were not able to reinvade. But the lack of trees does not mean that the climate is no longer suitable for their growth; as in the case of deserts, it is often difficult to appreciate the extent to which people limit the distribution of plants and animals.

Climate and fire also interact, as in the prairie region of central North America. Measurements of fire frequency in the past history of the prairies shows that the time between fires increases when the climate is cooler, so the composition of vegetation may be determined by climate but in an indirect way [39]. In the Mediterranean region of Europe, the extent of oak woodland was formerly greater than at the present day, but here the activities of humans have had a strong influence for many thousands of years, and climatic influences upon the vegetation are therefore difficult to discern.

The spread of forest in temperate areas during the early Holocene, caused by the increasing warmth, created heavily shaded, unsuitable conditions for many of the plants which had previously been widespread at the close of the glacial stage. Some of these, the arctic–alpine species, are also physiologically unsuited to high temperature. Many such plants, for example the mountain avens (*Dryas octopetala*), grow poorly when summer temperatures are high (above 23°C for Britain and 27°C for Scandinavia). The climatic changes which occurred during the postglacial therefore proved harmful to such species, and many of them became restricted to higher altitudes, especially in lower latitudes. Other plant species are more tolerant of high temperatures but are incapable of survival under dense shade. Low-latitude, low-altitude habitats which became covered by forests were unsuitable for the

continued growth of these species, and many of them also became restricted to mountains where competition from shade-casting trees and shrubs did not occur to the same extent. But for these species there were other opportunities. Lowland environments which for some reason bore no forest provided suitable places of refuge. Coastal dunes, river cliffs, habitats disturbed by periodic flooding, steep slopes, all provided sufficiently unstable conditions to hold back forest development and allow the survival of these plants.

The result of these processes was the production of relict distribution patterns (see Chapter 3). Sometimes the separation of species into scattered populations, even though it has lasted only about 10 000 years, has permitted genetic divergence, as in the case of the sea plantain (*Plantago maritima*), which has survived in both alpine and coastal habitats, but the different selective pressures of the two environments have resulted in physiological divergence between the two races. The coastal race is able to cope with high salinity but tends to be more frost-sensitive than the montane race.

Some of the less robust plant species, which are limited by competitive inadequacies rather than by climatic ones, have taken advantage of the disturbed conditions provided by human settlements and agriculture. These plants, which fared so poorly during forested times in the temperate latitudes, have become latter-day weeds and opportunists. Thus, climatic change and human habitat disturbance have interacted to provide different histories for each species.

Climatic deterioration

Although there has been an overall cooling of the climate during the past 5000 years, this has not taken place in a gradual way, but in a series of steps. Beginning about 3000 BC, a number of quite sudden temperature falls in the Northern Hemisphere had the effect of halting the retreat of glaciers and of increasing the rates of bog growth. One of the most pronounced of these steps, as far as north-west Europe is concerned, occurred about 500 BC and caused a sudden increase in the rate of growth of bogs over the entire area. In many bogs this has left a permanent mark upon the stratigraphic profile of the peat; a dark, oxidized peat typical of slow-growing bog surfaces is suddenly replaced by the almost undecomposed vegetable matter that typifies a fast-growing bog. The German botanists who first described this phenomenon called it the Grenzhorizont (boundary horizon) and the name is still occasionally used by palaeobotanists. Using such evidence of increased bog growth as an indicator of wetter or cooler climate is, however, fraught with problems. Often such changes are of only local significance and may be associated with local drainage patterns, human land use, or peculiarities of microclimate or bog hydrology. Only those changes which are synchronous over large areas can be considered as truly climatically induced.

Snow and ice deposited under warm conditions have higher proportions of the isotope [18]O than those accumulating in the cold (see p. 176). The changing proportions of isotopes in the ice cap therefore provide a detailed record of fluctuating temperature. In Fig. 10.15 the oxygen isotope record is compared with curves from Iceland and England which have been compiled from various types of indirect evidence, such as early literary sources. The combination of these pieces of information indicates that there have been considerable variations in climatic patterns even during recent, historical times.

Evidence of climatic change derived from the growth rings of trees also serves to emphasize regional variations in climate. For example, work on tree rings in pine from Fennoscandia, forming a continuous record back to AD 500, shows a great deal of variation and suggests that the so-called Little Ice Age in that region was a relatively brief event, lasting only from 1570 until 1650. In England it lasted from 1300 to 1700 (Fig. 10.15). Studies of lake sediments in the northern Great Plains of the United States have shown that drought events leading to elevated salinity were much more frequent prior to AD 1200 than they have been subsequently (Fig. 10.16). While Europe was suffering its Little

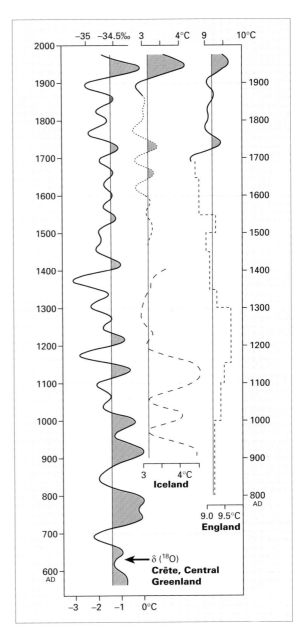

Fig. 10.15 Oxygen isotope curve from an ice core taken in central Greenland (left), and the projected temperature curve for Iceland (centre) and for England (right). From Dansgaard *et al.* [40].

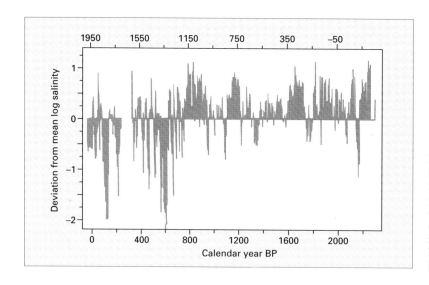

Fig. 10.16 Changes in the salinity in the sediments of Moon Lake, North Dakota (Great Plains), determined from analysis of fossil diatoms. Saline episodes, caused by drought and evaporation, are more frequent prior to AD 1200 than they have been since. From Laird *et al.* [41].

Ice Age, continental North America was receiving more precipitation [41].

Some biological consequences of the generally deteriorating climate over the past 5000 years are apparent from changing distribution patterns. In Europe, such species as the hazel (*Corylus avellana*) extended further north and up to higher altitudes in Scandinavia than they do today, and the pond terrapin (*Emys orbicularis*) had a wider occurrence in north-western Europe. It is very difficult, however, to be sure that such contractions in range have not resulted from human destruction of either the species or their habitats during this time period, because the impact of human cultures on the natural landscape through these millennia has become increasingly severe. In North America one might regard the resurgence of spruce in the north-west (see Fig. 10.4a) as a product of climatic deterioration and an indication that the current interglacial is entering its latter days.

The environmental impact of early human cultures

As the postglacial climatic optimum passed and conditions in the temperate regions became generally cooler and wetter, the agricultural concept continued to spread into higher latitudes, and with agriculture came the incentive to modify the environment to make it more suitable for enhanced productivity of domestic animals and plants. Temperate forests are unsuitable for the growth of domesticated plants, most of which have a southern origin and a high demand for light. Similarly, domestic animals such as sheep and goat are not at their most efficient in a woodland habitat, preferring open grassland conditions. Cattle and pigs, on the other hand, can be herded within forest, but even they can be managed more efficiently in a habitat that is more open. Even in pre-agricultural times the Mesolithic people of northern Europe had discovered that the opening of forest and burning to retain open glades provided a higher productivity of red deer (*Cervus elephas*). Like some of the Palaeoindian tribes of North America, who became closely dependent on the bison, the European people of the Middle Stone Age were often reliant on the red deer (known in North America as elk).

The intensification of forest clearance with the coming of agriculture in northern Europe is very apparent from the pollen diagrams (see Fig. 10.3), where the pollen of open habitat species (such as grasses, plantains and heathers) rises and the proportion of tree pollen falls. The precise pattern of forest clearance and the development of heathland, grassland, moorland and blanket bog as consequences of this activity varies from one area to another, depending on the local conditions and the pattern of human settlement. By 2000 years ago the impact was severe through much of central and western Europe, although the forests of the far north had been little influenced by humankind at that time. Some of the most severe deforestation, judging from the pollen record, had taken place in the north-west of Europe, including the British Isles. Perhaps it was in this region that the forest was least able to recover from human impact, and the maintenance of heavy grazing kept the area relatively open.

In North America, agriculture in the temperate zone in pre-European times was confined largely to the growing of maize and other weedy species, including purslane (*Portulaca oleracea*). This involved the clearance of small areas of forest, and the effect of these clearings can be detected in pollen diagrams [42]. Although these clearings seemed to recover, and there is little evidence for large-scale forest destruction of the European type, there are changes in the composition of the North American forests that may well have resulted from the activities of agricultural peoples. The burning and cutting of forests is often associated with a loss in certain species, such as sugar maple and beech, and an increase in the abundance of fire-resistant pines and oaks, together with a general increase in the frequency of birch. Intensive clearance in the eastern United States and Canada was usually delayed until the arrival of European settlers in the eighteenth and nineteenth centuries. It is often marked in the pollen diagrams by a marked rise in the ragweed (*Ambrosia*).

Recorded history

As soon as human beings appeared on the scene, they often inadvertently began to leave information about climate and its changes. Early records provide clues rather than precise information, such as the ancient rock drawings of hunting scenes discovered in the Sahara, which indicate that its climate was much less arid at the time they were made than it is today. With the development of writing, more accurate records of climatic changes began to be made. For example there are records of pack ice in the Arctic seas near Iceland in 325 BC, indicating the very low winter temperatures at the time. During the heyday of the Roman Empire, however, there was steady improvement in climate, allowing the growth of such crops as grapes (*Vitis vinifera*) and hemp (*Cannabis sativa*) even in such relatively bleak outposts as the British Isles. This warmer, more stable period reached its optimum between AD 1000 and 1200, after which it again grew colder.

In 1250 alpine glaciers grew and pack ice advanced in the Arctic seas to its most southerly position for 10 000 years. In 1315 a series of poor summers began in northern Europe, and these led to crop failures and famine. Climatic deterioration continued and culminated in the 'Little Ice Age' of AD 1300–1700, during which the glaciers reached their most advanced positions since the end of the Pleistocene glacial epoch [43]. During this time, trees on the central European mountains were unable to grow at their former altitudes, due to the increasingly cold conditions. The climate of Europe became somewhat warmer after 1700 and especially since 1850. There was a slight cooling after 1940, when winters in particular became colder, but since 1970 average temperatures globally have been rising again. An important question for climatologists is the extent to which human activities are influencing current trends in climate. For the biogeographer, the main task is one of detecting current changes in plant and animal responses and, as far as possible, predicting what the future may bring.

Summary

1 Climate change is not necessarily a smooth, steady process; it can alter abruptly and very significantly over a matter of decades, as in the case of the cold Younger Dryas stadial between 11 000 and

10 000 years ago. Critical in such change is the failure of the oceanic conveyor to maintain its global heat transfer.

2 Pollen analysis of lake and peat sediments has permitted detailed reconstruction of the rates of movement of major tree genera and the directions they took during the warming of the climate in the current interglacial.

3 Of the many types of hominid, only *Homo sapiens* survived into the present interglacial. The extinction of many large vertebrate animals at the end of the last glacial had not occurred in former climatic cycles, and circumstantial evidence suggests that humans were involved in their demise.

4 The domestication of animals and plants provided new opportunities for human food production and hence population expansion.

5 Climatic changes over the past 10 000 years (the Holocene) are well recorded in lake sediments, ice sheets, marine deposits and in human historical records. There appears to have been a climatic optimum some 5000–3000 years ago, since which conditions have become cooler. Global sea level changes during these times have altered and created barriers to dispersal for species in the course of their spread.

Further reading

Beerling DJ, Chaloner WG, Woodward FI (eds) Vegetation–climate–atmosphere interactions: past, present and future. *Phil Trans R Soc London B* 1998; 353: 1–171.

Mannion AM. *Global Environmental Change*. London: Longman, 1991.

Pielou EC. *After the Ice Age: the Return of Life to Glaciated North America*. Chicago: University of Chicago Press, 1991.

Prentice IC, Webb T, III. BIOME 6000: reconstructing global mid-Holocene vegetation patterns from palaeoecological records. *J Biogeogr* 1998 (in press).

References

1 Moore PD, Chaloner B, Stott P. *Global Environmental Change*. Oxford: Blackwell Science, 1996.

2 Lowe JJ. Lateglacial and early Holocene lake sediments from the northern Apennines, Italy—pollen stratigraphy and radiocarbon dating. *Boreas* 1992; 21: 193–208.

3 Broecker WS, Kennett JP, Flower BP, Teller JT, Trumbo S, Bonani G, Wolfli W. Routing of meltwater from the Laurentide Ice Sheet during the Younger Dryas cold episode. *Nature* 1989; 341: 318–21.

4 Jones GA. A stop–start ocean conveyer. *Nature* 1991; 349: 364–5.

5 Fairbanks RG. A 17 000-year glacio-eustatic sea level record: influence of glacial melting rates on the Younger Dryas event and deep-ocean circulation. *Nature* 1989; 342: 637–42.

6 Smith JE, Risk MJ, Schwarcz HP, McConnaughey TA. Rapid climate change in the North Atlantic during the Younger Dryas recorded by deep-sea corals. *Nature* 1997; 386: 818–20.

7 Benson L, Burdett J, Lund S, Kashgarian M, Mensing S. Nearly synchronous climate change in the Northern Hemisphere during the last glacial termination. *Nature* 1997; 388: 263–5.

8 Dansgaard W, White JWC, Johnsen SJ. The abrupt termination of the Younger Dryas climate event. *Nature* 1989; 339: 532–4.

9 Bennett KD. Postglacial population expansion of forest trees in Norfolk, U.K. *Nature* 1983; 303: 164–7.

10 Jacobson GL, Webb T, Grimm EC. Patterns and rates of change during the deglaciation of eastern North America. In: Ruddiman WF, Wright HE, eds. *The Geology of North America*, Vol. K-3. New York: Geological Society of America, 1987: 277–88.

11 Andrews P. Evolution and environment in the Hominoidea. *Nature* 1992; 360: 641–6.

12 Hay RL, Leakey MD. The fossil footprints of Laetoli. *Sci Am* 1982; 246 (2): 38–45.

13 Kerr RA. Tracing the identity of the first toolmakers. *Science* 1997; 276: 32–3.

14 Semaw S, Renne, Pl. Harris JWK, Feibel CS, Bernor RL, Fesseha N, Mowbray K. 2.5-million-year-old stone tools from Gona, Ethiopia. *Nature* 1997; 385: 333–6.

15 Wood B. The oldest hominid yet. *Nature* 1994; 371: 280–1.

16 Swisher CC, Rink WJ, Anton SC, Schwarcz HP, Curtis GH, Suprijo A, Widiasmoro. Latest *Homo erectus* of Java: potential contemporaneity with *Homo sapiens* in southeast Asia. *Science* 1996; 274: 1870–4.

17 Bermudez de Castro JM, Arsuaga JLO, Carbonell E, Rosas A, Martinez I, Mosquera M. A hominid from the Lower Pleistocene of Atapuerca, Spain: possible ancestor to Neanderthals and modern humans. *Science* 1997; 276: 1392–5.

18 Ward R, Stringer C. A molecular handle on the Neanderthals. *Nature* 1997; 388: 225–6.

19 Rogers RA, Rogers LA, Hoffmann RS, Martin LD. Native American biological diversity and the biogeographic influence of Ice Age refugia. *J Biogeogr* 1991; 18: 623–30.

20 Gibbons A. The peopling of the Americas. *Science* 1996; 274: 31–2.

21 Meltzer D. Monte Verde and the Pleistocene peopling of the Americas. *Science* 1997; 276: 754–5.

22 Martin PS, Wright HE Jr. *Pleistocene Extinctions: The Search for a Cause.* New Haven: Yale University Press, 1967.

23 Grayson JE. Pleistocene avifaunas and the overkill hypothesis. *Science* 1977; 195: 691–3.

24 Stuart AJ. Mammalian extinctions in the Late Pleistocene of Northern Eurasia and North America. *Biol Rev* 1991; 66: 453–562.

25 Lewin R. What killed the giant mammals? *Science* 1983; 221: 1036–7.

26 Heun M, Schafer-Pregl R, Klawan D, Castgna R, Accerbi M, Borghi B, Salamini F. Site of einkorn wheat domestication identified by DNA fingerprinting. *Science* 1997; 278: 1312–14.

27 Hubbard RNLB. Development of agriculture in Europe and the Near East: evidence from quantitative studies. *Econ Bot* 1980; 34: 51–67.

28 Diamond J. The language steamrollers. *Nature* 1997; 389: 544–6.

29 Willis KJ, Bennett KD. The Neolithic transition—fact or fiction? Palaeoecological evidence from the Balkans. *Holocene* 1994; 4: 326–30.

30 Smith B. The initial domestication of *Cucurbita pepo* in the Americas 10,000 years ago. *Science* 1997; 276: 932–4.

31 Iltis HH. From teosinte to maize: the catastrophic sexual transmutation. *Science* 1983; 220: 886–94.

32 Guzman R, Iltis HH. Biosphere reserve established in Mexico to protect rare maize relative. *Diversity* 1991; 7: 82–4.

33 Van Der Merwe NJ. Carbon isotopes, photosynthesis and archaeology. *Am Scientist* 1982; 70: 596–606.

34 Vila C, Savlainen P, Maldonado JE, Amorim IR, Rice JE, Honeycutt RL, Crandall KA, Lundeberg J, Wayne RK. Multiple and ancient origins of the domestic dog. *Science* 1997; 276: 1687–9.

35 Davis S. The taming of the few. *New Scientist* 1982; 95: 697–700.

36 Rosignol-Strick M, Nesteroff W, Olive P, Vergnaud-Grazzini C. After the deluge: Mediterranean stagnation and sapropel formation. *Nature* 1982; 295: 105–10.

37 Lezine AM. Late Quaternary vegetation and climate in the Sahel. *Quaternary Res* 1989; 32: 317–34.

38 Moore PD. The origin of blanket mire, revisited. In: Chambers FM, ed. *Climate Change and Human Impact on the Landscape.* London: Chapman & Hall, 1993: 217–24.

39 Bond WJ, van Wilgen BW. *Fire and Plants.* London: Chapman & Hall, 1996.

40 Dansgaard W, Johnsen SJ, Reeh N, Gundestrup N, Clausen HB, Hammer CU. Climate changes. Norsemen and modern man. *Nature* 1975; 225: 24–8.

41 Laird KR, Fritz SC, Maasch KA, Cumming BF. Greater drought intensity and frequency before AD 1200 in the Northern Great Plains, USA. *Nature* 1996; 384: 552–4.

42 McAndrews JH. Human disturbance of North American forests and grasslands: the fossil record. In: Huntley B, Webb III T, eds. *Vegetation History.* Dordrecht: Kluwer, 1988: 673–97.

43 Crowley TJ, North GR. *Paleoclimatology.* New York: Oxford University Press, 1991.

CHAPTER 11: *Projecting into the future*

Biogeography is, or should be, a predictive science. We have shown in the preceding chapters how it has been possible to analyse the ecological and historical causes of current patterns of plant and animal distributions. It has become evident that we need to know much of their physiological and hence habitat requirements, their interactions with other species and also their evolutionary history, if we are to understand why organisms are found where they are, and why some parts of the globe are richer in species than others. The geologically recent history of the earth has been particularly turbulent in climatic terms, and has had a profound effect upon both distribution patterns of species and the associations between species that are found in nature—the communities.

The question that must be asked at the end of all these studies is whether an understanding of how these patterns have come about provides us with what we need in order to predict the future. Are we now better equipped to extrapolate from the past and the present into the future and to make informed projections about what will happen to the species, the ecosystems, the communities and the biomes in response to the global environmental changes that we observe accelerating around us?

The accuracy of biogeographical projection is, however, limited by our capacity to predict the direction and pace of the changes themselves. Will global warming continue, and at what speed? Will vegetation and soil responses create some kind of feedback mechanism, and will this be positive or negative? Will global patterns of temperature and precipitation change? Will ocean levels rise and, if so, how fast? Will patterns of forest clearance and human-induced desertification continue at their present rate? How will fragmentation of habitats affect species responses to change, interrupting their population movements? Will chronic pollution problems, such as the discharge of compounds of nitrogen and sulphur into the atmosphere, create new conditions within which plants and animals have to adapt and/or alter their distributions?

Most of these questions are currently being actively researched and have as yet no simple answers. Since future climates and conditions cannot be firmly predicted, it is obvious that the biogeographical responses of living organisms must remain a scientifically misty area, but it is possible to give a general indication of the direction that research is taking and the possible consequences of given climatic scenarios.

The changing climate

Meteorological records from around the world span back over the past century or more, and therefore trends in global climate can now be documented without recourse to proxy methods such as those described in previous chapters. We do not need to look to oxygen isotope profiles or to indirect records of climate in fossil animals and plants when we have plentiful direct sets of data. This means that we are in a position to determine whether global climate is changing and at what rate such alteration is taking place. We can also project these changes into the future on the basis of certain assumptions and also of past experience.

Figure 11.1 shows the record of global temperature since the mid-nineteenth century and it shows a clear overall upward trend. There is no doubt that the earth has become warmer (by about 0.6°C) over the past century. It is possible to resolve this increase in warmth into two main stages, first between 1900 and 1940, followed by a fairly stable plateau for around 30 years, then a further and

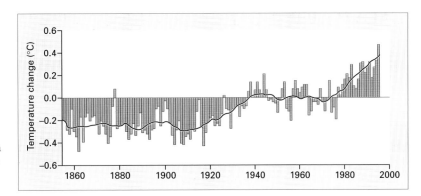

Fig. 11.1 Mean global combined sea surface and air temperatures for the past 130 years relative to the average for the period 1951–1980.

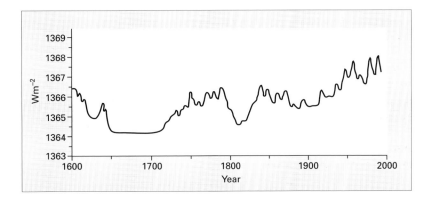

Fig. 11.2 Changes in solar irradiance over the past 400 years based on historical records of sunspot number. From Lean *et al.* [1].

continued rise since about 1970. Projecting on into the future demands that we understand the underlying causes of the rise. Perhaps it is part of a natural cycle, in which case we need to understand the detailed form of the cycle in order to predict its next move. Or perhaps we are observing a climate change in which human activity is playing a significant part, such as by contributing heat-absorbing gases into the atmosphere and creating a thermal blanket or 'greenhouse' effect.

Meteorologists are cautious about adopting the second of these explanations until alternatives have been adequately explored. If we look at changes in solar irradiance [1], which represents a possible explanation for climatic changes on this kind of time scale (Fig. 11.2), we find a pattern that generally follows the observed changes in global temperature. Solar irradiance is dependent on the abundance of sunspots (a high number of sunspots leading to

lowered irradiance) and reliable historical records of these go back almost 400 years. Sunspots are known to show a regular 11-year cycle and this is evident in the diagram, but of greater significance is the long-term variation which so closely follows the temperature curve. The precise mechanism by which solar irradiance might influence world climate remains obscure; possibly ozone production in the upper atmosphere is involved. Whatever the mechanism, predicting future climate must take into account the possibility of solar forcing.

Many climatologists, on the other hand, believe that the much-debated greenhouse warming is now taking place as a result of human injection of carbon dioxide, ozone, methane, chlorofluorocarbons, oxides of nitrogen and other infrared absorbing gases into the atmosphere. If this is indeed so, then predictions must allow for future trends in the production of these gases. But the picture will be

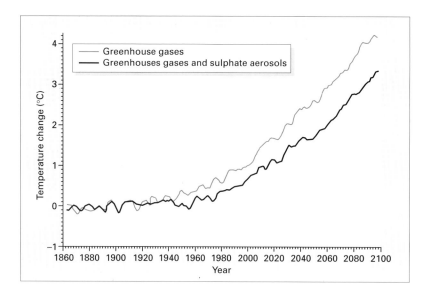

Fig. 11.3 Calculated global surface air temperature modelled to take into account the production of greenhouse gases alone and the combination of greenhouse gases and sulphate aerosols. Observed values of temperature follow the lower curve, which should therefore be regarded as a more reliable forecast of future developments.

complicated by human generation of other waste gases, such as sulphur dioxide, that are released into the atmosphere and that have the reverse effect and may lead towards some degree of cooling. Thus, predictions need to take into account the whole range of gaseous pollutants that we generate and how they may increase or decrease in the future. An international group of scientists examining the evidence for climate change and the human role in this process (the Intergovernmental Panel on Climate Change, IPCC) has, after much careful debate and analysis, eventually come to the conclusion that 'the balance of evidence suggests that there is a discernible human influence on global climate'. This is quite a strong statement for a collection of scientists, and it has been confirmed by detailed studies of differences in temperature rises in the Northern and Southern Hemispheres in relation to their respective output of gases and aerosols [2]. These results imply that any predictions about the future of the earth's climate must take human activities into account.

Given all the physical and human variables involved, it is not surprising that the future remains somewhat unpredictable. It is possible, however, to set up a number of scenarios based on certain assumptions and then to check which of these is proving most robust as time and climate change proceeds. Figure 11.3 gives an example of two scenarios extrapolating global temperature into the future. One is based simply on the continued projected input of greenhouse gases into the atmosphere, and the lower curve adds to this the additional impact of other pollutants, such as sulphur dioxide, that will ameliorate the temperature rise. At present, the lower curve is proving a better model in matching the observed temperature rise.

Even when global temperature changes have been predicted, there remain many further questions, such as the geographical variations in the temperature rise, the consequent influence on cloud-cover patterns and precipitation, and the impact on storm frequencies, all of which will be of biogeographical significance. Current projections suggest that the high latitudes will warm up faster than most other regions [3], which will impact strongly on tundra habitats. Computer simulation models have shown that a rise in sea-surface temperature will increase the intensity of hurricanes in the north-west Pacific basin [4]. An important question that will affect much of the world's climate is whether global temperature change will influence the frequency and intensity of the El Niño phenomenon in the equatorial Pacific. Recent years have shown how this periodic build-up of warm water in the eastern Pacific can have widespread repercussions, sometimes with catastrophic consequences from Indonesia to California and Peru, and perhaps even wider afield. The 1997–1998 El Niño was very carefully studied and modelled, but the future of these events is still uncertain [5].

Global warming, especially if it is concentrated in the higher latitudes, may be expected to cause the melting and breakup of the ice sheets around the Antarctic ice cap. Observations suggest that this process has indeed begun. Over the last 50 years there has been considerable ice-shelf retreat around the Antarctic Peninsula [6]. Workers from the British Antarctic Survey team have come to the conclusion that a monthly average temperature of –2.5°C is the threshold for melting; above that temperature, the melt is quite rapid. The number of summer days in which melting occurs has been increasing at the rate of about one extra day per year since the late 1970s and consequential ice-shelf collapse has been rapid, as evidenced by a series of satellite photographs [7]. An obvious possible result of such melting is a rise in global sea level, and many models have been developed to predict just how great a rise we might expect and how rapid it is likely to be. The problem is that this will be dependent on models of global temperature rise which, as we have seen, are themselves still open to question. One recent estimate [8], based upon a rigorous consideration of all the variables, suggests that between 1990 and 2100 we can expect a global sea-level rise of 20 cm, resulting from the melt of glaciers and the Greenland ice cap. In this model, the Antarctic ice cap will not contribute to global sea levels because increased atmospheric temperature will cause additional precipitation over the area and this will accumulate as new ice in the ice cap. It is possible that the Antarctic increase in ice load will actually compensate for the melting of the Greenland ice cap, leaving the sea to rise by only about 13 cm as a consequence of the melting of glaciers. Thus, the outcome as far as sea levels is concerned is still uncertain.

Perhaps a more serious potential problem associated with the oceans is the possibility of disruption in the oceanic thermohaline circulation—the conveyor belt—that transports heat around the world and which depends for its circulatory mechanism on subtle changes in sea-water salinity and density. Changes in global temperature and their resulting inputs of additional fresh water from melting ice could disrupt these movements of water masses, just as happened during the Younger Dryas event in the early stages of the current warming (see Chapter 10). The North Atlantic is a key area in this balance, and the melting of the Greenland ice cap and the glaciers of northern Europe could prove critical in switching off the conveyor movement and leading to a southward extension of the polar front of icy waters from the Arctic. There is even a chance that this could precipitate another Ice Age.

It is difficult to predict exactly how far we have to go along the current road of greenhouse gas production and global warming to reach the point of crisis for the thermohaline circulation, but a model developed by Thomas Stocker and Andreas Schmittner [9], from the University of Bern in Switzerland, suggests that if the atmospheric level of carbon dioxide were to rise to 750 p.p.m. (from its current level of about 360 p.p.m.) over the next 100 years, it would lead to a complete shut-down of the ocean conveyor. A slower rise in carbon dioxide would slow the ocean circulation but might not shut it down. Models of this sort will be measured against observations over the coming years, but the possibility that global warming could lead to extreme levels of localized cold as a result of oceanic changes cannot be ignored.

Nitrogen and sulphur overload

Carbon is not the only elemental cycle to be disrupted by human activity, with potential global consequences. Nitrogen compounds have been released into the environment both deliberately with the intention of fertilizing crops, and inadvertently as a byproduct of industrial processes and fossil fuel combustion. As in the case of global

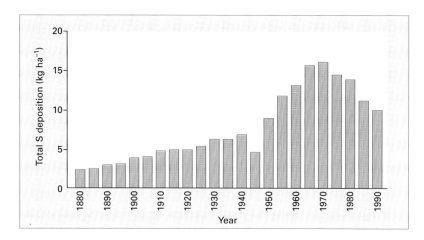

Fig. 11.4 The calculated deposition of sulphur from the atmosphere over the past 120 years from an area in southern Sweden. Data from Mylona, cited by Lee [10].

climate change, the current overload of compounds of nitrogen in the atmosphere and in the earth's soils and waters could lead to extensive biogeographical responses. Oxides of sulphur are also produced as waste products in the combustion of organic materials and these too have a high possibility of impact [10]. An indication of the increasing load of sulphur in the atmosphere is given in the example of southern Sweden shown in Fig. 11.4.

Sulphur has long been known as a potent fungicide, copper sulphate being the basis of 'Bordeaux Mixture' long used in the treatment of fungal pathogens of domestic plants. An increase in sulphur in the atmosphere and the rainfall will therefore impact upon microbial populations in soils and on plant surfaces. One of the incidental effects of sulphur dioxide pollution in urban areas over the last few decades has been a decrease in the abundance of leaf spot and mildew diseases of roses. Recent reductions in sulphur emissions (see Fig. 11.4), a consequence of clean air legislation, are already resulting in a return of these familiar fungal garden blights in temperate zone cities.

The effects of sulphur on decomposition processes are not always easy to observe, but lichens, which are formed by a combination of algal and fungal components, have reacted to raised sulphur levels in ways that are apparent. Epiphytic species (those growing on the twigs and branches of trees and shrubs) have proved particularly sensitive, and

surveys of lichen distributions have provided reliable information relating to atmospheric sulphur in regions where no chemical data are currently available. Lichen biodiversity has even been shown to correlate closely with the incidence of lung cancer in a studied area in the Veneto region of Italy [11], but interpreting correlations of this type is fraught with difficulties, and much research remains to be done to determine precisely how this kind of linkage can be explained. An understanding of the physical demands and limits of particular species of plants and animals can thus enable us to use them as bio-indicators of environmental conditions.

Injury by sulphur has also been recorded from plants other than lichens. The impact of sulphur on the bog mosses (*Sphagnum* species) has been severe in heavily polluted areas, leading to the disruption of vegetation and consequent erosion of peatlands. The southern part of the Pennine Mountains in Britain is a particularly severe example of this, with large expanses of eroding peats and limited capacity for recolonization and 'healing' in the absence of bog mosses. A confirmation that this is indeed the consequence of sulphur pollution over long periods of history was provided by an experiment in which bisulphite ions in dilute solution were sprayed onto *Sphagnum* growing on an unpolluted bog in North Wales; this led to the complete destruction of the bog moss cover within a year [12].

One might suppose that the impact of nitrogen

compounds on vegetation, on the other hand could be beneficial to plant growth, enhancing it where natural supplies are poor. In general terms this is probably true, but it does mean that those ecosystems which are naturally poor in nitrogen are being greatly modified by the additional inputs. Heathland and acid grassland ecosystems usually have low nitrate contents in their soils, and nitrogen additions will favour nitrogen-demanding plant species, enhancing their growth and producing a higher biomass ecosystem in which robust and competitive species may supplant the distinctive heathland flora and fauna. Heathland losses are already apparent over the Netherlands, where the dominant plant, heather (*Calluna vulgaris*), is adversely affected. Burning of heathland is a traditional management method which serves to keep nitrogen levels low (much is lost in smoke as oxides of nitrogen) but the aerial input of nitrogen compounds acts antagonistically to this management practice.

In the Subarctic zone of Sweden, the aerial input of nitrogen compounds may also disrupt the plant communities, stimulating grasses such as *Calamagrostis lapponica*, and leading to a tall grass canopy that overtops and shades out the characteristic dwarf shrub community of the tundra [10]. In the high Arctic, where nitrogen-demanding grasses are not present, the mountain avens (*Dryas octopetala*, see Fig. 10.1, p. 198) is initially favoured by the aerial inputs and grows faster [13] but continued enhanced supply of nitrates eventually leads to winter injury in the plant.

In temperate forests, excess nitrogen may have caused increased tree growth in recent years and resulted in other elements, such as magnesium, becoming limiting to further growth. It is possible that some of the degenerative symptoms observed in forests and broadly blamed on 'acid rain' could have resulted from this type of stimulated growth and then element limitation leading to leaf yellowing and other problems [14]. Such nitrogen uptake could be taken directly from the air and the water falling on leaves, leading to rapid growth flushes in forests, in much the same way that water bodies

receiving inwashed nitrates become the site of rapid algal growth, followed by death, decay and deoxygenation [15]. Eutrophication is becoming a global economic and conservation problem in all kinds of ecosystems.

Biogeographical consequences of global change

If biogeography is, indeed, to become a predictive science, then we need to develop methods of experimentation, analysis and modelling that will allow us to estimate changes in the range of individual species, changes in the composition of communities, and changes in the geographical extent of biomes resulting from a given set of climatic or other alterations. This is not an easy task, especially since the physical predictions of future conditions, as we have seen, are themselves so tentative.

The simplest approach is to begin with a single species, perhaps even a single population of a species, and experiment upon its responses to artificially induced changes. This type of work is essentially an exercise in applied physiology, but is the starting point for understanding biogeographical response processes.

A simple experiment, for example, would be to grow a particular plant species under conditions of double the normal carbon dioxide levels and compare its growth with those achieved at current levels. This type of experiment has been carried out for many plant species, both C_3 species and C_4 species. Take a common C_3 weed species such as *Chenopodium album* (fat hen) [16]. In a growth experiment under present levels of carbon dioxide in the atmosphere (360 p.p.m.), this plant was able to increase its weight by 0.22 g per day for every gram of plant tissue present over the course of its growth period. This type of measure is called relative growth rate and is a convenient way of measuring growth in terms that allow comparison between different sized plants. When carbon dioxide is raised from 360 p.p.m. to 700 p.p.m., the relative growth rate of fat hen becomes 1.08 g per day for every gram of plant tissue present. This is a fairly typical figure

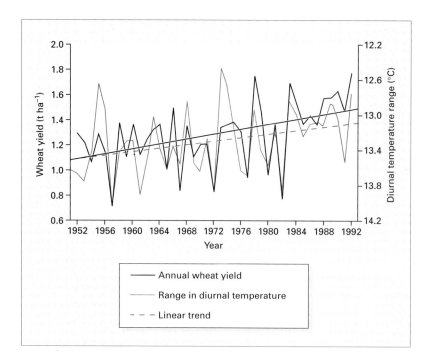

Fig. 11.5 Historical record of average Australian wheat yield (black solid line), together with the linear trend (dashed line). Also shown is the record of diurnal temperature variation over the same period (blue solid line), with the axis inverted to allow easier comparison. Much of the increase in yield is probably due to reduced diurnal temperature variability with fewer frosts. From Nicholls [18].

for annual C_3 plants and is explained by the fact that photosynthesis is limited by how fast the plant can accumulate carbon through its pores (stomata). Elevated carbon dioxide in the atmosphere means that it can fix carbon faster and hence grow more rapidly. C_4 species (see Chapter 3) are more efficient at collecting carbon dioxide at low atmospheric concentration, so they generally do not respond as strongly to raised levels. This means that we might hazard a prediction that C_3 species will gain advantages over C_4 species in a greenhouse world and extend their ranges into those currently occupied by C_4 species. Perhaps the balance line of the two groups of species in the Great Plains region (see Fig. 3.21, p. 55) will shift southwards. On the other hand, C_4 species tend to perform better than C_3 at higher temperatures, so this may compensate to some extent.

Some of our most important crop plants, especially in the temperate zone, are C_3 and C_4 grasses, including wheat, barley and maize. Is it possible that the productivity of these plants will increase with more carbon dioxide and higher temperatures?

Higher temperature will probably favour the C_4 species, like maize and sugar cane, but the higher carbon dioxide, as we have seen, will be particularly advantageous to the C_3 species, such as wheat and barley. At present, maize can be grown in Britain only as a silage crop for the production of cattle food. Grain maturation is poor in the relatively cool moist summers of the area, because ripening requires 850 degree-days above a base temperature of 10°C (number of degrees above 10°C multiplied by the number of days experiencing these temperatures). If the average temperature were to rise by 3°C, only the extreme north of Scotland would be unsuitable for growing grain maize [17].

In the United States, an increase in temperature of 3°C and an increase in precipitation of 8 cm would cause a shift in the grain belt to the northeast, around the Great Lakes [17]. In Australia, there is already indication that the productivity of wheat has improved with warmer conditions, and consequently fewer frosts, in the eastern part of that continent [18]. This is shown in Fig. 11.5,

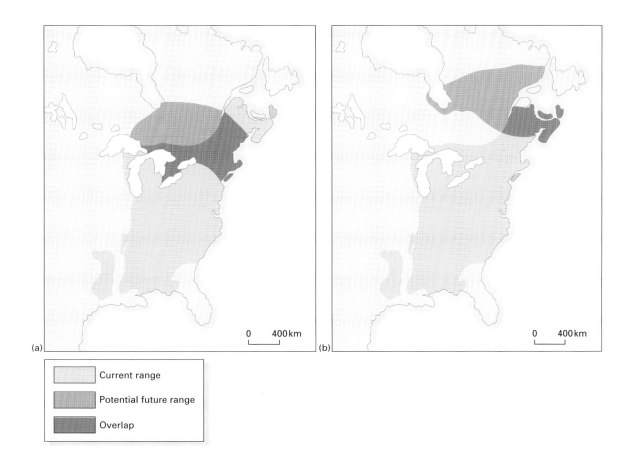

Current range	
Potential future range	
Overlap	

Fig. 11.6 Projected change in the distribution of beech (*Fagus grandifolia*) in North America under two greenhouse scenarios: (a) less severe and (b) more severe. From Roberts [19].

where yield is shown compared with the decrease in diurnal temperature range, an expression of more uniform conditions with less variation between high day and low night temperatures (note the inverted scale for temperature range). Not only grain production and maturation, however, but also grain quality could be affected by rising temperatures. Some additional research from Australia shows that when wheat grain is produced at high temperature (above 30°C) it is less suitable for the production of bread because the strength of the dough is reduced.

This could have serious implications for the quality of food production.

Similar experimental information relating to temperature optima could also lead to certain predictions about future species ranges. Such is the likely case, for example, with the beech tree (*Fagus grandifolia*) in North America [19] (Fig. 11.6). Two possible scenarios are considered, but both show a northward shift in distribution pattern in response to higher temperature. At the other end of the process, the retreat of species from areas where they have formerly succeeded is already apparent in some locations. The red spruce (*Picea rubens*), for example, is declining over much of its range in the eastern United States, and is believed to have been in decline since the early part of last century.

It has been proposed [20] that this change is a response to a general increase in both mean annual and summer temperatures.

This argument, however, may be too simple. It assumes that plants and animals inhabit 'climate envelopes', and that when the climate shifts the organism's distribution will alter accordingly. The real world is likely to be more complex, because we have to take into account the responses of other species, some of which may prove more competitive under the new set of conditions. It is also possible that spread into a new area will be prevented by barriers of some kind, whether artificial, such as the construction of agricultural land across which the species cannot travel, or natural in the form of mountain or water barriers. Some attempts have been made to set up experimental systems in miniature that will reflect these interactions, and they confirm that any climate envelope approach to biogeographical predictions is likely to fall well short of reality [21].

We can, of course, look at current changes in species range and observe what is really happening in nature as conditions change. The problem here is that we cannot always be sure about the precise causes in such range changes. Is it a response to climate or to some other unseen factor (such as human impact in one form or another)? An example of this is Edith's checkerspot butterfly from the western coastal area of North America. Camille Parmesan, from the University of California Santa Barbara [22], has studied the range and the fortunes of this butterfly and has found distinct patterns of range alteration even in the absence of any marked human impact. Given the climatic range of the species as a whole, one would expect that, as temperature increases, populations in the south and at low altitude would be more likely to become extinct than those in the north or at high altitude. As shown in Fig. 11.7, this is precisely what is found. Populations in Mexico are four times as likely to have become extinct than populations in Canada.

A more complicated situation arises in the case of migratory organisms. A bird that breeds in the

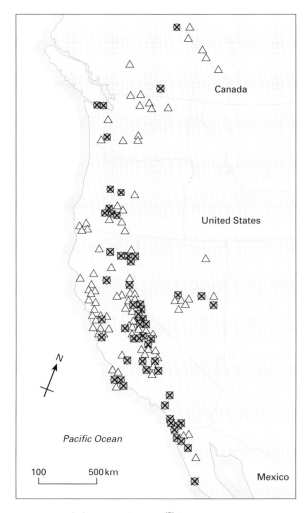

△ Occupied sites   ⊠ Abandoned sites

Fig. 11.7 Western North America showing populations of Edith's checkerspot butterfly (*Euphydryas editha*) recorded in a census taken between 1992 and 1996 (white triangles). Sites recorded in the past but no longer containing the butterfly are shown by the crossed circles. Most local extinctions have taken place in southerly and low altitude sites and probably reflect the impact of recent climate change. From Parmesan [22].

Northern Hemisphere but spends its non-breeding period in the Southern Hemisphere, for example, may be sensitive to environmental change in either of its residential sites or on the route in between.

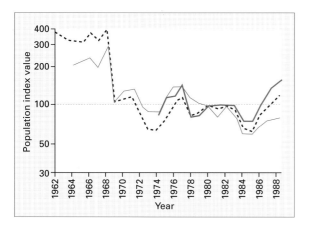

Fig. 11.8 Variations in the population density (an index is expressed on a logarithmic scale) for the whitethroat (*Sylvia communis*), an insectivorous migratory bird, in the British Isles between 1962 and 1988. The three lines represent three different censuses, but are in good general agreement. Note the population crash in the late 1960s. Data from the British Trust for Ornithology.

An example is the whitethroat (*Sylvia communis*), a small insectivorous warbler that breeds in Europe and spends its winters in West Africa having crossed the Sahara Desert on its way. Regular recording of its abundance in breeding sites by the British Trust

for Ornithology showed a collapse in populations in the late 1960s that continued until the mid-1980s, since when it has begun to recover (Fig. 11.8). Perhaps its most hazardous activity is migrating through the Sahara on the way to and from its wintering ground, and examination of the rainfall records of the Sahel region of the southern Sahara (Fig. 11.9) shows a marked decline in rainfall through the late 1960s and some recovery in the late 1980s. Although it is dangerous to argue from correlations, this does provide quite strong circumstantial evidence that the population decline was related to the climatic changes resulting in greater stress for the bird during its migrations.

Climate may not always be the cause of such population changes in migratory organisms, however. Many North American migrant birds, for example, have declined in recent years. The golden-winged warbler is down by 46% over the past 25 years; the wood thrush is 40% down and the orchard oriole has seen a 29% reduction. These changes have been related to an alteration in habitat management in their wintering grounds in Mexico, Central America and the Caribbean. In the past the traditional agriculture of these regions has been growing coffee under a canopy of trees, but

Fig. 11.9 Rainfall trends in the Sahel region of the southern Sahara expressed as departure from the long-term mean (represented as 0). Note the general trend towards drought conditions, particularly in the late 1960s resulting from the failure of monsoon rains to penetrate this region. Compare the pattern of this record with that of the whitethroat population in Fig. 11.8.

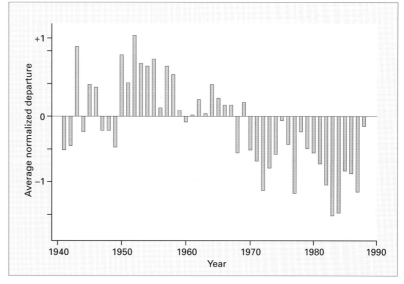

high-yielding coffee varieties demand more open conditions, and the varied habitat formerly available has been much modified. Thus, the loss of migrants in this case may have more to do with human land use than climatic change [23].

Changing communities and biomes

Studies of the past history of plants and animals lead us to believe that when communities are subjected to rapid environmental change they alter their species composition, some species becoming locally extinct (or emigrating) and others immigrating from elsewhere. The individualistic (Gleasonian) model of the community usually applies, with each species reacting in its own manner and according to its own requirements, rather than shifting as an integral member of a particular association. Our projections concerning a future world would therefore anticipate a shuffling of current assemblages of species as each reacts in different ways to a new set of conditions.

Many experimental manipulations of whole communities have been conducted to supplement the individual species studies described earlier in this chapter. In one experiment on hay meadows and pastures in the Swiss alpine foothills, for example, elevation of atmospheric carbon dioxide levels within plastic containers from 330 p.p.m. to 660 p.p.m. led to some species gaining advantages over others [24]. The outcome was a new balance in the assemblage of species, with some becoming more abundant and others less so. If we scale up from experiments of this type, we can begin to predict alterations in community composition over wide areas.

If, instead of considering the impact of climate change upon individual species, we consider the functional types of organisms present (see Chapter 4), then we can begin to predict the outcome of change upon biome distribution. There have been several attempts to model the changes that would be brought about by a range of possible climatic changes, and one of them is shown in Fig. 11.10 [25]. Here the current state of affairs in Europe

is compared with predicted vegetation characteristics, given a rise of 5°C together with an increase in average precipitation of 10%. Effectively, the boreal forests of Scandinavia would be replaced by deciduous forest, and the zone currently bearing deciduous forest would develop a Mediterranean type of vegetation. In practice it is questionable whether these changes could actually take place, because of the rapid rate of climatic change, the relatively slow response rate of vegetation (due to slow breeding and spread, especially among trees) and the fact that so much of the available land area is already greatly modified by human activity. Perhaps the most immediate response would be seen among invertebrate animals, especially those with the capacity for flight, and among those plants that have a short generation time and an efficient system of seed dispersal (i.e. weed species).

The tundra biome may well be particularly vulnerable in a greenhouse world. Not only is it found in a situation where there is little opportunity for retreat, but there is also a very real danger that the greatest rise in temperature will be experienced in the high latitudes. However, in their studies of the ecosystems of the high Arctic, Philip Wookey of the University of Uppsala, Sweden, and Clare Robinson of King's College, London [26], have discovered that these tundra communities have a high level of inertia. Their soils are very poor in nutrients, especially nitrogen and phosphorus, and therefore immigration by more demanding plants may be held back. Also, there are many physical barriers to plant spread, since many tundra ecosystems are found fragmented on remote islands. Even in the event of prolonged warming, the spread of more temperate species would prove difficult, and refugia of tundra fragments would survive, just as they have survived previous very warm interglacials. There is also a remarkable genetic diversity remaining in the arctic flora which would render it resilient in the face of climatic change and able to adapt to the new conditions. The tundra biome might well shrink, but would probably survive.

A very serious threat to biodiversity conservation in the greenhouse world is the fragmentation

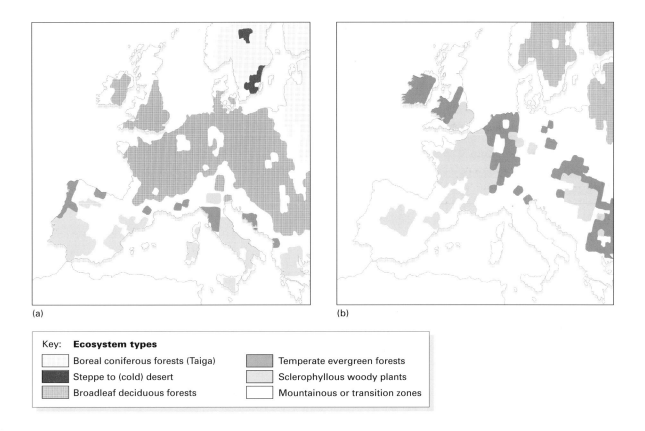

(a)   (b)

Key: **Ecosystem types**

	Boreal coniferous forests (Taiga)		Temperate evergreen forests
	Steppe to (cold) desert		Sclerophyllous woody plants
	Broadleaf deciduous forests		Mountainous or transition zones

Fig. 11.10 Model of the potential natural vegetation of Europe based on: (a) the current average temperature and precipitation and (b) an increase of +5°C in average temperature together with +10% average precipitation, which may be the outcome of a greenhouse scenario. Note the northerly shift in vegetation zones and the increase in the area of temperate evergreen forest. From de Groot [25].

of habitats as a result of human activity. The wholesale migration of biomes is conceivable only if species are free to spread from one area to another, and the fragmentation of habitats by agricultural, urban and highway development may prevent this process. The history of climatic change described in Chapters 9 and 10 shows that the earth's biota has survived some very rapid climatic changes in the past without too great a loss of biodiversity

(the megafauna being a notable exception). But conditions have altered since the last rapid climatic shift, about 10 000 years ago, for there is no longer a continuity of habitats. Human activity has so modified the face of the planet and has erected so many new biogeographical barriers that the former successful responses to climatic change may not be repeated this time round.

The future undoubtedly holds changes for the earth's biosphere, only some of which are predictable. Perhaps it is the unpredictable ones which are most to be feared. The changes in climate and in the physical and chemical conditions of our planet will result in a modification of its biogeography, inevitably involving the extinction of some, probably many, species. The study of the requirements and the interactions of plants and animals, including their past distribution patterns and evolution, provides an important means of understanding the complexity of nature and also of our own role within it. The story of our planet has been one of constant change and adjustment, and the biosphere has proved remarkably resilient to all the changes so far experienced. It may seem complacent to have faith in the ability of the natural world to cope with any stresses, including those directed at it by our own species, but the probability is that this is the case. The most important question facing humanity, however, is whether we are equally resilient, and whether we are equally able to adapt socially and technologically to changing conditions fast enough to permit the survival of *Homo sapiens*. Nature, in some form, will undoubtedly survive the next few centuries, but will we?

Summary

1 Taking the past and the present as the key to the future, it should be possible to make biogeography into a predictive science and to make projections concerning the future of the earth.
2 We are limited in this by our capacity to make predictions about future climate and nutrient cycling changes which will vary both in response to external forces, such as solar irradiance, and as a consequence of less predictable human activity.
3 A knowledge of the physiological requirements of individual species provides a basis for prediction in biogeography, but it does not take into account the reactions of other species, nor the problems of dispersal to new sites.
4 Functional types provide a better base for projection, so the mapping of biome distribution for a given set of climatic conditions is possible, but allowance must still be made for modifications due to human impact.

Further reading

Huggett RJ. *Environmental Change: The Evolving Ecosphere*. London: Routledge, 1997.
Moore PD, Chaloner B, Stott P. *Global Environmental Change*. Oxford: Blackwell Science, 1996.
Solomon AM, Shugart HH. *Vegetation Dynamics and Global Change*. London: Chapman & Hall, 1993.
Walker B, Steffen W (eds) *Global Change and Terrestrial Ecosystems*. Cambridge: Cambridge University Press, 1996.

References

1 Lean J, Beer L, Bradley RS. Reconstruction of solar irradiance since A.D. 1600: implications for climate change. *Geophys Res Lett* 1995.
2 Kaufman RK, Stern DI. Evidence for human influence on climate from hemispheric temperature relations. *Nature* 1997; 388: 39–44.
3 Moore PD, Chaloner B, Stott P. *Global Environmental Change*. Oxford: Blackwell Science, 1996.
4 Knutson TR, Tuleya RE, Kurihara Y. Simulated increase of hurricane intensities in a CO$_2$-warmed climate. *Science* 1998; 279: 1018–20.
5 Webster PJ, Palmer TN. The past and the future of El Niño. *Nature* 1997; 390: 562–4.
6 Vaughan DG, Doake SM. Recent atmospheric warming and retreat of ice shelves on the Antarctic Peninsula. *Nature* 1996; 379: 328–30.
7 Rott H, Skvarca P, Nagler T. Rapid collapse of Northern Larsen Ice Shelf, Antarctica. *Science* 1996; 271: 788–92.
8 Gregory JM, Oerlemans J. Simulated future sea-level rise due to glacier melt based on regionally and seasonally resolved temperature changes. *Nature* 1998; 391: 474–6.
9 Stocker TF, Schmittner A. Influence of CO$_2$ emission rates on the stability of the thermohaline circulation. *Nature* 1997; 388: 862–5.
10 Lee JA. Unintentional experiments with terrestrial ecosystems: ecological effects of sulphur and nitrogen pollutants. *J Ecol* 1998; 86: 1–12.
11 Cislaghi C, Nimis PL. Lichens, air pollution and lung cancer. *Nature* 1997; 387: 463–4.
12 Ferguson NP, Lee JA. Some effects of bisulphite and sulphate on the growth of *Sphagnum* species in the field. *Environ Pollution A* 1979; 21: 59–71.
13 Wookey PA, Robinson CH, Parsons AJ, Walker JM, Press MC, Callaghan, TV, Lee JA. Experimental constraints on the growth, photosynthesis and reproductive development of *Dryas octopetala* to simulated environmental change in a high arctic polar semi-desert. *Oecologia* 1995; 102: 478–89.
14 Schulze E-D. Air pollution and forest decline in a spruce (*Picea abies*) forest. *Science* 1989; 244: 776–83.
15 Moffat AS. Global nitrogen overload problem grows critical. *Science* 1998; 279: 988–9.

16 Bunce JA. Variation in growth stimulation by elevated carbon dioxide in seedlings of some C3 crop and weed species. *Global Change Biol* 1997; 3: 61–6.

17 Parry M. *Climate Change and World Agriculture.* London: Earthscan, 1990.

18 Nicholls N. Increased Australian wheat yield due to recent climate trends. *Nature* 1997; 387: 484–5.

19 Roberts L. How fast can trees migrate? *Science* 1989; 243: 735–7.

20 Hamburg SP, Cogbill CV. Historical decline of red spruce populations and climatic warming. *Nature* 1988; 331: 428–31.

21 Davis AJ, Jenkinson LS, Lawton JH, Shorrocks B, Wood S. Making mistakes when predicting shifts in species range in response to global warming. *Nature* 1998; 391: 783–6.

22 Parmesan C. Climate and species' range. *Nature* 1996; 382: 765–6.

23 Tangley L. The case of the missing migrants. *Science* 1996; 274: 1299–30.

24 Leadley PW, Stocklin J. Effects of elevated CO_2 on model calcareous grasslands: community, species and genotype level responses. *Global Change Biol* 1996; 2: 389–97.

25 De Groot RS. *Assessments of potential shifts in Europe's natural vegetation due to climatic change and implications for conservation.* Report to International Institute for Applied Systems Analysis, Luxembourg, 1987.

26 Wookey PA, Robinson CH. Responsiveness and resilience of high Arctic ecosystems to environmental change. *Opera Bot* 1997; 132: 215–32.

CHAPTER 12: *Drawing lines in the water*

Life clothes the land; it merely stains the seas.

Introduction

The biogeography of the land and of the sea are similar: both involve the analysis of the biotas of vast areas of the surface of the globe. However, because their environments are very different, the marine biotas are very much less complex than those of the land, yet also much more difficult to study. As a result, we know far less about the composition, structure and ecology of marine organisms than we do about those of the land. In many areas of marine research we are therefore still at the stage of constructing and evaluating hypotheses at a comparatively basic level. This fact is of more than merely academic and technical importance, in view of our desire and need to conserve the world's present diversity of organisms. To do this, we must first understand the fundamental patterns of distribution, both of the ecosystems and of the organisms they contain. Only then can we identify those that are threatened because of their rarity or because of their vulnerability to ecological change—whether natural or the result of human activities.

The terrestrial biogeographical regions are, effectively, the different continents. The interplay of the topography of the land and of the seasonal cycles in climate within each region also produces a considerable variety of physical environments. These regions are usually separated from one another by barriers of ocean, mountain or desert that make it difficult for organisms to disperse from one to another. The geographical boundaries of the regions are therefore easy to define. Their inhabitants live in air, which has a very low density. As a result, it is impossible for terrestrial organisms to be permanently airborne, and it does not provide much help

in their long-range dispersal. On the other hand, it allows plants to become structurally complex. In most parts of the continents (excluding the tundra, steppe and desert) the plants therefore dominate the environment. In addition to the variety of physical environments, the plants therefore add their own living architecture (grasssland, woodland, forest, etc.) within which the animals exist. These habitats provide a framework for biogeographical analysis at a finer level of detail, while our investigations are helped by the fact that we, too, are terrestrial.

The liquid world of the oceans is quite different. The major oceans are all interconnected, so that their geographical boundaries are less clear than those of the continents. As a result, their biotas cannot show such clear differences as those on land. The oceans are also far larger then the continents, for they make up 71% of the surface of our planet. The oceans themselves are continually moving, because the water within each ocean basin slowly rotates. These moving waters carry marine organisms from place to place, and also help the dispersal of their young or larvae. Furthermore, the gradients between the environments (and therefore between the different faunas) of different areas of sea floor or of ocean water mass are very gradual, and often extend over wide areas that are inhabited by a great variety of organisms of differing ecological tolerances. There are no firm boundaries within the seas.

However, because of the density and power of the waters, there are no large, complex plants to provide the equivalent of the terrestrial biomes. Photosynthesis in the sea is carried our by tiny, single-celled organisms known as the phytoplankton. The density of plant life in the ocean is therefore far less than on land, and the primary productivity per unit area is only one-fifteenth of that in tropical

238

rainforest, so that much less solar energy is being fixed in the system.

Nevertheless, there is one way in which the oceans are more complex than the land: they have an important extra dimension, that of depth. The physical conditions of light, temperature, density and pressure, and often also of the concentrations of nutrients and oxygen, change much more rapidly with depth in the seas than they do with altitude on land—and these lead to corresponding changes in the biotas. The resulting vertical patterns of distribution also interact with the horizontal patterns.

Because of these structural and productivity differences, the sea has provided much less opportunity for evolutionary diversification. Marine families contain fewer genera than terrestrial families, and these genera contain fewer species. As a result, there are many fewer species in the sea: only about 160 000 species of marine organism have been described, compared with about 1.8 million from the land. Similarly, there are probably over 250 000 species of land plant, but only 3500–4500 species of phytoplankton.

For all these reasons, marine organisms usually have a much wider distribution than those on land. Thus, while most families of mammal are found in a single zoogeographical region, most families of marine organisms are cosmopolitan or widespread throughout the world's oceans. Because of this, marine faunas differ from one another in containing different genera or species, rather than different families. These genera or species are not known by different English-language names, so we can only refer to them by their Latin names. As a result, although it is easy to explain that, for example, that the Neotropical zoogeographical region contains the endemic families of armadillo, ant-eater and sloth, one can only explain the differences between the biotas of marine regions by giving long lists of Latin names.

Finally, because we ourselves are terrestrial and air-breathing, it is difficult for us to study and to census the life of the sea, even in the near-shore or surface-water regions, and even more difficult in its depths. Our knowledge of the fauna of the deep-ocean sea bed, which covers an area of 270 million km^2, is derived from cores totalling only about 500 m^2, together with the areas sampled from dragging a number of trawls and deep-sea sleds over the bottom!

Despite all these differences, the great ocean basins are similar to the continents as biogeographical units. Even though their patterns are more diffuse, they exhibit the same phenomena, and raise the same problems of explanation by dispersal or vicariance, or of the extent to which differences are due to different histories of enlargement, fusion or subdivision, or to different evolutionary events. But, since the marine faunas are still much less well known, we need to be much more cautious in coming to conclusions, or in assuming that particular deductions and generalizations associated with continental biogeography are necessarily valid for marine biogeography.

Zones in the ocean and upon the sea floor

Because the physical conditions alter as one moves downwards, away from the light and warmth of the surface, a number of different depth zones can be distinguished in the waters of the sea, while the shape of the sea floor similarly defines several different regions there (Fig. 12.1a).

The surface layer of the sea, to about 200 m in depth, is the most liable to extremes of temperature. This can be as low as –1.9°C (the freezing point of sea water) below the high-latitude ice cover, or over 30°C in enclosed low-latitude waters such as the Red Sea. Below the 200 m level lies the thermocline, a transitional zone through which the temperature falls rapidly until, at a depth of 0.8–1.3 km, it reaches about 4°C. At greater depths, the temperature decreases slowly until, at most latitudes, it reaches a minimum of 2–3°C at around 3000 m. In polar seas, however, it may be as low as –0.5°C in the Norwegian Sea, or –2.2°C in parts of the Southern Ocean. Below 3000 m, there is only a very slight temperature gradient.

However, although the deep ocean makes up just over 50% of the marine environment, there are also

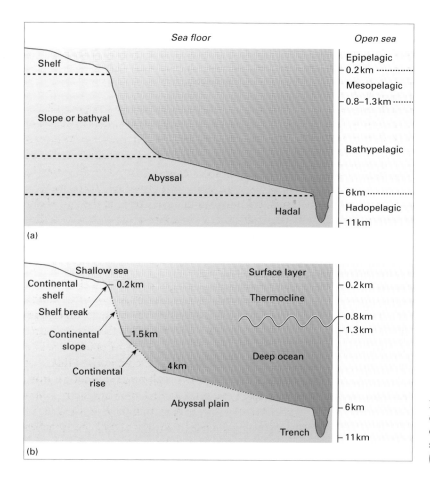

Fig. 12.1 (a) Diagram of the vertical divisions of the sea floor (left) and open sea (right). (b) Life zones on the sea floor (left) and in the open sea (right). (Not to scale.)

areas of shallow sea that are adjacent to the continents. The distinction between the two results from the basic structure of our planet. The continents rise above the level of the floor of the oceans, because the rocks of which they are made are less dense and therefore lighter than those of the ocean floor. Seas cover the lower-lying parts of the continents, which are known as the continental shelves (Fig. 12.1b). The depth of the water in these seas, which are known as 'epicontinental seas' or shallow seas, varies according to how much of it has gone to form ice sheets and glaciers on the continents. At present, they have a maximum depth of 200 m (except in the Antarctic, where the great weight of ice has depressed the continent, so that the edge of the continental shelf lies at about 500 m).

At this point, known as the shelf break, the slight gradient of the sea floor increases sharply. From here onwards, the continental slope descends relatively steeply until it reaches the deep ocean floor, which is known as the abyssal plain. However, where the two meet, the margin of the plain is covered by the continental rise—a wedge of sediments derived from the adjacent continent. At an average depth of 4 km, the abyssal plain covers 94% of the area of the oceans and 64% of the surface of the world, and is therefore the most extensive of all environments. In the Indian and Pacific Oceans, the abyssal plain is also fringed by a system of trenches, reaching depths of up to 11 000 m, where old ocean floor disappears back into the depths of the earth (see p. 113).

The basic biogeography of the seas

The shape of the ocean basins is the cause of the most basic divide within marine biogeography—between the shallow-sea (or 'neritic') realm and the open-sea realm. The major difference between these two realms is that of scale. Of course, the shallow seas occupy a smaller area than the oceans, and each of them is also far smaller than any ocean. But the individual shallow seas also differ more from one another than do the oceans, and each shallow sea contains within itself a greater variety of conditions than any equivalent area of ocean. As we shall see, the patterns of life in the oceanic realm are the results of external factors, such as the earth's rotation and the heat and light of the sun. The resulting patterns can be seen at the largest scale, for example at the level of the North Pacific or the South Atlantic Ocean. Although these oceanic patterns also affect the shallow seas, they are less obvious there than local influences such as the nature of the sea floor, the contribution of sediments and fresh water by local rivers and streams, or the pattern of tides. In addition to these physical differences, the various shallow seas are also isolated from one another by wide stretches of ocean. As a result, their biotas are similarly isolated, giving the opportunity for independent evolution and endemicity.

A subsidiary divide within marine biogeography is between the patterns of distribution of pelagic organisms, which swim or float within the waters themselves, and those of benthic organisms, which live on or in the sea floor. (Pelagic organisms that swim are known as nektonic, while those that float are known as planktonic.) The distribution of benthic organisms depends on the local characteristics of the sea floor, while that of the pelagic organisms is quite independent of the depth or nature of the sea floor. However, pelagic and benthic species are often dispersed by similar processes, and therefore tend to show similar patterns of distribution. This is also because both depend on the patterns of productivity in the surface waters, from which organic material falls to the lower levels and, eventually, to the sea bottom.

The biogeography of the open-sea and shallow-sea realms will now be dealt with in turn.

The open-sea realm

There is not enough sunlight to support photosynthesis below the upper few tens of metres of water, which is called the euphotic zone. The heating effects of the sunlight are also mainly restricted to the upper, wind-mixed layer, which varies in thickness from a few tens of metres to 200 m. As a result, this superficial layer, termed the epipelagic zone, contains the highest concentration of living organisms (Fig. 12.1b). Deeper than this, in the twilight zone below the thermocline, lies the mesopelagic zone, which extends to depths of approximately 1000 m. Many of the fish that live in this zone during the day move up towards the surface at night, to feed on the richer fauna there. Below the mesopelagic zone lies the bathypelagic zone, which extends to a depth of 6000 m; it is a zone of total darkness and almost unchanging cool temperatures. The fish of this zone are too far from the surface to migrate there daily. Finally, the hadopelagic faunas of the deep trenches are thought to be mainly endemic, because many of the trenches are isolated from one another.

The biogeography of the open-sea realm is best approached in two stages. First, one must understand the geographical/geological histories of the different oceans, for they have caused some of the faunal differences between them. Secondly, one needs to appreciate the patterns of circulation within the oceans, for these have led to differences in their nutrient concentrations, which in turn have affected the patterns of distribution of life there. Much of this has recently been reviewed by the British oceanographer Martin Angel [1,2].

The history of the ocean basins

Over the last 175 million years, the patterns of land and sea have been greatly modified by the breakup of the old supercontinent of Pangaea and the dispersal of the resulting fragments (see Figs 6.3

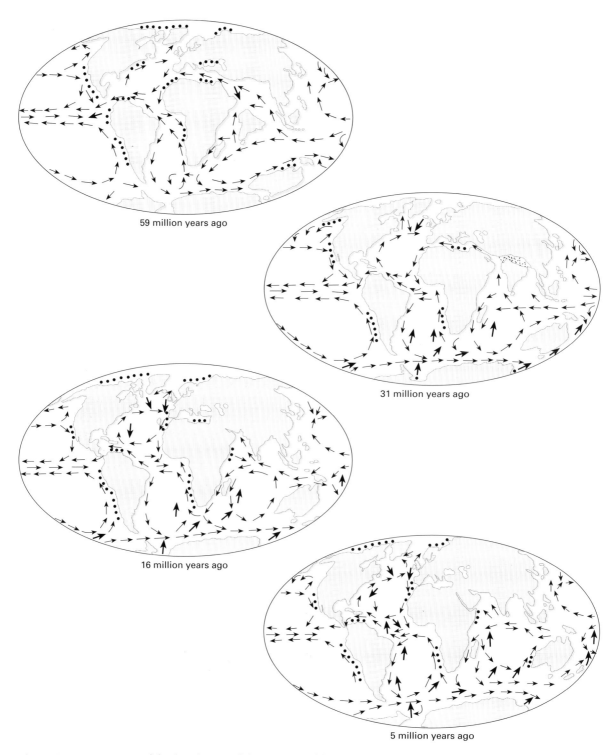

Fig. 12.2 Reconstruction of the distribution of the continental land masses and the inferred pattern of circulation of the ocean currents over the last 59 million years. The large dots indicate regions of upwelling, and the large arrows indicate possible regions of bottom-water formation. (In many cases, the formation of bottom water is by the sinking of warm saline water.) From Angel [1] with permission, based on Haq [37].

and 6.4). Whatever patterns existed in the single worldwide ocean (sometimes known as 'Panthalassa') that surrounded Pangaea were obliterated by the impact event at the end of the Cretaceous Period, 65 million years ago (see p. 126). As a result of the climatic changes at that time, the whole system of water circulation and of its gradients of nutrients appears to have been disrupted for something like half a million years, and marine productivity was reduced by 50%. Some 65–70% of the species that lived in the upper levels of the seas became extinct, but the distribution of microfossils suggests that those pelagic species that did survive became widely spread through the world's oceans. The differences between the faunas of the different oceans of today therefore appeared after this time.

Subsequent events led to major changes in the patterns of the oceans, and of ocean circulation in the Antarctic region and in the equatorial regions (Fig. 12.2). The separation of Australia from Antarctica approximately 45 million years ago and the deepening of the passage between Antarctica and South America about 34 million years ago had profound effects. They enabled the establishment, and later intensification, of the current of water that flows eastwards around Antarctica (see p. 150). This ultimately led, not only to the glaciation of Antarctica and a drastic change in the climate of Australia (see p. 149), but also to the eventual establishment of the Antarctic contribution of cold, oxygen-rich bottom water to the thermohaline circulation (see below, p. 244).

The other changes were in the equatorial region, where there had long been a tropical ocean corridor (sometimes called the Tethys Sea) extending from the western end of the Indo-Pacific Ocean between Africa and Eurasia, and across the then-narrow Atlantic to the eastern Pacific. This sea was obliterated by several events. First, India and Africa moved northwards to contact Asia, followed by the approach of Australia to South-East Asia. This resulted in the formation of a separate Indian Ocean with its own pattern of circulation. Secondly and more recently, the tropical Pacific was severed

from the tropical Atlantic by the completion of the Panama Isthmus about 3.5 million years ago.

The evidence for these changes comes from the palaeontology of marine sediments, as well as from geophysical studies. For example, study of the fossil plankton faunas from either side of the Panama Isthmus provides detailed evidence of the progress of formation of the land link. Circulation of the intermediate depth of water between the Atlantic and the Pacific became reduced about 12 million years ago, and a similar reduction of the circulation of the surface waters gradually intensified from 7 to about 3.5 million years ago. From this latter date, which coincides with the date of completion of the land link, the plankton faunas became steadily more different [3]. Similarly, there is also much evidence of vicariant evolution in the nektonic and benthic faunas on either side of the Isthmus: there are approximately 1000 fish species on both sides of the Isthmus but, apart from widely distributed species, only about 12 are identical, while the great majority of the shallow-water invertebrate species on each side differ from one another—although there are many closely related species-pairs across the Isthmus.

We can now turn to outline the factors that produce comparatively stable patterns of distribution within these oceans.

The dynamics of the ocean basins

The patterns of movement of the waters in the oceans are dominated by the stable gyres—huge masses of water, filling a large proportion of an ocean basin, which rotate horizontally, with a periodicity that is usually 19–20 years. Their rotation is caused by the patterns of wind, which themselves result from the uneven distribution of solar energy upon the surface of the earth, as well as from the eastwards rotation of the earth. Heat from the hot equatorial regions is distributed towards the poles by patterns of wind movement that rotate clockwise in the northern mid-latitudes and anticlockwise in the southern mid-latitudes (see pp. 84–8). These wind patterns create similar patterns of movement in the waters below, so that warm ocean currents flow away from the equator along the western edges of the oceans, and cold currents flow away from the poles along their eastern edges (see Fig. 12.5, below). The equatorial waters, in both the Atlantic and the Pacific Oceans, have a westwards surface current. Because the winds in the Indian Ocean show a seasonal reversal, the monsoon (cf. Fig. 4.9, p. 87), the current patterns in the Indian Ocean are similarly complex. The Arctic Ocean is almost totally enclosed by land, so that there is only a little southwards dispersal of its waters, mainly to the east and west of Greenland. The Antarctic continent is surrounded by a band of permanently cold, eastwardly flowing water.

In addition to these horizontal movements of the ocean waters, there is also a vertical circulation, driven by differences in water temperature and salinity. As sea water in the polar regions freezes into virtually salt-free ice, the excess salt enriches the water layer immediately below the ice. These waters are therefore unusually saline and dense, causing them to sink downwards. In two regions (along the eastern side of Greenland and near the Antarctic Peninsula) this water for some reason does not simply mix with the deeper ocean water, but sinks down to the ocean floor as a coherent body of water known as bottom water. Because it is cold, this water is also rich in dissolved oxygen and carbon dioxide, so that it can be recognized by these characteristics as well as by its salinity. As it spreads through the oceans, it therefore brings oxygen to even the deepest waters, far from the oxygen of the surface. This in turn displaces water upwards, producing what is known as a thermohaline circulation. The time taken for the ocean waters to complete a cycle of vertical circulation is estimated at 275 years in the Atlantic, 250 years in the Indian Ocean and 510 years in the Pacific [4].

A similar phenomenon arises from the fact that, near the centres of the great ocean gyres in the North Pacific and North Atlantic, where the weather is normally clear and sunny, the surface loses more water by evaporation than it gains by rainfall. It too, becomes more saline and dense so that, where

it meets neighbouring waters to the north and south, it sinks below their lighter, fresher waters along lines known as convergences (shown as dashed lines in Fig. 12.5 on p. 247). Convergences are also found where two sets of ocean currents converge, as in the North Pacific Polar Front Convergence and the Southern Subtropical Convergence (shown by dotted areas in Fig. 12.5).

Another cause of vertical water movement are the winds that blow off-shore along the western coasts of parts of the Americas, Africa and Australia. These winds blow the warm surface waters away from the coast, and they are replaced by an upwelling of deeper water. A similar upwards movement of water is found at regions known as divergences in the Atlantic and Pacific, where there is a shear between currents going in different directions. The Pacific Equatorial Divergence (Fig. 12.5 PEqD) lies to the south of the zone of meeting between the westwards equatorial current and the eastwards North Pacific Countercurrent (Fig. 12.5, NPEqC). The Antarctic Divergence (line of small circles in Fig. 12.5) lies between the eastwards Antarctic current and the narrow ribbon of westwardly directed Austral Polar Current that lies adjacent to the coastline of Antarctica.

The patterns of circulation of the ocean waters, both horizontally and vertically, also create patterns in the concentration of such major nutrients as nitrate, phosphate and silicate. The patterns of availability of these nutrients in space and time have profound effects upon the marine organisms; for example, silicate is vital for the production of the skeleton of the type of phytoplankton known as diatoms, which are the main source of the rain of organic material that descends into the deeper layers of the ocean. There is therefore a close correlation between the pattern of regions where waters rise to the surface at upwellings and divergences (Fig. 12.5), the availability of nutrients in the surface waters (Fig. 12.3), and the patterns of productivity in the oceans (Fig. 12.4). Thus, although upwelling areas account for only 0.1% of the surface of the oceans, these areas provide 50% of the world's fisheries catch! (It must be emphasized that

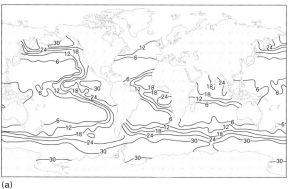

(a)

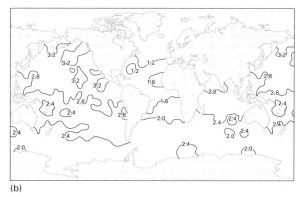

(b)

Fig. 12.3 (*above*) The distribution of nutrients in the oceans. (a) mean dissolved nitrate at a depth of 150 m; (b) mean dissolved phosphate at a depth of 1500 m. Levitus [38] with permission from Elsevier Science.

Fig. 12.4 (*below*) Patterns of annual primary production in the oceans. Numbered areas refer to annual productivity of carbon in g/m²/year, as follows: 1, 500–200 g; 2, 200–100 g; 3, 100–60 g; 4, 60–35 g; 5, 35–15 g. After Berger [39]

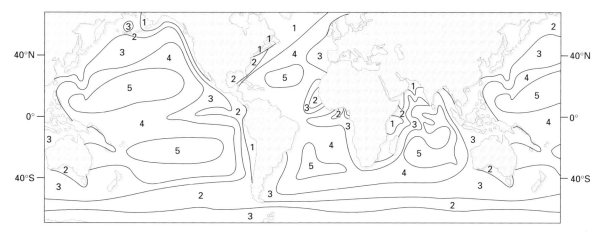

all the primary productivity in the oceans is confined to the uppermost few tens of metres as, below this level, the sunlight is absorbed and scattered by the water.) The extent to which nutrient-led productivity varies during the year depends upon local circumstances.

Patterns of life in the ocean waters: comparisons between the oceans

All the oceans interconnect at high southern latitudes, and the Panama Isthmus was only com-

pleted comparatively recently. It is therefore not surprising that very many open-sea species are widespread. For example, the little nektonic, epipelagic arrow-worm *Pterosagitta draco* is found in tropical and subtropical waters of all the oceans, and this is also true of polychaetes (bristle-worms) and of the pelagic crustaceans known as euphausids (which form the 'krill' upon which baleen whales, and many other marine animals, feed). Arrow-worms that live in the deeper mesopelagic and bathypelagic levels range from the subarctic to the subantarctic, because there is little latitudinal change in temperature at these greater depths [5], and this pattern is also found in fishes living there. Nevertheless, there is some endemicity in the oceans. For example, there are not only separate subspecies of the arrow-worm *Sagitta serratodentata* in the Atlantic and in the Pacific, but also a separate species, *S. pacifica*, in the southern Pacific. Furthermore, molecular evidence suggests that the uniform morphological characteristics of some of the widespread 'species' conceals considerable biochemical and physiological diversity, so that in reality they have probably diverged into several geographically defined species.

Another difference is that the cold-temperate marine fauna of the North Atlantic is much less diverse and has fewer endemic forms than that of the North Pacific. The fundamental cause may merely be that the Atlantic is a smaller ocean than the Pacific, and therefore provides a smaller area within which evolutionary novelties may appear and disperse to other parts of the ocean. Another factor that may have been important is the effect of the fluctuations in sea level during the Ice Ages. These caused major, and repeated, changes in the patterns of isolation and recombination of pelagic and benthic communities, particularly in the East Indies.

Patterns of life in the ocean waters: Domains and Provinces within the oceans

Although the patterns of movement of the surface waters have long been known, it is only since satellite observations became available that it has been possible to monitor their patterns of life comprehensively and continually. The chlorophyll content of the water can now be measured in this way, and this allows us to deduce the density of the phytoplankton, the depth of the euphotic zone, and the seasonal cycles in the balance between phytoplankton productivity and loss, which may or may not lead to a seasonal increase in phytoplankton biomass known as a 'bloom'. The British oceanographer Alan Longhurst has integrated this biological data with that of the movements of the ocean waters to identify and define three biogeographical

Domains in the oceans (the Polar, Westerly Winds and Trade Winds Domains), plus a Coastal Domain that comprises the shallow seas; these Domains are divided into ecological Provinces [6]. Figure 12.5 is a simplified, redrawn version of Longhurst's map, showing most of his 33 Provinces of the Oceanic Domains.

In being defined primarily by environmental features, Longhurst's marine Domains and Provinces are unlike the terrestrial biogeographical regions, which are characterized by differences in their faunas and floras—although this difference may disappear as further taxonomic work reveals more of the biotic differences between the Provinces. Another difference between the two systems is that the precise positions of the boundaries between these Provinces vary from year to year and from season to season—although the underlying pattern remains stable.

If the topography of the world were more uniform, each of the oceans would be occupied by two large gyres, one in the north and rotating clockwise, and another in the south rotating anticlockwise. The more-polar half of each would have westerly winds and ocean currents, and belong to Longhurst's Westerlies Domain, while the more-equatorial half would have easterly winds and currents and belong to his Trades Domain. This pattern does occur in the North Atlantic and North Pacific (Fig. 12.5), but elsewhere the pattern is more complicated. The winds and currents in the Indian Ocean, which Longhurst places in his Trades Domain, are variable because of the annual monsoon reversal; the winds blow from the north-east from November to March or April, and from the south-west from May to September. In the South Atlantic, the whole oceanic gyre is placed in the Trades Domain, because it is all under the influence of south-easterly winds, full westerlies only developing south of the latitude of Cape Horn. Both this and the South Pacific gyre are still very poorly known, and Longhurst therefore does not divide either of them into northern and southern Provinces.

The absence, occurrence or timing of planktonic blooms are an important part of Longhurst's sys-

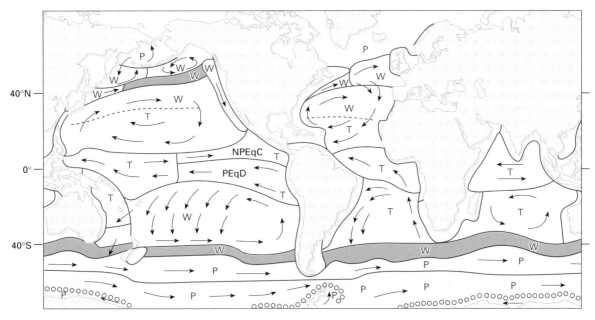

Fig. 12.5 The oceanic biogeographical Domains and Provinces, redrawn after Longhurst [6]. Letters in blue (P, T, W) indicate Provinces that are placed in the Polar, Trade Winds or Westerly Winds Domains, respectively. NPEqC, North Pacific Equatorial Counter-current Province; PEqD, Pacific Equatorial Divergence. North Pacific Polar Front Convergence and Southern Subtropical Convergence are shown as a dark tint. Antarctic Divergence shown as line of small circles.

tem. These blooms are best understood by imagining an ocean with stable conditions, in which the euphotic layer is of constant depth, and of the same depth as the thermocline. Within this euphotic layer, the light and heat of the sun allow continuous growth of the phytoplankton, but their total biomass is kept to a constant, low level by grazing by the zooplankton and by the supply of nutrients, which are normally used up by the phytoplankton as soon as they are available.

This is, in fact, more or less the pattern in most of Longhurst's Trades Domain. This mainly comprises the areas where winds and currents are always from the east, and where there is also enough year-round sunlight for the warm euphotic zone to be permanent and of approximately constant depth. Because of the stability of this layer, there is little vertical mixing of the surface waters, in which the phytoplankton exhausts its nutrient

supply, and the deeper, more nutrient-rich waters. There is, consequently, little seasonal variation in algal productivity, and therefore no algal bloom. However, the Province of the Trades Domain that lies across the equator in the Atlantic provides an exception to this, because in this Province there is an increase in the power of the trade winds in the summer. This leads to an accumulation of surface water in the western part of the basin, and a corresponding decrease in the eastern part. This in turn allows deeper, nutrient-rich water to rise upwards in this eastern region. As a result, it is now within the euphotic zone, so that its nutrients can fuel a summer algal bloom. (This phenomenon does not take place within the Pacific, because it is too broad for there to be enough time for such a seasonal wind-change to have such an effect.)

Longhurst's Westerlies Domain comprises Provinces in which there is an algal bloom in the spring. It includes Provinces with three rather different seasonal situations: those in high northern latitudes in the Atlantic, those in similar latitudes in the Pacific, and the rest, which are found in lower latitudes in both hemispheres. These three situations will be described in turn.

In the eastern North Atlantic, phytoplankton productivity is limited by both light and nutrients. The bloom therefore only takes place in the spring, when the amount of sunlight increases and the thermocline rises towards the surface. The rate of increase of the phytoplankton is extremely rapid and is unpredictable. As a result, it is not controlled by grazing zooplankton, whose increase lags behind that of the phytoplankton. Instead, within a few days, having exhausted all the available nutrients, the larger phytoplankton such as diatoms die or lose their buoyancy, and their remains fall like dense snow downwards through the water column to the abyssal levels. Their place in the upper waters is taken by tiny flagellates and cyanobacteria (the picoplankton). These are responsible for 80–85% of the subsequent primary production. However, because they are too tiny to sink or to be filtered out of the water by zooplankton, they do not form the basis for a classical food web leading to larger animals, such as is found in the Pacific (see below). There is often a second bloom in this area in the autumn. This may be because mixing of the waters by storms brings new supplies of nutrients to the surface waters, or because many of the zooplanktonic grazers have descended to greater depths to overwinter. This autumn bloom is short-lived, for the failing light intensity soon limits photosynthetic activity, and it declines rapidly when the nutrients in the upper level have been used up. Productivity remains at a very low level throughout the winter.

In the high latitudes of the North Pacific, the springtime bloom does not use up all the available nitrate. This may be because its timing is more predictable, so that the zooplanktonic grazers that have overwintered at depth, such as the copepod *Neocalanus*, rise to the surface at the appropriate time, ready to graze upon the phytoplankton and

control their numbers. However, it is also possible that the availability of iron is a limiting factor. Whatever may be the cause, the diatom bloom in the upper waters does not 'crash'. It is therefore available for a longer period of time, as the basis for a food chain upwards via zooplankton, fishes and larger crustaceans, and thence to sea birds, seals and whales.

The other Provinces of the Westerlies Domain (which together cover over half of the oceans) lie at lower latitudes, so that light is not a limiting factor. Phytoplankton productivity therefore increases during the winter as the progressive deepening of the mixed layer recharges the surface layers with nutrients, and declines in early summer as these nutrients are used up.

In Longhurst's Polar Domain it is light, rather than nutrients, that limits algal growth. This growth therefore rapidly increases in the spring, as the sunlight increases in duration and strength, and the ice (if any) melts. The peak of this bloom is near to midsummer, after which it declines because of grazing by zooplankton. However, there is a second peak in September, when the copepod grazers descend to greater depths to overwinter. This secondary peak is less developed in the Antarctic, where these copepods overwinter closer to the surface, among the phytoplankton.

Finally, it is worth noting that there are some areas where, for some reason, although the level of nitrates in the surface waters remains high, this does not lead to increased productivity—the north subpolar Pacific, the eastern tropical Pacific and the Southern Ocean around Antarctica. It has been suggested that this is because productivity is continually limited by grazing zooplankton, or that it is due to a lack of some other nutrient, perhaps iron.

If one reduces the scale of comparison from that of whole oceans to that of their component water gyres, there is some evidence for faunal distinctions. In particular, the American oceanographer Brian White [7] believes that one can find parallels between the sequence of appearance of different water masses in the Pacific Ocean, due to plate-tectonic changes, and the relationships between the different species of some fishes. It is likely that further work will expand on this, to give the faunas of Longhurst's different Provinces the same sort of taxonomic basis that has long been known for terrestrial zoogeographical regions. For example, the limits of distribution of many Pacific epipelagic organisms coincide with one another and seem to be related to the patterns of water masses [8]. The patterns of temperature, nutrient availability and phytoplanktonic production seen in Longhurst's Provinces are of fundamental importance to the marine organisms that inhabit them. Each population must therefore be physiologically and behaviourally adapted to the conditions in its own water mass. Such a set of adaptations can only evolve if the population is a separate species, physically and genetically distinct from others.

Although it is often stated that many marine taxa have very wide geographical ranges, it is very likely that this is more apparent than real, and that it results from inadequate taxonomy. There is little doubt that many marine 'species' as currently recognized are in fact complexes of subtly distinct species. For example, Gibbs [9] comments that the mesopelagic fish *Nominostomias*, which had been thought to contain only eight species, is now known to have over 100.

Patterns of life on the ocean floor

Apart from the intertidal zone between the high water mark and the low water mark, marine biologists recognize four different life zones (Fig. 12.1b) for organisms that live on the sea bottom. (It must be emphasized again that changes in conditions and in faunas are always gradual, so that the boundary between two zones is merely a level of more rapid change, rather than a level of abrupt change.) The shelf zone comprises the continental shelf below the low water mark. The other levels lie at successively greater depths: the bathyal zone (also known as the slope or archibenthal zone), the abyssal zone and the hadal zone. The hadal zone is easy to define, for it comprises the environment of the ocean trenches, at a depth of over 6 km. However, the depth of the boundaries between the shelf zone, the bathyal zone and the abyssal zone vary according to season, conditions and latitude.

The upper level of the bathyal zone is normally at the edge of the continental shelf, at about 200 m. Its lower level, and therefore the level of transition to the adjacent abyssal zone, varies considerably. Where the surface waters are cold, as in the Arctic region, the bathyal/abyssal transition is similarly at a shallow level, about 400 m, and the bathyal fauna itself extends to within 12 m of the surface. At lower latitudes, closer to the equator, where the water temperatures are in general higher, the bathyal/abyssal transition is at a deeper level, usually at the about 900 m base of the thermocline. At this level, the water temperature has dropped to 4°C, while below it drops to 1–2.5°C. The Swedish zoologist Sven Ekman [10], who was one of the founders of marine zoogeography, called this change in the depth of the bathyal/abyssal transition the principle of 'equatorial submergence'. The pattern of this variation strongly suggests that the depth at which the transition takes place is dependent on temperature.

To turn now to the nature of the faunas in these zones, one can distinguish two quite different environments. At one extreme, there is the fauna living on the continental shelf. Here, conditions vary in both time and space, but temperatures are higher and there is light, providing the energy basis for a rich ecosystem. At the other extreme, there is the fauna of the abyssal zone, adapted to lightless, cold waters in which nutriments are scarce. In each of these zones there live organisms that are specifically adapted to that environment. The bathyal zone is therefore an intermediate zone within which there is a gradual change with depth, from a mainly shelf-like fauna to a mainly abyss-like fauna. The precise pattern of faunal replacement with depth varies from place to place, according to the physiological tolerances and limitations of the individual species, and according to their interactions with the local competitors and predators.

The patterns of change with depth in the faunas of the bathyal and abyssal benthos have been intensively studied off the Atlantic coast of North America. As the continental shelf is the point at which the physical changes in the environment take place most rapidly, it is not surprising to find that this is the region where the rate of faunal change is most rapid; it then continues at a slower rate with increasing depth.

Another aspect of these faunas is the degree of diversity that they show. This, again, has been most intensively studied in the western North Atlantic, where the faunas show a very clear pattern [11]. In many groups, the faunal diversity is low on the continental shelf, high at a mid-bathyal depth (1000–2000 m), and low again on the abyssal plain. It seems likely that the ultimate cause of these changes is the gradual change in food and nutrient availability as one moves away from the high-productivity surface waters. The problem has been to identify how this change imposes itself upon the community structure of the sea bed. Various mechanisms have been suggested, such as changes in the intensity of predation, or competition, arising from differing rates of population growth. An interesting recent discovery is that there is a clear correlation between faunal diversity and the characteristics of the sediments of the sea floor, especially the diversity of particle size [12]. Because there is no primary production in the lightless deep sea, its economy is based upon the organic particles that settle upon the bottom, and the fauna that lives on or in the sea floor is dominated by detritivores. It may well be therefore that a greater range in particle size provides a greater range of niches for the bathyal benthic fauna.

Whatever the precise pattern (or range of patterns) of relative diversity in the deep-sea benthos, the total diversity on the abyssal plain is very great—even though, because of the low level of food supply, the density of the deep-sea biomass is very low [13,14]. The ecology of these faunas is so poorly understood that it is difficult to be sure of the reasons for this unexpected faunal richness. It is possible that the lack of barriers to dispersal on the abyssal plain makes it easy for many organisms, which have evolved at different locations on its enormous area, to become widely distributed. The coexistence of the resulting large number of species

might be facilitated by another phenomenon. The abyssal benthic environment is often disturbed by the settling of aggregations of plankton or of the remains of larger organisms. It has been suggested that this is so frequent that the community rarely has an opportunity to come to a final balance in which some species become locally extinct due to competition from rival species.

The amount of information on the composition of the fauna of the deep-sea benthos is still very limited and very geographically unbalanced. Most of it comes from the western Atlantic, and none is available from central oceanic regions, and it is also nearly all derived from soft-bottom communities, little being known of hard-bottom communities. It is therefore impossible to identify different biogeographical areas upon the deep-sea floor of a single ocean, or even to draw up lists of faunal differences between one ocean and another.

There is even less systematic information on the fauna of the hadal zone, which lies in the great submarine trenches, at depths of 6000–11 000 m. As might be expected from its great depth, its fauna is extremely sparse, even more impoverished than that of the abyssal plain, but it does also seem to be different—about 68% of the species, 10% of the genera and one family are endemic to the hadal zone. Most of the endemic species have a vertical depth range of less than 1500–2000 m, so there is steady faunal change with increasing depth. These faunas are confined to isolated patches along the deep-sea trenches (cf. Fig. 6.1, p. 113). This has permitted considerable independent evolution of endemic species within each trench fauna, and it is quite possible that their patterns of biogeographical relationship may be similar to that of the hydrothermal vents of the mid-oceanic rises (see below).

Unfortunately, it is still too early for marine biologists investigating these patterns to be able to make firm statements as to what patterns exist and in what variety, or as to how these patterns may vary according to geographical location or sea-floor topography.

There is, similarly, still not enough information on possible faunal differences between the faunas of the floors of the different ocean basins as a whole, although it seems extremely likely that these exist —earlier studies, mainly by Russian marine biologists, have suggested that only 15% of the deep-sea benthic species occur in more than one ocean, and only 4% in all of them. It is also possible that the pattern of mid-ocean ridges, where hot material rises from the depths of the planet to form new sea floor as the tectonic plates separate (see Fig. 6.1, p. 113), may similarly act as more minor barriers to faunal movement, and therefore delimit subsidiary faunal areas within the oceans. But the presence, nature and scale of such possible differences have yet to be established. In continental biogeography, larger-scale patterns of faunal change or difference are often found to have been caused by historical events of evolutionary innovation or extinction, followed by dispersal or vicariance arising from plate-tectonic events; smaller-scale differences are more likely to result from ecological factors. It will be interesting to see the extent to which marine biogeographical research reveals similar patterns in the deep sea. Even though the same processes are likely to be operative in the two environments, the very different scale, and probably very different rates, in the oceans may nevertheless lead to significantly different results.

The biogeography of hydrothermal vent faunas

Comparatively recently, in 1977, marine biologists discovered a dramatically different deep-sea environment, containing a fauna that shows a fascinating biogeographical pattern. This lies at the mid-oceanic ridges, which are at an average depth of 2.5–3.5 km. Although the ridges themselves are many hundreds of kilometres wide, they are split by a rift valley only about a kilometre wide, where hot lava is emerging. In some widely scattered areas known as hydrothermal vents, each of which covers only a few hundred square metres, the cold sea water penetrates fissures in the surrounding rocks. The water temperature there may reach 400°C (only the enormous pressures at this depth prevent it from turning to steam), and it reacts chemically with the rocks, so that it becomes rich in metals and sulphur. Where this superheated water emerges and is cooled by the surrounding waters of the ocean, these minerals precipitate out of the fluid. Some of them form solid 'chimneys', which can be many metres high, while others remain as distinct particles in the rising plume of water, which therefore looks like smoke emerging from the chimney.

Accompanying this extraordinary environment is a unique fauna, whose food web is not based on plants that have trapped the sun's energy, but instead on chemosynthetic bacteria that extract energy from the chemicals dissolved in the hot fluids. Some of these bacteria are consumed by grazing or filtering organisms, while others live symbiotically, rather as photosynthetic algae live in corals (see below, p. 256). They form the base of a fauna consisting mainly of worms, arthropods and molluscs.

The biogeographical problem that arises from these scattered vent communities is the method by which their organisms disperse from vent to vent. It is theoretically possible that their larvae disperse across the abyssal plain from one mid-oceanic ridge to another, but the highly unusual nature of the vent environment makes that seem unlikely. It is therefore no surprise to find that the degree of similarity between one vent fauna and another is not dependent on the direct distance between them across the abyssal plain, but instead upon the distance between them following a longer path along the pattern of mid-oceanic ridges and faults [15]. As a result, the vent communities in the Atlantic are quite different from those in the Pacific.

The shallow-sea realm

Some of the differences between the shallow-sea realm and the open-sea realm have already been considered (see above, p. 241), but some others need now to be noted. Even if one subdivides the open ocean into areas corresponding to the movements of the surface waters, each of these is immensely greater than any of the individual units of the

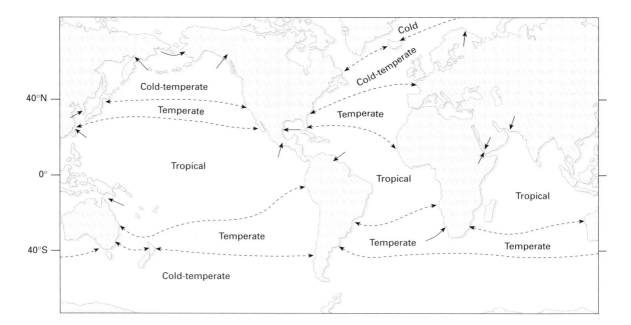

Fig. 12.6 The locations of major faunal changes in shallow-sea communities, shown by small arrows, and their relationships to the boundaries between areas with different surface-water temperatures. After van der Spoel & Heyman [40].

shallow-sea realm. Furthermore, each of these shelf seas is heavily influenced by the characteristics of the adjacent land, such as the nature of the coast and the presence of rivers, which may contribute both fresh water and a varying input of sediment. They are therefore also much more heterogeneous than the open ocean. Because of the relative shallowness of the sea, the nature of the sea bed also influences conditions through the whole of the overlying water mass, right up to the surface; there is no such interaction in the open sea.

Although the environment and biogeography of the shelf sea faunas themselves are quite different from those of the open sea, they can nevertheless be assembled into similar biogeographical units (cf. Figs 12.3 and 12.6). There are two reasons for this. One is that the waters of the open sea usually themselves traverse the shelf areas—and, even if they do not, they inevitably influence the temperature of

the shallow seas to which they lie adjacent. The other reason is that any faunal connection between individual shelf faunas can only be via the open ocean waters, which can carry their larvae from place to place. Thus, the shelf areas that are linked together as, for example, parts of the Northern Hemisphere warm-temperate region, share a particular temperature regime, because they are connected by an ocean surface current of that temperature range.

Faunal breaks within the shelf faunas

As we shall see, it is difficult for shelf organisms or their larvae to make the long crossing between the eastern and the western shores of an ocean. As a result, most of the similarities between the faunas of different shelf seas are between faunas at different latitudes on the same side of an ocean. These have been divided into units by the American marine zoologist Jack Briggs [16,17]. Developing a scheme originally proposed by Ekman [10] and using information on the patterns of endemicity in the coastal faunas, Briggs has identified over 60 locations where there appears to be a zone of more rapid faunal change (Fig. 12.6). Some of these lie in areas where there are changes in the ocean current that impinges upon the coast. This may result in both a change in the temperature of the water and also in the composition of the fauna that arrives within the current. These locations are therefore at the boundaries between different temperature regimes. The remainder lie within the influence of a single ocean-current temperature regime, and Briggs uses them to divide each regime into Regions, some of which are subdivided into Provinces. Excluding those that comprise only islands or island-groups, he defines a total of 23 Regions and 28 Provinces.

As would be expected, the clarity of these locations of faunal change varies according to the rapidity of the environmental change that causes them. For example, the sea-temperature change that results from an alteration in the origin of the ocean currents is inevitably diffused over a considerable extent of coastline, and the faunal change is therefore similarly gradual. A change of this kind takes place just north of Santa Barbara in southern California, where Point Conception projects westwards and guides the cold, southerly directed California Current away from the coast. From here southwards to near the southern end of Baja, California, there is a gradual reduction in the domination of the Northern Hemisphere cold-temperate fauna, and a progressive increase in the faunal element derived from the tropical fauna. About 40% of the species of shallow-water fishes of this transitional, warm-temperate fauna are endemic.

In some other areas, the boundary points between adjacent faunas are even less definite. For example, parts of the west coast of southern Africa are bathed by tropical waters from the north, while others receive cooler waters from the south. As a result, there is a change in the temperature of the surface water from Cape Santa Marta at 14°S to Cape Frio at 18°30′S—a distance of some 480 km (300 miles). Studies of the faunas of jellyfish, polychaete worms, molluscs, crustaceans and shore fishes indicated a variety of points of faunal change, so that Briggs felt unable to decide on any one firm location. However, we cannot expect that there will always be a convenient point at which we can confidently draw lines on the map. Different groups, with different physiologies, will often show slightly different points of faunal transition, even if they are all faced with the same sequence of changes in environmental conditions (which themselves may also differ somewhat from year to year). It is fruitless to seek constancy within the changing waters.

At the other extreme, one of the best-defined points of faunal change lies at the entrance to the Red Sea, which is only approximately 32 km (20 miles) wide and is also partially blocked by a shallow sill only 125 m deep. In addition to this geographical barrier, there is an ecological change; because the waters of the Red Sea evaporate in the dry climate, and there is no significant contribution of fresh water from the land, the Red Sea is unusually saline. As a result, there is considerable endemicity in its fauna: corals 25%, crustaceans 33%, cephalopods 50%, echinoderms 15%, fishes 17%.

A different type of break in the nature of the shallow-water fauna is where there is a change in the nature of the bottom sediments. For example, as one travels eastwards along the northern coastline of South America, there is an ecological transition at the delta of the Orinoco River in Venezuela. From here eastwards, almost to the north-eastern corner of Brazil, the bottom of the continental shelf is covered with mud brought down by the great tropical rivers, whose fresh waters also greatly reduce the salinity of the waters of the shallow sea. The coral reefs so characteristic of the Caribbean region to the west are absent, together with their associated fish fauna, which is replaced by such groups as the sea catfishes and croakers.

Trans-oceanic links and barriers between shelf faunas

Although currents are potentially capable of carrying the larvae of shelf organisms from one side of an ocean to the other, many larvae live for so short a time that there are comparatively few examples of east–west linkages between shelf faunas. The deep ocean therefore acts as a very effective barrier. For example, the approximately 6500-km wide, almost island-free expanse of the East Pacific has long been recognized as a barrier to the dispersal of shelf organisms across the Pacific Ocean. Of the shore fishes that are found either in the Hawaiian Islands or between Mexico and Peru at the eastern end of the Pacific Ocean warm-temperate region, only 6% are found in both. Similarly, the Swedish biologist Sven Ekman [10], who recognized and named the East Pacific Barrier, showed that only 2% of the 240 species and 14% of the 11 or 12 genera of echinoderms found in the Indo-West Pacific area had been successful in reaching the west coast of the Americas. (The greater success of the genera is because, on average, different genera are likely to have diverged from one another earlier than the different species of which they are composed, and they therefore have had a greater length of time in which to cross the barrier.)

The great antiquity of the East Pacific Barrier has been shown by Richard Grigg and Richard Hey of the University of Hawaii [18], who studied the zoogeographical affinities of fossil and living genera of coral. They found that those of the East Pacific are more closely related to those of the West Atlantic than to those of the West Pacific. This is true even of corals living as long ago as the Cretaceous Period, before the Panama Isthmus formed (see Fig. 7.8, p. 154), proving that the Barrier was effective in inhibiting dispersal across the Pacific as long ago as that time. Shelf faunas also provide evidence on the progressive widening of the Atlantic Ocean. Calculation of the degree of similarity of the shelf faunas on either side of the North Atlantic from the Early Jurassic onwards, using the coefficient of faunal similarity (see the caption to Fig. 7.5, p. 143) shows a steady decrease in their similarity as they are gradually separated by the widening ocean [19].

The deep waters of the Atlantic form a similar Mid-Atlantic Barrier to the dispersal of shelf organisms between the African tropics and the South American tropics. However, as the Atlantic is narrower than the Pacific, this barrier is less effective, so that in most groups there is a greater proportion of species that are found on both sides of the ocean. For example, in the shore fishes, there are about 900 species on the western shelf and about 434 species on the eastern; of this total, about 120 species (9%) are common to both faunas [16]. Although the trans-Atlantic currents are predominantly from east to west, most of these dispersals appear to have been from South America to Africa. Perhaps the greater richness of the South American shelf fauna, in terms of numbers of both species and of individuals, makes it more likely that they will succeed in dispersing.

An example of the appearance of a new link between shelf faunas is that which took place between the Arctic and the North Pacific about 3.5 million years ago, after submergence of the Bering Strait. This led to an exchange of cold-water species, in which the majority (125 species) dispersed from the Pacific to the Arctic Basin (in this case, in the direction of the current flow), and only 16 species dispersed in the reverse direction [20].

Latitudinal patterns in the shelf faunas

Marine organisms show a general latitudinal trend of decreasing diversity as one moves away from the tropics. However, much of that is due to the distribution of coral reefs, which have a faunal diversity unparalleled elsewhere in the shallow seas. Their distribution is centred upon the tropics (see below), and therefore heavily distorts the underlying pattern. In addition to this, recent work by Crame [21] has shown that Antarctic and sub-Antarctic shelf faunas are far richer (and more ancient) than had previously been thought, further emphasizing the fact that this pattern, and its significance, must now be reconsidered. (For a general discussion of latitudinal patterns of diversity, see p. 14.)

Many marine organisms also provide examples of the phenomenon known as bipolar distribution (also known as an antitropical distribution). This term is used to describe a situation in which related organisms are to be found in temperate or polar environments in both the Northern and the Southern Hemisphere, but not in the intervening equatorial region. Whatever may be the reason for terrestrial examples of this pattern, its occurrence in marine faunas has prompted suggestions of specifically marine mechanisms. Charles Darwin suggested that the cooling of the equatorial waters during the Ice Ages had allowed these genera to pass through waters that are now once again too warm for them to inhabit. Brian White [22] has argued that it was instead the warmer temperatures earlier in the Cenozoic that made it impossible for these genera to live in the equatorial waters, so that they now show a relict distribution on either side of this zone. This is supported by Gordon Howes in his useful analysis of the biogeography of the gadoid fishes [23]. Yet another explanation is that of Jack Briggs [24], who links it with his theory that the Indo-West Pacific region is a centre of evolutionary origin (see below), and suggests that bipolar genera have become extinct in the equatorial regions because of competition from genera that have newly evolved there. Perhaps the simplest suggestion is that the organisms living in cool waters on either side of the equator have been able to disperse beneath it via cooler waters at greater depth (cf. the principle of equatorial submergence, p. 249) [25].

In his review of this topic, Crame [26] pointed out that most theories had tacitly assumed two premises: first, that these patterns had originated within the last five million years, and were possibly related to the climatic changes of the Ice Ages, and secondly that the two now-separate taxa had achieved this pattern as a result of dispersal, rather than of vicariance (see p. 161). Concentrating on the distribution of marine molluscs, which have a good fossil record extending over 245 million years, Crame identified three main periods of bipolar distributions. The first was in the Jurassic–Cretaceous, and seems to have been caused by vicariance resulting from the breakup of Pangaea. The second was in the Oligocene–Miocene; this may have been caused by vicariance resulting from the cooling temperatures at that time, which allowed temperate taxa to spread across the Equator, only to become extinct there when temperatures rose again. The third period was during the Plio-Pleistocene Ice Ages which, together with the closing of the Panama Isthmus, caused increased cooling and upwelling along the equatorial divergences in both the Pacific and the Atlantic, allowing dispersal of temperate forms from one hemisphere to the other. Our now steadily increasing understanding of past patterns of climatic change on both land and sea are likely to lead to further understanding of the problem of bipolar distributions.

Coral reefs

For many reasons, corals provide fascinating and unique aspects of marine biogeography. One of the most complex and diverse environments on earth, they are clearly definable in nature, with limits of distribution that are simply explained by fundamental aspects of their biology. They therefore provide a good example of the extent to which marine patterns can be explained when the taxonomic and environmental aspects are more simple than elsewhere in the sea. On the other hand, interpretation of their patterns of diversity raises fundamental problems. Finally, because they are easily recognizable in the fossil record, historical biogeography can contribute more to our understanding of patterns of coral distribution than to most other aspects of marine biogeography. Two recent books, one by the Australian marine biologist Charlie Veron [27] and the other edited by the American Charles Birkeland [28], have dealt with many aspects of the biology and history of corals (although Veron's theories of coral evolution are controversial [29]).

Coral reefs provide a complex, three-dimensional environment that is home for an immense diversity of other marine organisms [30]. To date, 35 000–60 000 different species of reef-dwelling organism have been described, and this is probably only a fraction of the total number—between 1950 and 1994, the number of species of fish, molluscs, echinoderms and corals known from the Cocos (Keeling) Islands in the Indian Ocean tripled. The coral reef fauna includes at least 4000 species of fish, almost one-third of the total number of species of marine fishes, and comprising the greatest diversity of species of vertebrate per square metre known on earth.

The types of coral that form reefs are known as hermatypic corals, and they are a spectacular example of animal–plant symbiosis. The animals are colonial organisms called hydrozoans, in which the individuals, or polyps, resemble tiny sea anemones and feed on zooplankton. The reefs result from the activities of these polyps in secreting a basal calcium carbonate skeleton, to which the individual polyp is attached, and which is continuous with that of its neighbours. The plants are a type of algae known as zooxanthellae, which live within the tissues of the hydrozoans.

The biology of the corals limits their distribution to particular circumstances of nutrient levels, temperature and light. Corals are found in areas where the nutrient levels are so low that there is too little primary productivity from free-living algae or phytoplankton to provide the basis for a diverse ecosystem. Corals can flourish in this environment because their zooxanthellate algae are living within the hydrozoans, where nutrient levels are high. Of the other two factors, temperature is more important than light, as is shown by the fact that some corals can grow in deeper water as long as the temperature level is adequate. Coral reefs are only found where there is a minimum sea-surface temperature of at least 18°C, sustained over long periods of time, with a maximum of 30–34°C. As a result, relatively diverse coral reef assemblages are found up to about 30° of north and south latitude, with extremes in Japan at 35°N, Lord Howe Island at 32°S and Bermuda at 32°N. Hermatypic corals also cannot flourish where there is significant sedimentation, for this impedes the light that is vital for their photosynthetic algae.

Within these limits, the biogeography of corals shows a very interesting pattern (Fig. 12.7), involving changes in diversity according to both latitude and longitude, which have been discussed by the American marine biologist Gustav Paulay [31]. One of the most obvious features is the comparative poverty of the eastern ends of the oceans, compared with the western. There are several different reasons for this. One major influence is the pattern of ocean currents, for the upwellings of cool, nutrient-rich waters that take place along the eastern margins of the oceans (see Fig. 12.5) inhibit coral growth in those regions. In addition, most of the warm, equatorial ocean currents are directed westwards; when they reach the edge of the continent, they diverge both northwards and southwards, bringing warmer waters to higher latitudes. Another contributory factor is that the continental shelves, on which most coral reefs lie, are much narrower along the western margins of Africa and the Americas than along the eastern margin of Asia and in the Caribbean region. However, not all corals grow on the continental shelves; some grow around the margins of volcanic islands. Since these are most common in the older, western parts of the Pacific Ocean, this provides another increment to the greater reef area of that region. As a result of all of these factors, 85% of the area of coral reef lies in the Indo-Pacific Ocean, and only 15% in the Atlantic Ocean. Similarly, the reefs of the eastern Pacific are only a few metres thick, while those of the Indian Ocean and West Pacific (Briggs' Indo-West Pacific, or IWP, Region) are up to more than a kilometre in thickness.

A different factor causes the gap in the distribution of coral reefs along the tropical northern coastline of South America. The westerly winds of the equatorial Atlantic bring heavy rains to the lowlying river basins of tropical South America. These drain back to the sea via the great rivers (cf. above, p. 253), whose sediments blanket the sea bed, making it inimical to the growth of corals. (The impact of this phenomenon has recently been increased by the deforestation of the Amazon Basin, which has resulted in the waters of the great rivers bearing even more sediment and also having an increased concentration of nutriments, while fires in Central America have recently had a similar effect on corals in the neighbouring part of the Caribbean.)

Most corals, then, are to be found in the IWP region. Within this area, there is a clear latitudinal gradient in their diversity at the generic level (Fig. 12.7). However, it is difficult to distinguish the roles of temperature and of reef area in this, for both are at a maximum in the equatorial, island-rich

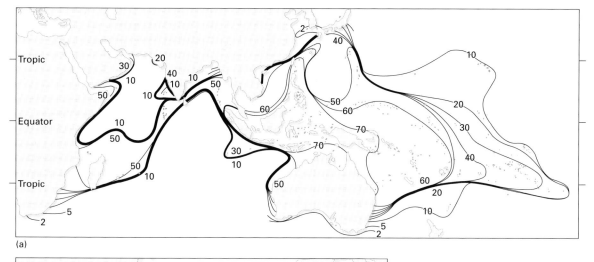

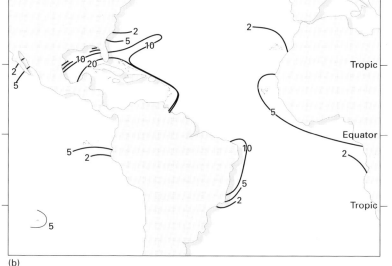

Fig. 12.7 Contours of generic diversity in corals relative to the Equator and Tropics, combining the distribution ranges of all the genera. (a) the Indo-West Pacific Region; (b) the Eastern Pacific, West Atlantic and East Atlantic Regions. The contours indicate the maximum generic diversity for the regions as a whole; remote or small areas within those regions may have a lower diversity. After Veron [27].

area of the East Indies, and decline both north and south of that area.

More interesting is the similar latitudinal gradient in the diversity of coral genera, which is not only paralleled in the pattern of diversity of their species, but also in that of many other reef organisms, such as molluscs and fishes [32] (Fig. 12.8). Jack Briggs [16,33], has suggested that the East Indies region is a centre of origin of new species, which gradually disperse away from there to produce the

pattern of diversity we see today. The fact that the ages of the genera involved also seemed to decrease away from the centre of the IWP towards its periphery also seemed to support this suggestion. However, the British workers Moyra Wilson and Brian Rosen have studied the fossil record of zooxanthellate corals, and found no evidence of extensive coral reefs in South-East Asia before the early Miocene, about 21 million years ago [34]. They suggest that the modern coral diversity of the IWP

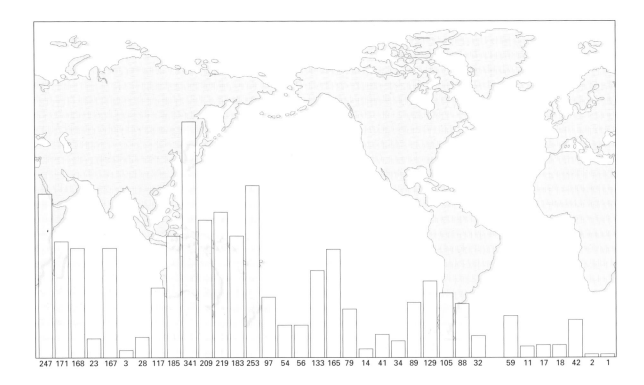

247 171 168 23 167 3 28 117 185 341 209 219 183 253 97 54 56 133 165 79 14 41 34 89 129 105 88 32   59 11 17 18 42 2   1

Fig. 12.8 Longitudinal gradients in fish species richness. The columns represent the total numbers of fish species (from a sample of 799 species) that occur in each 10°-wide extent of longitude. After McAllister *et al.* [41].

resulted from the collision of the Australian and South-East Asian plates, which commenced at that time (see p. 151). This led to the decreased isolation of the region, and to the appearance of numerous shallow-water areas around the emerging islands. This would have encouraged the evolution of many new taxa in these areas, a process that would have been further facilitated by the changes in sea level during the Ice Ages, which would have led to fragmentation of the larger islands each time sea levels rose. The Ice Age glacial periods also seem to have produced a seasonal cold-water barrier to the east of the South-East Asian coastline, subdividing the IWP region and producing vicariant speciation in the pelagic zooplankton [35], and this might have had similar effects on the reef faunas.

Crame, too, does not support the Centre of Origin hypothesis, arguing that the diversity of the tropical marine taxa is simply because they are older taxa, which therefore have had a longer period in which to diversify, rather than having any inherently greater ability to do so [21]. Many other suggestions have been put forward, such as that of Jokiel and Martinelli [36], who mathematically modelled the fate of the faunas of a grid of equally spaced islands set in a pattern of surface currents similar to that found in the Pacific today. This model eventually produced a pattern of diversity very similar to that which we see today, heavily skewed towards the western side of the ocean at equatorial latitudes. Ormond and Roberts [32] supplemented this method by factoring in the pattern of available reef area, and comment that the result shows that these two factors alone can account for 90% of the longitudinal variation in species richness. (It is interesting and instructive that this simple pattern has generated so many different explanations!)

Another aspect of coral biogeography is the pattern of relationships between the coral faunas of the IWP, eastern Pacific, western Atlantic and eastern Atlantic (see Fig. 12.7). Although most of the coastal marine groups of the eastern Pacific are highly diverse, are endemic to that region at the species level, and are closely related to the biota on the eastern side of the Panama Isthmus, the coral reef biota shows a very different pattern. It is much less diverse than that of the IWP, but is closely related to it—all of the east Pacific coral genera and over 90% of the species are shared with the IWP. This pattern of relationship in the corals is the result of the loss of many coral taxa in the eastern Pacific after the Panama Isthmus closed, followed by dispersal of some of the coral taxa from the IWP to the eastern Pacific.

The coral faunas of the eastern Pacific, western Atlantic and eastern Atlantic, and their interrelationships, have been greatly affected by the continental movements that closed the old Tethys Sea during the Cenozoic (see above, p. 243). In the Late Oligocene, about 25 million years ago, Africa had not yet met Europe (cf. p. 145) and South America was still separate from North America, so that tropical waters extended from the western end of the Indian Ocean to the eastern Pacific. As a result, the coral fauna of the proto-Caribbean gap between the Americas had 25 genera in common with that of the Mediterranean seaway; four genera were endemic to the proto-Caribbean, and nine to the Mediterranean. But the closure of the Mediterranean seaway about 12 million years ago, followed by the completion of the Panama Isthmus 3.5 million years ago, caused major changes in the environment of this whole stretch of tropical waters, as well as separating the fauna of the eastern Pacific from that of the Caribbean. Many genera became extinct in the western Atlantic, but most of these are still to be found in the IWP region. There has also been the evolution of new taxa in each of the now-separate areas, further reducing the similarity of their coral faunas. The corals of the IWP region and of the western Atlantic now differ greatly; only eight of their combined total of 107 genera are found in both, and they share none of their species of zooxanthellae. Similarly, only one-third of the genera and none of the species of the eastern Pacific are shared with that of the Caribbean region on the other side of the Panama Isthmus. The coral fauna of the Eastern Atlantic also suffered badly from extinction after the closure of the Mediterranean seaway, losing some 85% of its coral genera, but long-distance dispersal has led to some exchange of taxa with the Caribbean region.

Summary

1 The composition, structure and ecology of marine organisms are still relatively poorly understood, and their environment differs greatly from that of terrestrial organisms. Nevertheless, it is becoming clear that marine biogeography is basically similar to that of the land, although the boundaries between the units are more gradual and are not fixed in their locations.
2 The major subdivision is between the open-sea realm and the shallow-sea realm.
3 The open-sea realm is divided vertically into zones based on light, temperature and nutrient availability. The surface waters are divided into Domains and Provinces, which are related to the rotating patterns of ocean circulation and the patterns of productivity.
4 The units in the shallow-sea realm are much smaller than those in the open-sea realm, because they are dependent on the characteristics of the local land, rivers, sediments, sea bed, tides and ocean currents.
5 Coral reefs provide the most diverse environments within the seas, and the clearest examples of gradients of marine diversity, but these patterns seem to be of comparatively recent origin.

Further reading

Gage JD, Taylor PA. *Deep Sea Biology*. Cambridge: Cambridge University Press, 1991.
Longhurst A. *Ecological Geography of the Sea*. London & New York: Academic Press, 1998.

Nybakken JW. *Marine Biology. An Ecological Approach*, 4th edn. Menlo Park, CA: Addison Wesley Longman, 1997.

Angel MV (ed.) Reports of the Scientific Committee on Oceanic Research. 1, The ecology of the deep-sea floor. 2, Pelagic biogeography. *Prog Oceanogr* 1994; 34: 79–284.

Ormond RFG, Gage JD, Angel MV (eds) *Marine Biodiversity. Patterns and Processes.* Cambridge: Cambridge University Press, 1997.

References

1 Angel MV. Spatial distribution of marine organisms: patterns and processes. In: Edwards PJR, May NR, Webb NR, eds. *Large Scale Ecology and Conservation Biology*. British Ecological Society Symposium no. 35. Oxford: Blackwell Science, 1994: 59–109.

2 Angel MV. Pelagic biodiversity. In: Ormond RFG, Gage JD, Angel MV, eds. *Marine Biodiversity Patterns and Processes.* Cambridge, Cambridge University Press, 1997: 35–68.

3 Duque-Caro H. Neogene stratigraphy, paleoceanography and paleobiogeography in northwest South America and the evolution of the Panama seaway. *Palaeogeogr Palaeoclimatol Palaeoecol* 1990; 77: 203–24.

4 Stuiver M, Quay PD, Ostlund HG. Abyssal water carbon-14 distribution and ages of the world's oceans. *Science* 1983; 139: 572–6.

5 Pierrot-Bults AC. Biological diversity in oceanic macrozooplankton: more than merely counting species. In: Ormond RFG, Gage JD, Angel MV, eds. *Marine Biodiversity Patterns and Processes.* Cambridge: Cambridge University Press, 1997: 69–93.

6 Longhurst A. Seasonal cycles of pelagic production and consumption. *Prog Oceanogr* 1995; 36: 77–168.

7 White BN. Vicariance biogeography of the open-ocean Pacific. *Prog Oceanogr* 1994; 34: 257–82.

8 McGowan JA. The biogeography of pelagic ecosystems. In: *Pelagic Biogeography.* UNESCO Technical Papers in Marine Science no. 49, 1985: 191–200.

9 Gibbs RH. The stomioid fish genus Eustomias and the oceanic species concept. In: *Pelagic Biogeography.* UNESCO Technical Papers in Marine Science no. 49, 1985: 98–103.

10 Ekman S. *Zoogeography of the Sea.* London: Sidgwick & Jackson, 1958.

11 Rex MA, Etter RJ, Stuart CT. Large-scale patterns of species diversity in the deep-sea benthos. In: Ormond RFG, Gage JD, Angel MV, eds. *Marine Biodiversity*

Patterns and Processes. Cambridge: Cambridge University Press, 1997: 94–121.

12 Etter RJ, Grassle JF. Patterns of species diversity in the deep sea as a function of sediment particle size diversity. *Nature* 1992; 360: 576–8.

13 Grassle JF. Deep sea benthic diversity. *Bioscience* 1991; 41: 464–9.

14 Grassle JF, Maciolek NJ. Deep-sea species richness: regional and local diversity estimates from quantitative bottom samples. *Am Naturalist* 1992; 139: 313–41.

15 Tunnicliffe V, Fowler MR. Influence of sea-floor spreading on the global hydrothermal vent fauna. *Nature* 1996; 379: 531–3.

16 Briggs JC. *Marine Zoogeography.* New York: McGraw-Hill, 1974.

17 Briggs JC. *Global Biogeography.* Amsterdam: Elsevier, 1995.

18 Grigg R, Hey R. Paleoceanography of the tropical Eastern Pacific Ocean. *Science* 1992; 255: 172–8.

19 Fallaw WC. Trans-North Atlantic similarity among Mesozoic and Cenozoic invertebrates correlated with widening of the ocean basin. *Geology* 1979; 7: 398–400.

20 Vermeij GJ. Anatomy of an invasion: the trans-Arctic exchange. *Paleobiology* 1991; 17: 281–307.

21 Crame JA. An evolutionary framework for the polar regions. *J Biogeogr* 1997; 24: 1–9.

22 White BN. The isthmian link, antitropicality and American biogeography: distributional history of the Atherinopsidae (Pisces; Atherinidae). *Syst Zool* 1986; 35: 176–94.

23 Howes GJ. Biogeography of gadoid fishes. *J Biogeogr* 1991; 18: 595–622.

24 Briggs JC. Antitropical distribution and evolution in the Indo-West Pacific Ocean. *Syst Zool* 1987; 36: 237–47.

25 Boltovskoy D. The sedimentary record of pelagic biogeography. *Prog Oceanogr* 1994; 34: 135–60.

26 Crame JA. Bipolar molluscs and their evolutionary implications. *J Biogeogr* 1993; 20: 145–61.

27 Veron JEN. *Corals in Space and Time.* Sydney: University of New South Wales Press, 1995.

28 Birkeland C. (ed.) *Life and Death of Coral Reefs.* New York: Chapman & Hall, 1997.

29 Paulay G. Circulating theories of coral biogeography. *J Biogeogr* 1996; 23: 279–82.

30 Kohn AJ. Why are coral reef communities so diverse? In: Ormond RFG, Gage JD, Angel MV, eds. *Marine Biodiversity Patterns and Processes.* Cambridge: Cambridge University Press, 1997: 201–15.

31 Paulay G. Diversity and distribution of reef organisms. In: Birkeland C, ed. *Life and Death of*

Coral Reefs. New York: Chapman & Hall, 1997: 298–353.

32 Ormond RFG, Roberts CM. The biodiversity of coral reef fishes. In: Ormond RFG, Gage JD, Angel MV, eds. *Marine Biodiversity Patterns and Processes.* Cambridge: Cambridge University Press, 1997: 216–57.

33 Briggs JC. The marine East Indies: centre of origin? *Global Ecol Biogeogr Lett* 1992; 2: 149–56.

34 Wilson MEJ, Rosen BR. Implications of paucity of corals in the Paleogene of SE Asia: plate tectonics or Centre of Origin? In: Hall R, Holloway JD, eds. *Biogeography and Geological Evolution of SE Asia.* Leiden: Backhuys, 1998: 165–95.

35 Fleminger A. The Pleistocene equatorial barrier between the Indian and Pacific Oceans and a likely cause for Wallace's Line. In: *Pelagic Biogeography.* UNESCO Technical Papers in Marine Science no. 49, 1985: 84–97.

36 Jokiel P, Martinelli FJ. The vortex model of coral reef biogeography. *J Biogeogr* 1992; 19: 449–58.

37 Haq BU. Paleoceanography: a synoptic overview of 200 million years of ocean history. In: Haq BU, Milliman HD, eds. *Marine Geology and Oceanography of Arabian Sea and Coastal Pakistan.* New York: Van Nostrand Reinhold, 1984: 201–31.

38 Levitus, S. Conkright ME, Reid JL, Najjar RG, Mantyla A. Distribution of nitrate, phosphate and silicate in the world oceans. *Prog Oceanogr* 1993; 31: 245–74.

39 Berger WH. Global maps of ocean productivity. In: Berger WH, Smetacek VS, Wefer G, eds. *Productivity of the Ocean: Past and Present.* London: John Wiley & Sons, 1989: 429–55.

40 van der Spoel S, Heyman RP. *A Comparative Atlas of Zooplankton: Biological Patterns in the Oceans.* Utrecht: Bunge, 1983.

41 McAllister DE, Schueler FW, Roberts CM, Hawkins JP. Mapping and GIS analysis of the global distribution of coral reef fishes on an equal-area grid. In: Miller R, ed. *Advances in Mapping the Diversity of Nature.* London: Chapman & Hall, 1994: 155–75.

CHAPTER 13: *Life (and death) on islands*

Most of this book is concerned with the patterns of life on the continents. We have seen how that has developed, both over approximately 300 millions of years of evolution, and also over the last two million years of drastic climatic change and (more recently) of human intervention. Those patterns of life are robust and long term, because organisms that live on the continents can gradually change their area of distribution in response to changing environmental conditions. Although as yet still less well documented, the patterns of life in the seas are basically similar.

Island biogeography instead concentrates upon the organisms themselves. Little is known of their history, and their fate is inevitably played out within the small confines of the island they inhabit—there is no adjacent territory to which they can retreat if the island environment becomes less congenial. But the biota of an island is much simpler than that of a continent, so that its composition and the interaction of the different components are easier to analyse. This comparative simplicity invites the biologist to ask two different types of question.

The first type of question is concerned with analysis of the island biota and environment in general. In what ways is the island biota different from that of the mainland? How did it arrive? How does the island community gradually establish itself and mature? What can we learn about the processes and rates of evolutionary change upon islands, some of which can now be accurately dated?

The second type of question is an attempt at synthesis. How much can we learn from comparisons between the biotas of islands of different age, history, climate, size or topography, or of islands that lie at different latitudes or at different distances from their source of colonists? Is it possible to deduce any general, underlying regularities that may con-

trol the diversity of the island biota, and perhaps enable us to predict this? In particular, this research attempts to identify and quantify the factors that control three phenomena: the rate at which new species reach an island, the rate at which species become extinct on an island, and the number of species that an island can support.

These are the interlocking, fascinating themes that this chapter will explore.

Types of island

First of all, it is important to realize that there are three types of island, for the differences between them also affect the nature of the biota that they are likely to contain.

The first type of island was originally a part of a nearby continent, but became separated from it by rising sea levels (e.g. Britain, Newfoundland, the larger islands of the East Indies that lie upon the continental shelf of South-East Asia, see Fig. 7.7), or by tectonic processes that split them away from an adjacent continent (e.g. Madagascar, New Zealand).

A second type of island is part of a volcanic island arc. These arcs form along trenches (see p. 113), where old ocean crust is disappearing into the depths of the earth. As the ocean crust is consumed, the position of the trench, and of the island arc, gradually moves across the floor of the ocean. As a result, the trench and island arc may eventually collide with an adjacent land mass, the collision forming volcanic islands and mountain ranges; the volcanic island Krakatau (see p. 273), and the mountains of southern Java, Sumatra and northern New Guinea are examples of this.

Finally, other islands form as a result of the activity of what geologists call 'hotspots'. These are scattered, but fixed, locations over 700 km deep within the earth, from which plumes of hot material rise to form volcanoes at the Earth's surface. Where this volcano lies within an ocean rather than within a continent, it may either remain submerged as a 'sea mount', or grow to rise above the surface as a volcanic island. However, because the sea floor is part of a moving tectonic plate (see p. 113), the island is gradually moved away from the plume of hot material. Volcanic activity in the island then ceases, and the surrounding sea floor cools and contracts, so that the island gradually disappears below the surface of the ocean. Meanwhile, a new volcano is developing in that part of the sea floor which now lies above the hotspot. Over millions of years, repetition of this process causes the appearance of a chain of volcanoes (extinct, except for the youngest), the orientation of which clearly shows the direction of movement of the underlying sea floor. The Hawaiian chain of islands is the clearest example of this phenomenon.

The characteristics of the biota of these three types of island are different. Those of former continental fragments would originally have been a subset of the biota of the continent itself, but will have changed because of independent evolution and extinction within the new island. Some of that evolutionary change will have been a response to the different conditions of life upon an island compared with those on the mainland (see p. 266).

The biota of island arcs and of hotspot island chains, on the other hand, originally arrived by trans-ocean dispersal. Like that of the continental fragments, the biota of these islands will subsequently have become changed by evolution, but they may also show a progressive ecological change as the island ecosystem matures and offers new opportunities (see pp. 277–9, below).

The islands of an arc all appear more or less simultaneously, while those of a hotspot chain appear (and disappear) in turn, but in both cases a number of islands exist at the same time. This provides the potential for inter-island dispersal, and for a more complex pattern of evolutionary change —a pattern that is sometimes called archipelago speciation. Where an island arc has collided with a larger land mass, the resulting biota may be complex, containing the descendants of the original continental fragment as well as the descendants of the island-arc biota that had arrived by trans-ocean dispersal. (The possible importance of moving island arcs in interpreting complex biogeographical patterns in the Pacific Ocean is discussed on p. 152.)

Getting there: problems of access

Oceans are the most effective barrier to the distribution of land animals. Because few terrestrial or freshwater organisms can survive for any length of time in sea water, organisms can only reach an island if they possess special adaptations for transport by air or water. Dispersal to islands is therefore by a sweepstakes route (see p. 37), the successful organisms sharing adaptations for crossing the intervening region rather than for living within it. This greatly restricts the diversity of life that is capable of dispersing to an island.

Some flying animals, such as birds and bats, may be capable of reaching even the most distant islands unaided, using their own powers of flight, especially if, like water birds, they are able to alight on the surface of the water to rest without becoming waterlogged. Smaller birds and bats and, especially, flying insects may reach islands by being carried passively on high winds. These animals may, in their turn, carry the eggs and resting stages of other animals, as well as the fruits, seeds and spores of plants.

Most land animals cannot survive in sea water for long enough to cross oceans and reach a distant island, but it seems possible that some may occasionally make the journey on masses of drifting debris. Natural rafts of this kind are washed down the rivers in tropical regions after heavy storms, and entire trees may also float for considerable distances. Such a floating island could carry small animals such as frogs, lizards or rats, the resistant eggs of other animals, and specimens of plants not adapted to oceanic dispersal. It is rare for such raft-aided dispersal to be seen and documented, but there has been a recent example of this in the West Indies [1]. Soon after two hurricanes had passed through the Caribbean in September 1995, a mass of logs and uprooted trees, some over 10 m long, was found on the beach of the island of Anguilla, and at least 15 individuals of the green iguana lizard, *Iguana iguana*, were seen on logs offshore and on the beach. They included both males and females in reproductive condition, and specimens were still surviving on the island (where the species was previously unknown) over 2 years later. Judging by the track of the hurricanes, the lizards probably originated on the island of Guadeloupe, some 250 km (160 miles) away, and the journey probably lasted for about a month.

It is potentially far easier for a plant to adapt to long-distance dispersal. Very many plants show some adaptation to ensure that the next generation is carried away from the immediate vicinity of the parent [2]. It requires little elaboration of some of these dispersal devices to make it possible for them to traverse even wide stretches of ocean. In addition to this, successful colonization requires only a few fertile spores or seeds, whereas in most animals it requires the dispersal of either a pregnant female or a breeding pair. The spores of most ferns and lower plants are so small (0.01–0.1 mm) that they are readily carried considerable distances by winds. Some plants have seeds that are specially adapted to being carried by the wind. Orchid seeds, for example, are surrounded by light, empty cells, and some have been known to travel over 200 km. *Liriodendron* and maple seeds have wings, and the seeds of many members of the Asteraceae (daisies and their relatives) have tufts of fluffy hairs; those of thistles have been carried by the wind for 145 km. Many fruits and seeds have special sticky secretions or hooks to make them adhere to the bodies of animals. Examples are the spiny fruits of burdocks and beggarticks, and the berries of mistletoe, which are filled with a sticky juice so that the seeds that they contain stick to birds' beaks. The seeds of many plants can germinate after passing through a bird's stomach, and those of some (e.g. *Convolvulus*, *Malva*, *Rhus*) can germinate after up to 2 weeks there.

A few plants have developed fruits and seeds that can be carried unharmed in the sea. For example, the coconut fruit can survive prolonged immersion, and the coconut palm (*Cocos nucifera*) is widespread on the edges of tropical beaches. However, since the beach is as far as most seaborne fruits or seeds are likely to get, only species that can live on the beach are able to colonize distant islands in this way. The fruits or seeds of plants that live inland would be less likely to reach the sea and, even if they were able to survive prolonged immersion and were later cast up on a beach and germinated, they would be unable to live in a beach environment.

In most cases, the biota is strongly affected by the degree of isolation of the island. However diverse the habitats that it offers, the variety of the island life depends, in the short term, very much upon the rate at which colonizing animals and plants arrive. This, in turn, depends largely upon how far the island is from the source of its colonizers, and upon the richness of that source. If the source is close, and if its biota is rich, then the island in its turn will have a richer biota than another, similar island which is more isolated or which depends upon a source with a more restricted variety of animals and plants. Each sea barrier further reduces the biota of the next island, which in turn becomes a poorer source for the next. For example, the data provided by Van Balgooy [3] make it possible to map the diversity of conifer and flowering plant genera in the Pacific island groups (Fig. 13.1). This clearly shows that diversity is much lower in the more isolated island groups of the central and eastern Pacific.

However, in several of the more westerly island groups, the diversity is much higher than their geographical position alone would lead one to predict. A logarithmic graph of the relationship between the number of genera and the area of the islands (Fig. 13.2) clearly shows that, in most cases, the generic diversity is simply dependent on island area. (The fact that some islands therefore have more genera than do the islands from which most of their flora is derived suggests that other genera were once also present in these latter islands, but have since become extinct there.) Nearly all of the

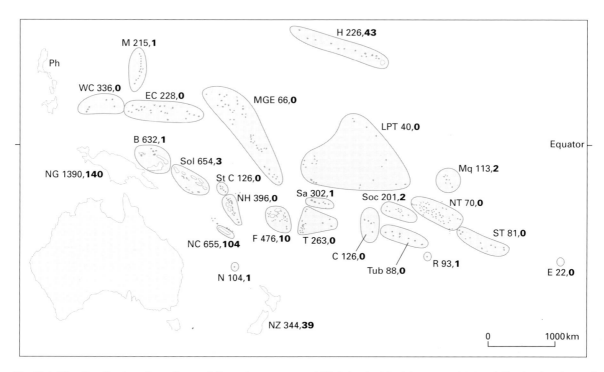

Fig. 13.1 The distribution of conifers and flowering plants in the Pacific Islands. The first number beside each island group is the total number of genera found there; the second is the number of endemic genera found there. B, Bismarck Archipelago; C, Cook Islands; E, Easter Island; EC, East Carolines; F, Fiji Islands; H, Hawaiian Islands; LPT, Line, Phoenix and Tokelau Island groups; M, Marianas; MGE, Marshall, Gilbert and Ellis Islands; Mq, Marquesas; N, Norfolk Island; NC, New Caledonia; NG, New Guinea; NH, New Hebrides; NT, Northern Tuamotu Islands; NZ, New Zealand; Ph, Philippines; R, Rapa Island; Sa, Samoa group; Soc, Society Islands; Sol, Solomon Islands; ST, Southern Tuamotu Islands; StC, Santa Cruz Islands; T, Tonga group; Tub, Tubai group; WC, West Carolines. Data from Van Balgooy [3].

Fig. 13.2 The relationship between island area and the diversity of conifer and flowering plant genera in the Pacific Islands. The more isolated islands are indicated by triangles. The data from the other islands lie very close to a straight line (the regression coefficient), suggesting that generic diversity in these islands is almost wholly controlled by island area—the correlation coefficient is 0.94, indicating a very high degree of correlation. For abbreviations, see legend of Fig. 13.1, plus Loy, Loyalty Islands. Data from Van Balgooy [3].

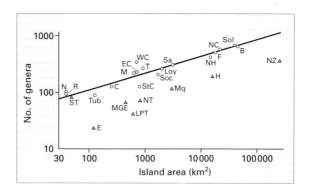

	Area (km²)	Angiosperm genera	Bird genera
Solomon Islands	40 000	654	126
New Caledonia	22 000	655	64
Fiji Islands	18 500	476	54
New Hebrides	15 000	396	59
Samoa group	3 100	302	33
Society Islands	1 700	201	17
Tonga group	1 000	263	18
Cook Islands	250	126	10

Table 13.1 The relationships between island area and the diversity of bird genera and nonendemic flowering plant genera in some Pacific islands. Data from Van Balgooy [3]; Mayr [51]; MacArthur & Wilson [52].

more isolated island groups (shown as triangles in Fig. 13.2) have, as would be expected, a much lower diversity than would be predicted from their areas alone.

The number of land and freshwater bird species in each island shows a similar relationship to island area (Table 13.1), but this is probably due, not to island area directly, but instead to the resulting higher floral diversity.

Dying there: problems of survival

Like any other population, the island population of a species must be able to survive periodic variations in its environment. But island life is more haz-ardous than that on the mainland, for several rea-sons. Catastrophe, such as volcanic eruption, has longer-lasting effects in an island situation, for there is little opportunity for a species to vacate the area and return subsequently, nor is reinvasion easy should extinction take place. On the mainland, by contrast, chance extinction of a species in a partic-ular area can soon be made good by immigration from elsewhere. An island will therefore contain a smaller number of species than a similar mainland area of similar ecology. For example, study of a 2-ha plot of moist forest on the mainland of Panama showed that it contained 56 species of bird, while a similar plot of shrubland contained 58 species. The offshore Puercos Island, 70 ha in area and inter-mediate between the two mainland plots in its ecology, contained only 20 of these species [4].

Since its success and survival are the only meas-ures of an organism's degree of adaptation to its

environment, the fact that a species has become extinct also demonstrates that it was not able to adapt to the biotic or climatic stresses to which it was exposed. The island environment is inevitably different from that of the mainland, which was the source of the colonists, and adaptation to it is not easy. First, if the colonists are few in number, they can include only a very small part of the genetic variation that provided the mainland population with the flexibility to cope with environmental change; this is sometimes known as the 'founder principle'. Secondly, small populations are also far more susceptible to random non-adaptive changes in their genetic make-up (see p. 103). Since it is less likely to be closely adapted to its environment, a small population is therefore also more liable to chance extinction.

The extent of this added risk to survival in islands is shown by the fact that, although only 20% of the world's species and subspecies of bird are found only on islands, they contributed 155 (90%) of the 171 taxa that are known to have become extinct since 1600—an extinction rate about 50 times as great as on the continents. The influence of area on extinc-tion rate is underlined by the fact that 75% of the island extinctions took place on small islands [5].

A species which can make use of a wide variety of food is therefore at an advantage on an island, for its maximum possible population size will be greater than that of a species with more restricted food pre-ferences. The advantage of this will be especially great in a small island, in which the possible popu-lation sizes are in any case smaller. This is probably the reason why, for example, although on the larger islands of the Galápagos group in the eastern Pacific both the medium-sized finch *Geospiza fortis* and the small *G. fuliginosa* can coexist, on some of the smaller islands of the group there is only a single species, of intermediate size [6].

Chance extinction is also a particular danger for predators, since their numbers must always be far lower than those of their prey. As a result, island faunas tend to be unbalanced in their composition, containing fewer predators, and fewer varieties of predator, than a similar mainland area. This in turn reinforces the fundamental lack of variety of the animal and plant life of an island that is due to the hazards involved in entry and colonization. The complex interactions of continental communities containing a rich and varied fauna and flora act as a buffer that can cope with occasional fluctuations in the density of different species, and even with temporary local extinction of a species. This resilience is lacking in the simple island community. As a result, the chance extinction of one species may have serious effects, and lead to the extinction of other species. All these factors increase the rate at which island species may become extinct.

There are clearly several different possible reasons why a particular organism may be absent from a particular island. It may be unable to reach it; it may reach it but be unable to colonize it; it may have colonized it but later have become extinct, or it may simply as yet, by chance, not have reached the island [7]. It is often very difficult to decide which of these possible reasons was the cause in any particular case. In some instances, as for example where two species have complementary distributions in the same island group but are never found on the same island, the facts obviously suggest that they compete with one another so strongly that they cannot coexist.

Integrating the data: The Theory of Island Biogeography

As we have seen above, the number of species found on an island depends on a number of factors: not only on its area and topography, its diversity of habitats, its accessibility from the source of its colonists, and the richness of that source, but also on the equilibrium between the rate of colonization by new species and the rate of extinction of existing species. Many individual observations and analyses of such phenomena have been made over the past 150 years. However, scientists always attempt to synthesize a mass of isolated data of this kind into a unifying theory that will not only explain them, but will also enable predictions to be made. A quantitative theory of this kind was put forward in 1967 by the American ecologists Robert MacArthur and Edward Wilson in their book, *The Theory of Island Biogeography* [8]. This put forward two main suggestions: that the changing, and interrelated, rates of colonization and immigration will eventually lead to an equilibrium between these two processes, and that there is a strong correlation between the area of the island and the number of species it contains.

To explain the relationship between rates of colonization and of immigration, MacArthur and Wilson took the case of an island that is newly available for colonization. They pointed out that the rate of colonization will at first be high, because the island will be reached quickly by those species that are adept at dispersal, and because these will all be new to the island. As time passes, immigrants will increasingly belong to species that have already colonized the island, so that the rate of appearance of new species will drop (Fig. 13.3). The rate of immigration will also be affected by the position of the island, for it will be higher for islands that are close to the source of their colonists, and lower for those that lie further away (Fig. 13.4).

The rate of extinction, on the other hand, will start at a low level but gradually rise. This is partly because, since every species runs the risk of extinction, the more that have arrived, the more species there are at risk. In addition, as more species arrive, the average population size of each will diminish as competition increases—and a smaller population is at greater risk of extinction than a larger population.

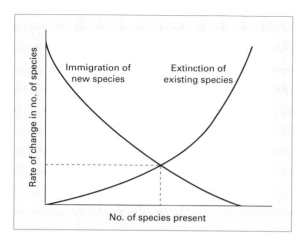

Fig. 13.3 Equilibrium model of the biota of an island. The curve of the rate of immigration of new species and the curve of the rate of extinction of species already on the islands intersect at an equilibrium point. The interrupted line drawn vertically from this point indicates the number of species that will then be present on the island, while that drawn horizontally indicates the rate of change (or 'turnover rate') of species in the biota when it is at equilibrium. After MacArthur & Wilson [52].

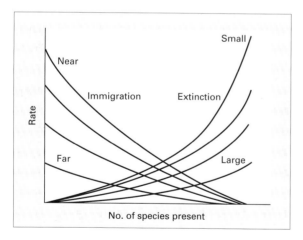

Fig. 13.4 The interrelationship between isolation and area in determining the equilibrium point of biotic diversity. Increasing distance of the island from its source of colonists lowers the rate of immigration (left). Increasing area lowers the rate of extinction (right). After MacArthur & Wilson [8].

At first, the few species present can occupy a greater variety of ecological niches than would be possible on the mainland, where they are competing with many other species. For example, in the comparison mentioned earlier (p. 266) between the Panama mainland and Puercos Island, the smaller number of bird species in the island were able, because of reduced competition, to be far more abundant: there were 1.35 pairs per species per hectare in Puercos Island, compared with only 0.33 and 0.28, respectively, for the two mainland areas [4]. This effect of release from competition was especially noticeable in the antshrike (*Thamnophilus doliatus*). On the mainland, where it competed with over 20 other species of ant-eating bird, there were only eight pairs of antshrike per 40 ha; on Puercos Island, where there was only one such competitor, there were 112 pairs per 40 ha.

This effect of release from competition will be reversed if the island is later colonized by a new species whose diet overlaps with that of one of the earlier immigrants. This may result in the extinction of one of them. This may be because they compete too closely with one another, so that they cannot coexist. Alternatively, it may be because the competition between them leads to a reduction in the population size of each—because each species has to become more specialized in its ecological requirements. This in turn renders the species more vulnerable to extinction (see p. 266). In either case, the rate of extinction upon the island will have increased.

In all these theoretical cases, the number of species present in the biota will obviously be the result of the balance between the rate of immigration and the rate of extinction. MacArthur and Wilson suggested that the biota will eventually reach an equilibrium, at which the rates of immigration and of extinction are approximately equal, and that this equilibrium level is comparatively stable.

The Theory of Island Biogeography was widely welcomed, for it gave biogeographers a theoretical background with which to compare their own individual results, and therefore encouraged a more structured and less *ad hoc* approach to biogeographical studies. Its methodology was also extended into types of isolation other than that of dry land surrounded by water. For example, mountain peaks [9], cave biota [10] and individual plants [11] were all interpreted in this way, and the theory was even extended to evolutionary time, with individual host-plant species being regarded as islands as far as their 'immigrant' insect fauna was concerned [12,13].

Second thoughts about the theory

Over the years that followed the publication of the Theory of Island Biogeography, many papers were written that interpreted individual biota in terms of the theory. These papers were in turn taken as providing such a wide measure of support for the theory that it became almost uncritically accepted as a basic truth. In turn, therefore, results which did not conform to expectations based on the theory were re-examined in search of procedural or logical faults, or for unusual phenomena that might explain this 'anomalous' result, or simply ignored, rather than being seen to cast doubts on the applicability or universality of the theory. Later, however, a number of criticisms were made, both of the nature of the theory itself and of the extent to which subsequent work could be taken to have 'proved' or supported it. For example, the theory treats species as simple numerical units, of equal value to one another. Their possible biological interactions, such as competitive or co-evolutionary effects, are therefore ignored or assumed to be trivial in comparison with the overall statistical effects, as is the possibility of an increase in the number of species by evolution rather than by immigration. The story of the rise of the theory and of the later mounting wave of criticism has been told in the fascinating book *The Song of the Dodo—Island Biogeography in an Age of Extinction*, by the American science writer David Quammen (see 'Further reading'). Its current status has also been sympathetically reviewed by one of MacArthur's former students, the ecologist Michael Rosenzweig [14].

Many studies that had been extensively quoted

as supporting the Theory of Island Biogeography are really far too imprecise, as pointed out by the American ecologist Dan Simberloff [15] and in a review by the British biologist Francis Gilbert [16], while the statistical procedures of many earlier studies were criticized by the American ecologists Edward Connor and Earl McCoy [17]. It has also been pointed out that the patterns of interaction predicted by the theory should be tested against an alternative 'null model' which suggests that the patterns are simply due to chance. However, as Robert Colwell and David Winkler have shown [18], it is in fact extremely difficult to design a null model that does not itself suffer from serious biases. It is equally difficult to use a real-world mainland biota as a comparison, because its composition has been conditioned by competition within a much more varied biota.

MacArthur and Wilson's first point, that larger islands contain more species, has met general acceptance. There is now considerable supporting evidence, and it is clear that the effects of greater area show themselves mainly via the resulting presence of a greater variety of environments and of larger populations, the latter reducing the rate of extinction [14].

In contrast, there has been considerable criticism of work that had seemed to support the theory's prediction that the number of species will come to, and remain at, an equilibrium. Jared Diamond [19] had compared the numbers of bird species found in the Californian Channel Islands in a 1968 survey, with those recorded in a 1917 review. He concluded that there had been an equilibrium in the number of species breeding there, even though the turnover had been as high as 20–60% of the species on each island. However, it was later pointed out by Lynch and Johnson [20] that Diamond had ignored the fundamental changes in the environment of the Channel Islands that had taken place between 1917 and 1968. These included the effects of human activities, such as hunting and the poisoning of birds of prey by pesticides, and the appearance of totally new species, such as European sparrows and starlings, in the islands. Given these changes in the environment, whether or not the number of species is similar or different in the two cases is irrelevant to the concept that the number of species will remain reasonably constant in an environment that does not change.

Lynch and Johnson found similar flaws in studies of bird faunas on an island near New Guinea, also carried out by Diamond [21], and on a West Indies island investigated by Terborgh and Faåborg [22]. Similarly, Simberloff [23] analysed the records of the bird faunas of two islands and three inland areas over 26–33 years, and found that none of them showed any evidence of regulation towards an equilibrium. (It is important to realize that these criticisms do not disprove the concept of biotic equilibrium. They merely show that the work investigated does not, after all, provide support for it.)

Apart from these specific studies, can we in any case be sure that any biota that we see today is in a state of stable equilibrium? The level of that equilibrium will alter if the environment changes—and the environment has changed a very great deal over both the comparatively distant past and in the recent past. The climatic changes of the last few millions of years not only caused the latitudinal movement of climatic belts, and so modified the environment of many islands, but they also caused changes in sea level. That in turn altered one of the fundamental parameters of the theory: island area. Changing sea level may either subdivide an island into two or more subordinate islands, or unite previously separate islands. In either case, there has been a change in the number and nature of the units (islands) that are being interpreted. It might be theoretically suggested that the island biota of a couple of thousand years ago might by now have made a full, complete adaptation to the results of these processes, and so have achieved a new biotic equilibrium. But the subsequent influence of our own species would have caused further major changes, causing the extinction of original endemic island taxa, and introducing new, exotic taxa—we have already seen that island populations are unusually vulnerable to extinction. Under

these circumstances, it is difficult to be confident that any island biotas are at stable equilibrium. If so, they cannot provide a database for estimates of the actual numerical equilibrium level in any existing situation, nor for any prediction of where such an equilibrium might emerge in the future.

Some biologists have attempted to integrate these climatic and geographical changes into the theory. For example, the American botanist Beryl Simpson [24] studied the flora of the high-altitude patches of 'paramo' vegetation on Andean mountains. She concluded that their diversity could be correlated better with the smaller areas that they would have occupied during the lower temperatures of the Pleistocene, than with their areas today. Similarly, Jared Diamond [25] suggested that the bird faunas of some Pacific islands could best be interpreted as in the process of reducing (or 'relaxing') towards a new, lower balance because rising sea levels had reduced the areas of the islands. The problem here is that the estimated required times for these biotic changes are so long that, almost inevitably, further geographical or climatic events will have changed the situation afresh. If so, this casts further doubt on the extent to which any of these island biotas can be thought of as being 'at equilibrium'.

When considering the rate of turnover of species, the longevity of the dominant species is also important. Case and Cody [26] have pointed out that the life spans of forest trees are so great that turnover is inevitably very slow. This is shown by the fact that the species structure of the forest at Angkor in Cambodia, which started to grow when the ancient Khmer capital was abandoned 560 years ago, has still not become identical to that of the surrounding older forest. Over even greater periods of time, the quality and adequacy of the data become even more unreliable.

It thus seems impossible at present to prove the theory's suggestion that, for any given set of circumstances, there is an equilibrium level for the biota of any island. But that does not, of course, prove that the suggestion is incorrect. Many observations show that the theory's basic proposals about the interrelationships between island area, position relative to its sources of colonists, and the rates of immigration and extinction, are correct. So these factors are leading each island biota towards a point of equilibrium, even though it may not yet have reached that point. The value of the theory is that it provides a unifying approach and standard format for analysis and comparison. These comparisons may, in their turn, reveal anomalies that stimulate and focus further research. But the theory cannot provide reliable predictions—of for example the equilibrium level for a given island under given circumstances.

The Theory of Island Biogeography and the design of nature reserves

It is not surprising that the Theory of Island Biogeography was warmly welcomed by those concerned with the management or design of nature reserves, for they could be analysed as islands amidst the sea of surrounding unprotected land. It seemed to promise an almost magical prescription for ensuring the ideal balance between effectiveness in terms of retaining the maximum number of species, and economy in their area (and therefore in their cost of acquisition and management). After the realization of the limitations of the theory, as reviewed above, it is now clear that it is no such remedy and reliable forecaster. (A valuable warning of the inadequacies of the theory, especially in applying it to the problems of conservation, is to be found in the book by Schrader-Frechette and McCoy [27].)

As Shafer [28] has commented, no aspect of the concept of an equilibrium level in species numbers can now be considered as established, and this would in any case be of limited value in an environment that is subject to both cyclical and occasional change. Only one of the predictions of the theory, that the number of species in an island will decrease directly with reduction in its area, now seems reliable. This is supported by the work of the American ecologist William Newmark [29], who studied the records of the species of large

mammal found in 14 National Parks of the western United States. He found that 42 separate populations of the species of lagomorphs, carnivorans and artiodactyls had disappeared from the parks since their foundation. Some of these losses had been the result of deliberate human intervention (e.g. the culling of all the grey wolves from Bryce and Yellowstone), but the patterns of disappearance of other species were correlated closely with the size of the park. Thus, the smallest (Bryce, Lassen and Zion) have each lost four to six species, while great Yellowstone, 20 times larger than Zion, has lost none. Furthermore, it seemed that area alone was the best predictor of the diversity of large mammals in each park, rather than any other factor such as the range of elevation found in the park, its latitude or the diversity of plants that it contained.

Large nature reserves should therefore retain more species and suffer fewer extinctions. But how large need they be? Very little is known about the area requirements of most species, and these requirements will depend on the degree of ecological specialization of the species and on their breeding habits. The larger mammals and birds that are a major preoccupation of many tropical game reserves require a very large home range to meet their energy needs. However, within the size range of the world's nature reserves, 97.9% of which are less than 10 000 km^2 in area, the variability of prediction of the Theory of Island Biogeography is so great as to be of no practical value [30]. Shafer comments [28] that it is probably impossible to separate the effects of the area of a nature reserve from other factors, even using sophisticated statistical techniques of multivariate analysis.

It is tempting to assume that a large nature reserve is simply better than a small one—and a larger reserve is certainly likely to support larger populations of the species that it contains, and therefore render them less vulnerable to chance extinction. Similarly, local ecological changes are also less likely to affect all the areas occupied by a particular species within a large park than within a small park. On the other hand, such random occurrences as fire, disease or the appearance of an over-competent predator or competitor might exterminate all of the individuals of a given species in one large park, but are unlikely to affect all of a number of small parks.

It has also been suggested that the disadvantageous effects of the isolation of nature reserves can be reduced by establishing corridors to interconnect them. To test this on the scale of large-mammal parks would be expensive and, in view of the long generation time of the animals concerned, would be so time consuming as to be impractical. However, Francis Gilbert and his colleagues [31] have recently carried out an analysis of the effect of corridors on the microarthropod fauna of isolated moss ecosystems over a period of 6 months—a time period equivalent to several generations of most of the fauna in the moss. This showed good evidence of a slower rate of extinction in the interconnected patches, which lost on average 15.5% of their species compared with 41% of the species in the case of unconnected islands. This was particularly marked in the case of predator species, which would be particularly vulnerable to extinction because of their smaller population size, but whose greater mobility would make them more likely to use the corridor. It is difficult to estimate how accurately these effects mirror what might happen at the larger scale of mammals, but Gilbert suggests that small mammals in patches of woodland might make a suitable larger-scale analogue of these experiments.

However, even a large park will not safeguard the future of all of the species that live within it, if it does not contain environments that are vital for some of them. This is shown by the work of the American zoologists Barbara Zimmerman and Richard Bierregaard [32], who studied the breeding requirements of 39 species of Amazon frog. Some of them gathered to breed in large streams, some in permanent pools of water, and others in temporary pools. Thus, a large park that omitted any of these crucial environments would be of less value in conserving a frog species than a smaller park that contained the appropriate breeding environment. The presence of the pig-like peccaries was also

important, for they created small wallows in which some of the frog species bred.

It is now clear that decisions on the size and location of nature reserves must be made of the basis of ecological analysis: which species do we wish to preserve, and what are their requirements? Furthermore, the ecological analysis must also include a study that will help to estimate the size of the minimum population that is likely to provide a reasonable chance of long-term survival for the species in question, together with an estimate of the area that is needed to support a population of that size. Finally, in some communities there are keystone species (see p. 78), that are crucial to the survival of many other species and whose disappearance would therefore fundamentally change the nature of the community; in the case of the frogs mentioned above, peccaries are important to their breeding. Any conservation area must therefore be large enough to support the minimum viable population of that species. It is considerations of this kind, rather than inferences from abstract curves of the numbers of species arriving, surviving or becoming extinct, that must lie at the heart of the design of nature reserves. There are no short cuts.

Irrespective of the usefulness of the Theory of Island Biogeography, islands are of great value in conservation, as their isolation provides a useful insulation from the factors that may be producing extinctions on the mainland. For example, the Australian conservationists Andrew Burridge, Matthew Williams and Ian Abbott [33] have shown that 35 islands around Australia protect 18 taxa of threatened mammals, and they emphasize the importance of preventing the introduction of exotic mammals, especially foxes. On the other hand, the low population sizes characteristic of islands make their biota more susceptible to extinction, as is shown by studies of human impact upon them. For example, of the original mammal species that were autochthonous to islands (i.e. that were not only endemic to them but had actually evolved there), 27% became extinct owing to human activities—and this figure rises to 35% if flying mammals are excluded from the calculation [34].

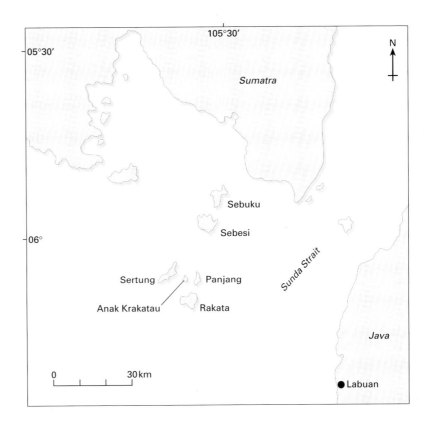

Fig. 13.5 Location of Rakata and neighbouring islands. After Whittaker *et al.* [53].

Starting afresh: the story of Rakata

In most cases, we have little knowledge of the history that lies behind the complex assemblage of animals and plants that inhabit a particular island. We may attempt to compare different islands, and to place their biota in a series, or several series, that might represent an historical process, but such an enterprise is fraught with the difficulties of subjective interpretation. We are on surer ground only where the history of the biota of a single island has been documented over an extended period of time—and we are now fortunate in having such a documentation for one island, Rakata, whose biota is in the process of being reassembled after total destruction. It has been studied since 1979 by the Oxford botanist Rob Whittaker and his associates, and since 1983 by the Australian zoologist Ian Thornton and his coworkers. Much of the infor-

mation in this section is taken from Thornton's enjoyable and stimulating book Krakatau [35], plus other data from Whittaker *et al.* [36]. The data accumulated (and still being accumulated) by these research programmes allows us to analyse the sequence of colonization of the island, and how different methods of colonization contribute to that.

Rakata lies in the East Indies, between the major islands of Java (40 km away) and Sumatra (35 km away), which act as the main sources of its colonists (Fig. 13.5). With an area of 17 km² and up to approximately 780 m high, Rakata is the largest remaining fragment of the island of Krakatau, which was destroyed by an enormous volcanic explosion in 1883. Two other islands, Sertung (13 km², 182 m) and Panjang (3 km², 147 m), are fragments of an older, larger version of Krakatau, and a new island, Anak Krakatau, appeared in 1930. All life on the

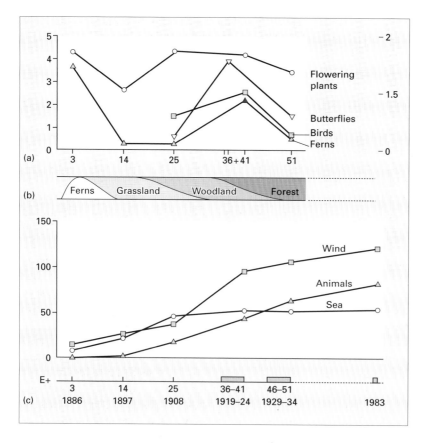

Fig. 13.6 The changing biota of Rakata, 1883–1934. The lower axis shows the dates, and the periods elapsed, since the eruption (E + *x*); the blocks indicate the periods at which collections were made. (a) Changes in the immigration rates of ferns and flowering plants (left-side scale) and of butterflies and resident land birds (right-side scale; these were not censused before 1908). After Thornton [35]. (b) A subjective interpretation of the rate and nature of the environmental changes, deduced from the descriptions given by scientific investigators. (c) Methods of initial colonization by ferns and flowering plants. Beyond the figures for 1934, the level has been extrapolated towards the figures from the 1983 census. After Bush & Whittaker [54].

islands was extinguished by the eruption, which covered them with a layer of hot ash 60–80 m deep on average, and up to 150 m deep in places. Surveys of the biota of Rakata were made intermittently from 1886, with a gap between 1934 and 1978 (apart from a little work in 1951), and intensive work has been carried out since the centenary of the eruption, in 1983. (In the following account, the dates of these surveys are indicated as the length of time that had elapsed since the 1883 eruption; thus 1908 = E + 25. See also the scale at the base of Fig. 13.6.) These surveys show that the patterns of colonization and extinction are not smooth, but are heavily influenced by the times of emergence of new ecosystems, and by the linkage between plants and animals due to food requirements or mechanisms of dispersal.

The coastal environment

The development of the biota of Rakata is best understood by analysing the beach and near-beach environment separately from the inland environment. There are two reasons for this. First, the beach environment is largely unchanged by the establishment of living organisms there, and is also itself ceaselessly being destroyed and recreated as ocean currents and storms erode one part of the beach and redeposit its materials elsewhere. This causes frequent local extinction of the biota, and simultaneous recolonization elsewhere. Secondly, these environments are primarily colonized by plants that have evolved methods of dispersal by sea, and for whom the ceaseless tides provide daily opportunities to colonize the beach.

There were already nine species of flowering plant on the beaches of Rakata by E + 3 (including two species of shrub and four species of tree). Eleven years later (E + 14), this had risen to 23 species of flowering plant, including three species of shrub and 10 species of tree. These trees and shrubs formed a coastal forest known as the *Terminalia* forest after its main component, the Indian almond tree, *Terminalia catappa*. All of the species in this forest are widely distributed on the beaches of South-East Asia and the western Pacific, showing that they are good at dispersal by sea. The number of trees in these early beach-arrival figures may at first seem surprising, but the larger size of the fruits or nuts of trees makes it easier for them to have flotation devices and not be overwhelmed by the waves. (Not all of these species necessarily arrived as solitary individuals; in 1986, the beach on the neighbouring island of Anak Krakatau bore a mass of vegetation 20 m², including complete palm trees 3–4 m tall.) Once they had grown, the early trees and shrubs also provided food in the form of fruit, as well as perching places, for birds and bats whose droppings were the probable source of other trees, such as two species of fig. Therefore, early colonists themselves provide a beach-head for other arrivals. The seeds and fruits of trees (such as *Terminalia*) that had arrived on the beach may also have been taken further inland by both fruit bats and crabs; the lack of mammals other than bats and rats on Rakata may explain why this inland spread seems to have been comparatively slow on the island.

By E + 25, 46 species of plant had arrived in Rakata by sea (Fig. 13.6c), but thereafter the seaborne component of its flora started to level off. Potential colonists are continually arriving by sea—a 2-month long survey of the beach of Anak Krakatau on 2 successive years found the fruits, seeds or seedlings of 66 species of plant. Because it is so easy for these species to colonize the beach, most of them soon do so, and thereafter there will be few arrivals of new seaborne species. So, for example, between E + 41 and E + 106, the diversity of the beach flora increased by only six new species (from 53 to 59

species), and the beach community is now relatively stable, with few new gains or losses.

Life inland

The history of the colonization of the inland areas of Rakata is more complex, because the environment did not remain constant. Instead, the presence and activities of each wave of colonists not only changed the environment, but also produced a greater variety of habitats, some of which were suited to a new selection of colonists. Therefore, both the complexity and the variety of the inland ecosystems of Rakata steadily increased. Figure 13.6(b) is an attempt to give an impression of the timing and rate of these ecological changes, using the descriptions given in the early surveys, so that they can be compared with the changes in some of the animals and plants that were colonizing the island through this period of time (Fig. 13.6a).

At first, by E + 3, the devastated, ash-covered inland areas bore a gelatinous film of 'blue–green algae', or cyanobacteria. This had provided a moist environment in which the spores of 11 species of fern and two species of moss had already germinated and grown, as well four species of herb. (This early preponderance of ferns may be because their spores are lighter than the seeds of flowering plants.) By E + 14, many new species of flowering plant had arrived, so that the ferns were less dominant, although they still covered much of the upland areas. By that time, definite plant associations could be distinguished at different levels in the island. The interior hills and valleys were covered by a 'grass steppe', up to 3 m high, consisting of the wild sugar cane *Saccharum spontaneum*, together with other grasses and scattered trees. Interestingly, some of the creeping plants, and a shrub, that are normally confined to the beach had been able to extend inland in this still-impoverished Rakata flora.

Grassland also covered the higher slopes of Rakata, but here it was dominated by *Imperata cylindrica*, a grass that is usually the first to colonize fire-cleared areas in the East Indies, together with the bamboo grass *Poganotherum*. Although ferns were still present in these grassland floras, they contributed only 14 species, compared with 42 species of flowering plants.

By E + 25 the interior grasslands had already begun to be replaced by mixed woodland and forest, and they had almost completely disappeared by E + 45. The forest grew denser, and its canopy gradually closed between E + 36 and E + 51. This caused a progressive change in the physical habitat and microclimate of the forest floor: the wind velocity, light intensity and temperature all decreased, while the humidity increased. The graphs of the immigration rates of ferns, flowering plants, butterflies and birds (Fig. 13.6a) show interesting changes that appear to be results of these ecological changes. (The immigration rate is the rate of addition of new species to the biota, per year. The addition of 10 new species over the space of 5 years between one survey and the next would therefore be an immigration rate of two species per year.)

For both ferns and flowering plants, the grasslands that replaced the early fern phase do not seem to have provided an environment that encouraged a diversity of new colonists, and the immigration rate of both groups fell. The fall was greater for the ferns, because their immigration rate had earlier been particularly high, during the formation of the fern phase. The immigration rate of the flowering plants had improved when they were next sampled, at E + 25, but had fallen by E + 51; this may be because the increasing woodland at first provided a greater variety of habitats for them, but that the later closure of the forest canopy restricted the light. The immigration rate of the ferns, in contrast, continued to increase even during the early stages of canopy formation, probably because the moist forest provided an ideal environment for a second group of ferns, mainly shade-demanding species, many of which were epiphytes, living on the trunk and branches of the forest trees. In both butterflies and resident land birds, the immigration rate rose as the forest started to form, but had fallen by the time the canopy had closed. Thus, closure of the canopy took place at the same time as

a reduction in the immigration rate of all these groups. Nevertheless, because extinction rates still remained comparatively low, the total number of species increased slightly in the case of the flowering plants, and remained approximately constant in the other groups. (The fact that extinction rates remained low during the period of closure of the canopy suggests that patches of open ground or woodland must have remained within the forest, perhaps where trees had fallen and provided a continuing opportunity for the survival of species that preferred an open habitat.)

As noted above, the biota of the interior of Rakata also differs from that of the coast in that nearly all of it arrived by air, not by sea. That was not a very difficult journey, for winds from Java and Sumatra have an average speed of 20–22 kph (12–14 mph), so that wind-blown seeds could arrive in Rakata in about 2 h. However, although some of the airborne arrivals came on the wind, others came in or upon the bodies of animals. To begin with, the ash-covered interior of Rakata was totally uninviting to animals. Thus, as can be seen from Fig. 13.6(c), animal-aided dispersal only became a significant contributor to the biota from the time of the E + 25 survey. From then on, the graphs of the arrival of wind-dispersed species and of animal-dispersed species move in parallel. But they also rise more steeply, because of a positive-feedback effect between the plants and the animals. The woodlands that had developed by E + 25 provided an environment that other species of flowering plant could colonize (Fig. 13.6b). The increasing diversity of these plants in turn provided food for an increasing diversity of animals, as can be seen from the fact that the immigration rates of both butterflies and birds increased at that time. This effect became especially evident as the forest canopy formed. But the increasing numbers and diversity of animals arriving in the growing forests also brought the seeds of other new species of plant, either within their alimentary canal or adhering to their bodies. The resulting increase in plant diversity in turn encouraged more animal diversity, and so on.

The animal-dispersed component has been the most important one ecologically, because the seeds of nearly all the tree species of the inland forests arrived in this way. Wind dispersal, on the other hand, was particularly important in the addition of new forest species. This method of dispersal not only provided all of the forest ferns, but many of the herbs and shrubs. Of these, 17% belong to the Compositae and 13% to the Asclepiadaceae (milkweeds, whose seeds bear silky hairs rather like those of the Compositae and are easily carried in the wind), while over 50% of these species belong to the Orchidaceae, some of which are dispersed by wind and others by animals. (Orchid seeds have no food reserve and need root fungi in order to germinate and grow, which suggests that the original colonists, at least, may have arrived on the legs of birds, in mud that also contained the fungus.) For the flora as a whole, species with small, wind-dispersed seeds form a much higher proportion of the flora of Rakata than they do on neighbouring Java.

Rakata is the largest and highest of the three islands that were the surviving, initially lifeless, fragments of Krakatau (see Fig. 13.5). It might have been expected that the forests that eventually appeared on these three islands would be similar to one another and to those on neighbouring Java and Sumatra—but they are not. The lowland forests of Rakata are unique, because they are dominated by the wind-dispersed tree *Neonauclea* (Rubiaceae), which grows up to 30 m high. The forests of Panjang and Sertung, in contrast, are instead dominated by the animal-dispersed trees *Dysoxylum* (Meliaceae) and *Timonius* (Rubiaceae). Why should the forests of these islands be unique and also different from one another? Various theories have been put forward.

One of the differences between the trees is that *Neonauclea* is less tolerant of shade than are the other two species. It may therefore be 'shaded out' if all three species arrive on an island at about the same time. However, *Neonauclea* arrived on Rakata in 1905, nearly 25 years before the others, and this may have given it a head start and allowed it to become dominant on that island. It also seems that

it requires overturned soil or fresh ash in order to establish itself. Few scientific visits were made to the other two islands, so the history of their forests is not well known—only that all three species were present by 1929. The Japanese ecologist Hideo Tagawa and his colleagues have suggested that, if they all arrived at about the same time, the shade produced by the growing *Dysoxylum* and *Timonius* trees might have made it more difficult for *Neonauclea* to flourish. Rob Whittaker and his colleagues initially pointed out that, unlike Rakata, the other islands had been partially covered by up to 1 m of ash during the eruption of Anak Krakatau in 1930, and that this might well have affected the floras of the islands. More recently [36], they have suggested that there is a very strong chance element determining which species happens to form the dominant element, its success depending on such factors as the time of year, the prevailing climate, what elements of the previous vegetation had survived and which of they were fruiting, which dispersal agents were available, etc. (Fig. 13.7).

All of this is a good illustration of how difficult it is to interpret ecological biogeography, even in the apparently simple situation of the colonization of a lifeless island environment.

Fig trees, which are a particularly important part of the flora of tropical forests, provide another interesting problem in the colonization of Rakata. They are a major component of the forests; by E + 40, the 17 fig species found on Rakata, Panjang and Sertung made up nearly two-thirds of the total number of tree species there. Figs are also important because they are used as food by many animals. (In Malaya, one individual fig tree was visited by 32 species of vertebrate, and 29 species of fig tree were used by 60 species of bird and 17 species of mammal.) Just as important, however, is the fact that few of these animals eat only figs, and therefore those fig-eating animals arriving in Rakata were quite likely to bring, within their digestive system, the seeds of other trees. But figs also provide a problem in colonization, for each species of fig requires the services of its own species of pollinating wasp in order to produce fertile seeds—and

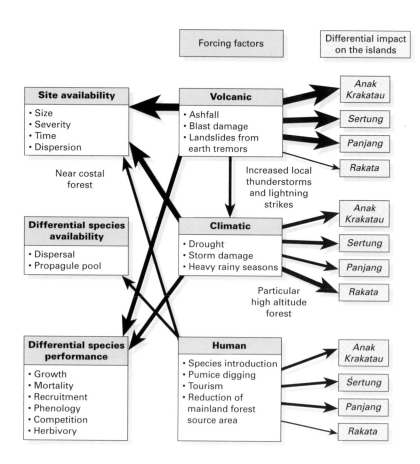

Fig. 13.7 The relative importance, and hierarchy, of the different factors affecting plant succession in the Krakatau Islands. From Schmitt & Whittaker with permission [37].

the fig is similarly necessary for the life cycle of the wasp. Therefore, for this symbiotic system to become effective and established, there has to be a sufficiently large population of both the fig tree and the wasp, each component having arrived in Rakata independently, the figs having been borne by animals and the wasps having arrived by air.

Finally, one can learn something about the processes and difficulties of island colonization by noting what types of animal or plant have so far *not* been able to colonize Rakata. Small, non-flying mammals such as cats, monkeys and most rodents are absent; they are incapable of crossing sea barriers of more than 15 km (9 miles). The exception is the country rat *Rattus tiomanicus*, which has been known to swim 35 km and which had colonized Rakata by E + 45. Because there is no supply of

running or standing fresh water on the island, there are no mangrove trees, no freshwater birds, no insects that have aquatic larvae, and no freshwater molluscs. There is as yet no mature forest on Rakata, and therefore none of the birds that require that environment, such as trogons, parrots, nuthatches, hornbills, pittas and leafbirds. Finally, although some trees (e.g. dipterocarps) and bushes have winged seeds, these do not normally travel in the wind for more than a mile (1.6 km), and these species are absent from Rakata.

There are also some interesting interactions at a more detailed level. For example, the bird fauna of Rakata includes the flowerpecker *Dicaeum*, which distributes the seeds of the plant family Loranthaceae, which are epiphytic plant parasites of the trees of the forest canopy. However, the

Rakata forest is not yet old enough to contain the mature and dying trees that the parasite can attack. As a result the Loranthaceae are absent, together with the butterfly *Delias* that feeds upon these plants, even though the butterfly itself is highly migratory and a competent potential colonist.

The complexity of all these ecological and successional changes shows that the colonization history of an island will not follow the simple path predicted by the Theory of Island Biogeography. MacArthur and Wilson themselves pointed this out in their book, in which they gave an example of a single-peaked colonization curve and suggested that the sequence of invasion might affect the nature of colonization and the equilibrium number [8]. The replacement of one plant community by another causes pronounced irregularities in the graphs of immigration and extinction, not only of the plants themselves but also of the associated fauna. The integrated nature of the successional ecosystems therefore makes it likely that the colonization history of an island like Rakata will show pronounced waves of change. The simple monotonic curves predicted by the Theory of Island Biogeography therefore apply only to the situation after the pattern of communities in an island has already become firmly established.

Evolving there: opportunities for adaptive radiation

Colonists may encounter many difficulties when they first enter an island, but there are rich opportunities for those species that can survive long enough for evolution to adapt them to the new environment. These opportunities exist because of the lack of many of the parasites and predators that elsewhere would prey upon the species, and of many of the other species with which it normally competes. Like Darwin's finches on the Galápagos Islands (see p. 95), it may be able to radiate into ways of life not formerly available to it.

A good example of this can be found in the Dry Tortugas, the islands off the extreme end of the Florida Keys, which only a few species of ant have

successfully colonized [6]. One species, *Paratrechina longicornis*, on the mainland normally nests only in open environments under, or in the shelter of, large objects; but on the Dry Tortugas it also nests in environments such as tree trunks and open soil, which on the mainland are occupied by other species. Not every species, however, is capable of taking advantage of such opportunities in this way. On the Florida mainland, the ant *Pseudomyrmex elongatus* is confined to nesting in red mangrove trees, occupying thin, hollow twigs near the tree top. Although it has managed to colonize the Dry Tortugas, it is still confined there to this very limited nesting habitat.

Such opportunities for alterations in behavioural habits or in diet provide in turn the opportunity for the organism to become permanently adapted, by evolutionary change, to a new way of life. This process requires a longer period of time, and is therefore unlikely to take place except on islands that are large and stable enough to ensure that the evolving species does not become extinct. But if an island does provide these conditions, then remarkable evolutionary changes may take place as colonizing species become modified to fill vacant niches. Instances of such evolution in isolation are provided by the Great Lakes of Africa, large enough to provide a great diversity of environments, and old enough for these ecological opportunities to have become realized through evolutionary change. The cichlid fishes, in particular, have been able to take advantage of this, and have undergone a rapid evolutionary change, producing 37 genera and 126 species in Lake Tanganyika, and 20 genera and 196 species in Lake Malawi [38]. Most other lakes are probably too small, too impermanent, or too recent in origin, for colonization or evolutionary change to have been able to replenish the losses due to extinction, and are faunally impoverished. In particular, it probably takes a considerable time for the deeper parts of lakes to become exploited by evolutionary change of the normally shallow-water fishes that are the only possible colonists.

Another tendency is for island species to lose the dispersal mechanisms that originally allowed them to reach their new home. Once on the restricted area of the island, the ability for long-distance dispersal is no longer of value to the species: in fact it is a disadvantage, for the organism or its seeds is now more likely to be blown out to sea. Seeds tend to lose their 'wings' or feathery tufts, and many island insects are wingless. The loss of wings by some island birds may be partly for this reason, and partly because there are often no predators from which to escape. A few out of many examples are the kiwi and moa of New Zealand, the elephant birds of Madagascar, and the dodo of Mauritius (the last three are extinct, but only because they were killed by humans).

Islands that are treeless provide an opportunity for other plants to fill this ecological niche. The seeds of trees are usually much larger and heavier than those of other plants, and therefore do not often cover long distances. As a result, other plants may develop to fill this vacant niche [39]. The modifications needed to produce a tree from a shrub which already possesses strong, woody stems are comparatively slight—merely a change from the many-stemmed, branching habit to concentration on a single, taller trunk. For example, although most members of the Rubiaceae are shrubs, this family has produced on Samoa the 8-m-tall tree *Sarcopygme*, which has a terminal palm-like crown of large leaves. Although more comprehensive changes are needed to produce a tree from a herb, many islands show examples of this. In many cases the plants involved are members of the Compositae, perhaps because they have unusually great powers of seed dispersal, are hardy, and often already have partly woody stems. To this family belong both the lettuces, which have evolved into shrubs on many islands, and the sunflowers. On the isolated island of St Helena in the South Atlantic can be found five different trees, 4–6 m high, which have evolved there from four different types of immigrant sunflower (Fig. 13.8). Two of these (*Psiadia* and *Senecio*) are endemic species of more widely distributed genera, while the other three (*Commidendron*, *Melanodendron* and *Petrobium*) are recognized as completely new genera. These latter genera are

therefore both endemic (i.e. known only in St Helena) and also autochthonous.

It is not unusual to find that island species are different in size from their mainland relatives, and this phenomenon has been discussed by the American ecologist Ted Case [40]. Sometimes an island lacks a particular type of predator, because the size of the population of its prey is not large enough to provide a reliable source of food. This may decrease the death rate of the prey species, and allow it to grow more rapidly, for it can feed at times and places that were previously dangerous. The predator might also have preferred to catch larger specimens of the prey species. For all these reasons, the average size of island herbivores, such as some rodents, iguanid lizards and tortoises, may become greater than that of their mainland relatives.

Another reason for a change in size in island species is if some of its competitors are absent. If that competitor was smaller than the island species, then the latter may itself become smaller to colonize that vacant niche. The converse may be true if its competitor was larger, as in the case of the Komodo dragon (*Varanus komodoensis*), a giant lizard which lives on Komodo Island and nearby Flores Island, in the East Indies. These animals increased in size to occupy niches which on the mainland are filled by much larger animals.

All these organisms evolved in islands to fill habitats normally closed to them. But other evolutionary changes frequently found on islands are the direct result of the island environment itself, not of the restricted fauna and flora. We have seen how serious the effect of a small population may be. But the same island will be able to support a larger population of the same animal if the size of each individual is reduced. This evolutionary tendency on islands is shown by the find of fossil pygmy elephants that once lived on islands in both the Mediterranean and the East Indies.

The Hawaiian Islands

As has been seen, there are many aspects of island life that are unique, and many others that differ

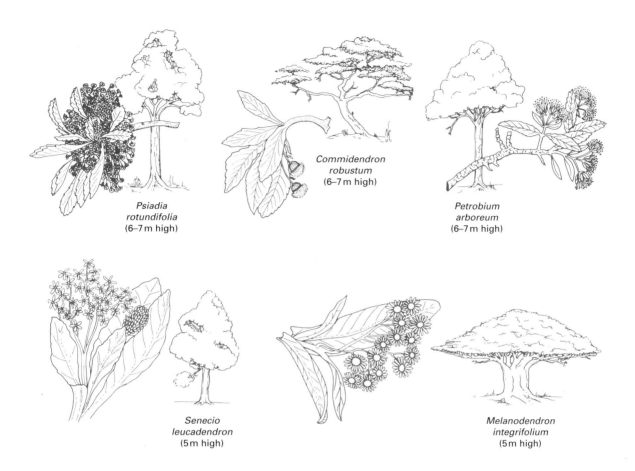

*Psiadia
rotundifolia
(6–7 m high)*

*Commidendron
robustum
(6–7 m high)*

*Petrobium
arboreum
(6–7 m high)*

*Senecio
leucadendron
(5 m high)*

*Melanodendron
integrifolium
(5 m high)*

Fig. 13.8 The varied trees that have evolved from immigrant sunflowers on St Helena Island. From Carlquist [38].

only in degree from life on the continental land masses. The result of the action of all these different factors can be seen by examining the flora and fauna of one particular group of islands. The Hawaiian Islands provide an excellent example, for they form an isolated chain, 2650 km long, lying in the middle of the North Pacific, just inside the Tropics (Fig. 13.9). Sherwin Carlquist has provided an interesting account of the islands and of their fauna and flora, pointing out the significance of many of the adaptations found there [41]. More recently, a collection of papers on Hawaiian biogeography [42] contains many fascinating contribu-

tions, and others are to be found in a collection of papers on Pacific island biotas in general [43].

The Hawaiian Islands are a part of the most outstanding example of a volcanic island chain generated by a geological hotspot (see p. 262). They rise from a sea floor that is 5500 m deep, and that is moving north-westwards at 8–9 cm per year. The whole chain comprises 129 volcanoes; 104 of these were originally high enough to reach the surface and form islands (or add to an existing island). The youngest and biggest island, Hawaii, is only 700 000 years old and is made up of six volcanoes, two of which are still active. The most westerly island still visible above the sea, Kure, is nearly 30 million years old, but the Emperor Seamount chain, consisting of other submerged islands and volcanoes, extends further westwards and then

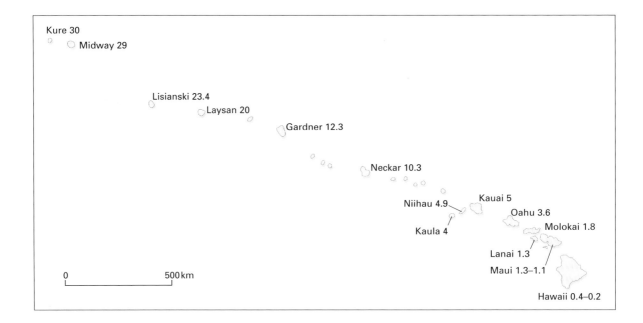

Kure 30

Midway 29

Lisianski 23.4

Laysan 20

Gardner 12.3

Neckar 10.3

Niihau 4.9

Kauai 5

Oahu 3.6

Molokai 1.8

Kaula 4

Lanai 1.3

Maui 1.3–1.1

Hawaii 0.4–0.2

0   500 km

Fig. 13.9 The Hawaiian island chain.

northwards to near the Kamchatka Peninsula of Siberia. There, the Meiji Seamount is perched on the edge of the Kurile Trench and will eventually disappear back into the depths of the earth.

Meiji was formed by the hotspot 85 million years ago, at the present-day location of Hawaii, so islands and submerged volcanoes have been forming in the central North Pacific Ocean for at least that length of time. However, there have been periods in the past when there were no visible islands, or only low islands that were soon submerged. The most recent of these periods lasted for 18 million years, and ended with the appearance of Kure Island about 30 million years ago. Kure is therefore the oldest island that was a potential home for colonists that might be ancestral to the biota of the Hawaiian Islands of today. Although most later islands (until the last five million years) were small, there continued to be the occasional larger, higher island that might have provided a home for the descendants of the biota of Kure.

The rocks that form the Hawaiian Islands can

be accurately dated, and the modern technique of calculating the rates of molecular change in the lineages of animals and plants provides a similar system of dating the times of divergence of different genera and species. As the American workers Hampton Carson and David Clague [44] have shown, it is therefore now possible, in the case of the Hawaiian Islands, to produce a fascinating integration between the gears of those two great engines of our planet—the plate-tectonic engine that here leads to the appearance of new islands, and the evolutionary engine that responds to these opportunities by producing new forms of life. As will be seen below, this integration strongly suggests that the ancient islands, which appeared between 30 million years and 10 million years ago, were in fact the sites of evolution of some of the ancestors of today's Hawaiian animals and plants. Once they had arrived on any particular island, these animals and plants were able to diversify and, often, to spread to the newly formed additions to the chain as the older islands steadily eroded away and disappeared beneath the sea. The high volcanic peaks seem to be islands within islands, for it is difficult for their alpine plants to disperse from one island to another. Instead, they have evolved from the adjacent lower-lying flora independently within each island: 91% of the alpine plants of Hawaii are endemic, a far higher proportion than that of the island flora as a whole (16.5%) or even of the angiosperms alone (20%) [45].

The closest relatives of most of the Hawaiian animals and plants live in the Indo-Malaysian region. For example, of the 1729 species and varieties of Hawaiian seed plants, 40% are of Indo-Malaysian origin but only 18% are of American origin; also, nearly half of the 168 species of Hawaiian ferns have Indo-Malaysian relatives, but only 12% have American affinities. This is not surprising, for the area to the east is almost completely empty, while that to the south and west of the Hawaiian chain contains many islands, which can act as intermediary homes for migrants. Organisms adapted to life in these islands would also be better adapted to life in the Hawaiian Islands than would those from the American mainland.

Mechanisms of arrival

The way in which the Hawaiian birds reached the islands is obvious enough. One of the plants which probably came with them is *Bidens*, a member of the Compositae, whose seeds are barbed and readily attach themselves to feathers; about 7% of the Hawaiian non-endemic seed plants probably arrived in this way [3]. The Hawaiian insects, too, arrived by air. Entomologists have used aeroplanes and ships to trail fine nets over the Pacific at different heights and have trapped a variety of insects, most of which, as would be expected, were species with light bodies. These types also predominate in the Hawaiian Islands (an indication of their airborne arrival), although heavier dragonflies, sphinx moths and butterflies are also found there.

The influence of the winds in providing colonists is shown by the fact that, although angiosperms are far more common than ferns in the world as a whole, their diversity in Hawaii is more evenly balanced: 225 immigrant angiosperms, 135 immigrant ferns. The relatively greater success of the ferns is probably due to the fact that their spores are much smaller and lighter than the seeds of angiosperms. Of the non-endemic seed plants of the Hawaiian Islands, about 7.5% almost certainly arrived carried by the wind, while another 30.5% have small seeds (up to 3 mm in diameter) and may also have arrived in this way.

One of the most interesting plants that probably arrived as a windborne seed is the tree *Metrosideros*. It is unusual because its seeds are tiny compared with those of other trees, and this has allowed it to become widely dispersed through the Pacific Islands. It is a pioneering tree, able to form forests on lowland lava rubble with virtually no soil—a great advantage on a volcanic island. *Metrosideros* shows great variability in its appearance in different environments, from a large tree in the wet rainforest, a shrub on wind-swept ridges, to as little as 15 cm high in peatlands, and it is therefore the dominant tree of the Hawaiian forest. Although these differences are probably at least partially genetically based, the different forms are not distinct species, and intermediates are found where two different types (and habitats) are adjacent to one another.

Probably the single most important method of entry of seed plants to the Hawaiian Islands has been as seeds within the digestive system of birds that have eaten fruit containing them (e.g. blueberry, sandalwood); about 37% of the non-endemic seed plants of the islands probably arrived in this way. Significantly, many plants that succeeded in reaching the islands are those that, unlike the rest of their families, bear fleshy fruits instead of dry seeds (e.g. the species of mint, lily and nightshade found in Hawaii).

Dispersal by sea accounts for only about 5% of the non-endemic Hawaiian seed plants. As well as the ubiquitous coconut, the islands also contain *Scaevola toccata*; this shrub has white, buoyant fruits and forms dense hedges along the edge of the beach on Kauai Island. Another seaborne migrant is *Erythrina*; most species of this plant genus have buoyant, bean-like seeds. On Hawaii, after its arrival on the beach, *Erythrina* was unusual in adapting to an island environment, and a new endemic species, the coral tree *E. sandwichensis*, has evolved on the island. Unlike those of its ancestors, the seeds of the coral tree do not float—an example of the loss of its dispersal mechanism often characteristic of an island species.

The successful colonists of the Hawaiian Islands are the exceptions; many groups have failed to reach them. There are no truly freshwater fish and no native amphibians, reptiles or mammals (except for one species of bat), while 21 orders of insect are completely absent. As might be expected, most of these are types that seem in general to have very limited powers of dispersal. For example, the Formicidae (ants), which are an important part of the insect fauna in other tropical parts of the world, were originally absent. They have, however, since been introduced by humans, and 36 different species have now established themselves and filled their usual dominant role in the insect fauna. This proves that the obstacle was reaching the islands, not the nature of the Hawaiian environment.

Evolutionary radiations within the Hawaiian Islands

As ever, the absence of some groups has provided greater opportunities for the successful colonists. Several insect families, such as the crickets, fruitflies and carabid beetles, are represented by an extremely diverse adaptive radiation of species, each radiation derived from only a few original immigrant stocks. For example, the fruitflies, belonging to the closely related genera *Drosophila* and *Scaptomyza*, have undergone an immense radiation in the Hawaiian Islands; of the over 1300 species known worldwide, over 500 have already been described from the Hawaiian Islands, where there are probably another 250–300 species awaiting description. The abundance of species of fruitflies in the islands is probably due partly to the great variations in climate and vegetation to be found there, and also to the periodic isolation of small islands of vegetation by lava flows, each

island providing an opportunity for independent evolution of new species. But another major factor has been that the Hawaiian fruitflies, in the absence of the normal inhabitants of the niche, have been able to use the decaying parts of native plants as a site in which their larvae feed and grow. This change is probably also due to the fact that their normal food of yeast-rich fermenting materials is rare in the Hawaiian Islands.

Molecular studies indicate that the common ancestor of all the drosophilids of the Hawaiian Islands diverged from the Asian mainland type about 30 million years ago; comparison with the ages of the different islands (Fig. 13.9) suggests that this must have happened on Kure Island when that island was young and nearly 900 m high. Similarly, these studies indicate that *Scaptomyza* diverged from *Drosophila* 24 million years ago, which suggests that this event took place on Lisianski, which was at one time an island about 1220 m high. Detailed studies of the chromosome structure of the Hawaiian 'picture-winged' fruitflies are now making it possible to reconstruct the sequence of colonizations that must have taken place [44]. As might have been expected, the older islands to the west in general contain species ancestral to those in the younger islands in the east (Fig. 13.10a). The youngest, Hawaii itself, has 19 species descended from species in older, more westerly islands, but none of its species appears to be ancestral to those in the older islands. (The present-day islands of Maui, Molokai and Lanai together formed a single island until the postglacial rise in sea levels; their *Drosophila* species are therefore treated together as a single fauna.) Within Hawaii itself, molecular studies show that the species on the more southern part of the island, which is up to 200 000 years old, evolved from those on the more northern part, which is 400 000–600 000 years old.

The same phenomenon of a great adaptive radiation has taken place in other groups of animals and plants. In general, therefore, although the islands contain comparatively few different families, each contains an unusual variety of species, nearly all of which are unique to the islands. In fact, out of

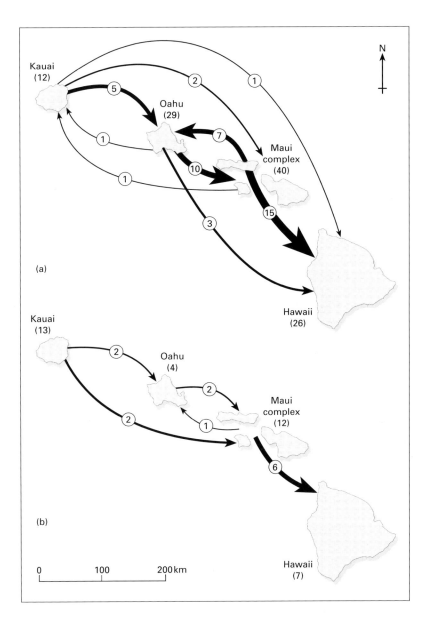

Fig. 13.10 The dispersal events within the Hawaiian Islands that are suggested by the interrelationships of the species of 'picture-winged' *Drosophila* flies (above) and of silverweed plants (below). The width of the arrows is proportional to the number of dispersal events implied, and the number of species in each island is shown in parentheses. From Carr *et al.* [46], by permission of Oxford University Press.

the whole of the Hawaiian flora, over 90% of the species are endemic to the islands.

There are many other examples of Hawaiian adaptive radiations, but three are of particular interest: the silverswords and lobeliads among the plants [41], and the honey-creepers among the birds. The silverswords are descended from the tar-

weeds of south-west North America (members of the family Compositae, which includes sunflowers and daisies), and probably arrived on the islands as sticky seeds attached to the feathers of birds. They have produced only three genera in the islands (*Dubautia*, *Argyroxiphium* and *Wilkesia*), but these have colonized a variety of habitats. For example,

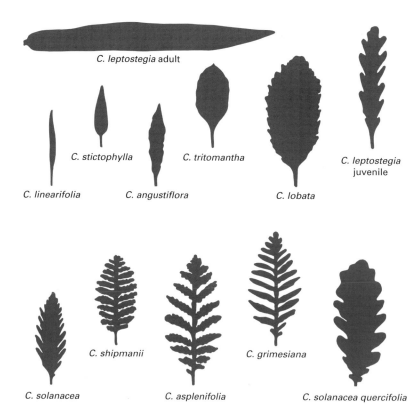

C. leptostegia adult

C. stictophylla

C. tritomantha

C. linearifolia

C. angustiflora

C. lobata

C. leptostegia juvenile

C. shipmanii

C. grimesiana

C. solanacea

C. asplenifolia

C. solanacea quercifolia

Fig. 13.11 The leaves of different species of the lobeliad genus *Cyanea.*

on the bare cinders and lava of the 3050 m high peak of Mt Haleakala on the island of Maui, two of the few plant species that can survive are silverswords. *Dubautia menziesii* is adapted to this arid environment by its tall stem and stubby succulent leaves, while *Argyroxiphium sandwichense* is covered by silvery hairs that reflect the light and heat. A few hundred metres below the bare volcanic peaks, conditions are at the other extreme, because most of the rain falls at heights of from 900 to 1800 m; these regions receive from 250 to 750 cm of rain per year. The upper regions of 1770 m high Mt Puu Kukiu on Maui are covered by mire in which thrives another silversword, *A. caliginii.* On the island of Kauai the heavy rainfall has led to the development of dense rainforest, in which *Dubautia* has evolved a tree-like species, *D. knudsenii,* with a trunk 0.3 m thick, and large leaves to gather the maximum of sunlight in the dim forest. Kauai bears another silversword which shows the tendency for island

plants to become trees. In the drier parts of this island grows *Wilkesia gymnoxiphium,* with a long stem which carries it above the shrubs that compete with it for light and living space. This species also shows another example of the loss of the dispersal mechanism that first brought the ancestral stock to the island: the seeds of *Wilkesia* are heavy and lack the fluffy parachutes usually found among the Compositae. The pattern of dispersal of the 28 species in the three silversword genera within the Hawaiian Islands is very similar to that of the drosophilid flies (Fig. 13.10) [46]. Molecular evidence suggests that these three genera had a common ancestor that arrived on Kauai.

Lobeliads are found in all parts of the world, but they have undergone an unusual adaptive radiation in the Hawaiian Islands, because their normal competitors, the orchids, are rare. The Hawaiian lobeliads include 150 endemic species and varieties, making up six or seven endemic genera.

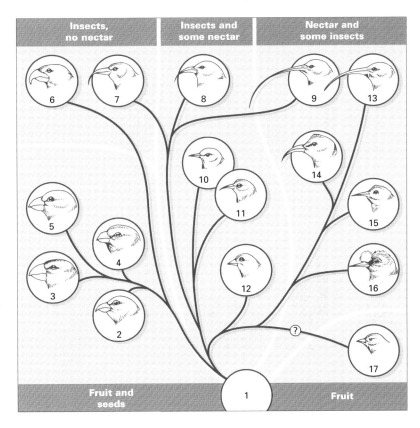

Fig. 13.12 The evolution of dietary adaptations in the beaks of Hawaiian honey-creepers. 1, Unknown finch-like colonist from Asia; 2, *Psittirostra psittacea*; 3, *Chloridops kona*; 4, *Loxioides bailleui*; 5, *Telespiza cantans*; 6, *Pseudonestor xanthophrys*; 7–11, *Hemignathus wilsoni, H. Iucidus, H. procerus, H. parvus, H. virens*; 12, *Loxops coccineus*; 13, *Drepanis pacifica*; 14, *Vestiaria coccinea*; 15, *Himatione sanguinea*; 16, *Palmeria dolei*; 17, *Ciridops anna*. Taxonomy after Pratt *et al.* [55].

Over 60 species of the endemic genus *Cyanea* are known, showing an incredible diversity of leaf form (Fig. 13.11). The plants range from the tree *C. leptostegia*, 9 m tall (similar in appearance to the tarweed *Wilkesia*) to soft-stemmed *C. atra*, only 0.9 m tall. The species of another genus, *Clermontia*, are less varied in overall size, but are very varied in the size, shape and colour of their flowers. These are mainly tubular and brightly coloured, a type of flower that is often associated with pollination by birds. On isolated islands such as Hawaii, the adaptation of larger flowers to bird pollination may be because the large insects that would normally pollinate such flowers on the mainland are absent. It is no coincidence that the adaptive radiation of the Hawaiian lobeliads has been accompanied by the adaptive radiation of a nectar-eating type of bird, the honey-creepers [47].

The ancestors of these birds were probably finch-like immigrants from Asia [48] which fed on insects and nectar. From the original immigrants, adaptive radiation has produced 11 endemic genera comprising the endemic family Drepanididae (Fig. 13.12). Many of the genera, such as *Himatione, Vestiaria, Palmeria, Drepanidis*, many species of *Loxops*, and one species of *Hemignathus* (*H. procerus*) are still nectar eaters, feeding from the flowers of the tree *Metrosideros* and the lobeliad *Clermontia*. Since insects, too, are attracted to the nectar, it is not surprising to find that many nect-areating birds are also insect-eaters, and it is a short step from this to a diet of insects alone. *Hemignathus wilsoni* uses its mandible, which is slightly shorter than the upper half of its bill, to probe into crevices in bark for insects, and *Pseudonestor xanthophrys* uses its heavier bill to rip open twigs and branches in search of insects. Other types have heavy, powerful beaks, which they use for cracking open seeds, nuts or beans. The light bill of the recently extinct *Ciridops* was used for eating the soft flesh of the fruits of the Hawaiian palm *Pritchardia*.

Recent studies of the Hawaiian avifauna also show the unreliability of the modern biota as a basis for estimates of rates of biotic change or of the relationship between island area and the number of species. It has long been known that about a dozen Hawaiian bird species became extinct after the arrival of Europeans and the animals that they introduced. But studies by the American biologists Storrs Olson and Helen James [49,50] revealed at least 50 now-extinct species of Hawaiian bird— more than the entire avifauna of the islands today. These included flightless types of ibis, rail and goose-like ducks, six hawks and goshawk-like owls, and nearly two dozen species of drepanidid finches, mostly of insectivorous type. Most of these species were still alive when the Polynesians arrived in the Islands in about AD 300, but had become extinct before Europeans arrived 1500 years later. Similar evidence for bird extinctions has been reported from many other Pacific islands, so that the patterns of distribution, endemism and numbers of species on individual islands that are seen today are wholly unreliable as a basis for generalizations about bird populations on islands.

Living as they do on fruits, seeds, nectar and insects, it is not surprising that none of the drepanidids shows the island fauna characteristic of loss of flight. However, both on Hawaii and on Laysan to the west, some genera of the Rallidae (rails) have become flightless (a common occurrence in this particular family of birds). The phenomenon of flightlessness is also common in Hawaiian insects: of the endemic species of carabid beetle, 184 are flightless and only 20 are fully winged. The Neuroptera or lacewings are another example—their wings, usually large and translucent, are reduced in size in some species, while in other species they have become thickened and spiny.

Summary

1 Islands provide a unique opportunity to study evolution, for their impoverished faunas and floras are the ideal situation for rapid evolutionary modification and adaptive radiation.
2 Island life is unusually hazardous, so that there is a complex interaction between the processes of immigration, colonization and extinction.
3 Quantitative analysis of this has recently commenced, and is proving extremely valuable in providing insights into the history and structure of ecosystems, on the continents as well as on islands. However, attempts to construct a predictive theory of the numbers of species that would be found on islands of different sizes and locations have proved to be unreliable, and these theories also provide little help in the design of nature reserves.

Further reading

Whittaker RJ. *Island Biogeography: Ecology, Evolution and Conservation.* Oxford: Oxford University Press, 1998.
Quammen D. *The Song of the Dodo—Island Biogeography in an Age of Extinction.* London: Pimlico/Random House, paperback, 1996.

References

1 Censky EJ, Hodge K, Dudley J. Over-water dispersal of lizards due to hurricanes. *Nature* 1998; 395: 556.
2 van der Pijl L. *Principles of Dispersal in Higher Plants*, 3rd edn. Berlin: Springer-Verlag, 1982.
3 Van Balgooy MMJ. Plant-geography of the Pacific as based on a census of phanerogam genera. *Blumea* 1971; Suppl. 6: 1–222.
4 MacArthur RH, Diamond JM, Karr J. Density compensation in island faunas. *Ecology* 1972; 53: 330–42.
5 Diamond JM. Historic extinctions: a Rosetta Stone for understanding prehistoric extinctions. In: Martin PS, Klein RG, eds. *Quaternary Extinctions: a Prehistoric Revolution.* Tucson: University of Arizona Press, 1984.
6 Lack D. Subspecies and sympatry in Darwin's finches. *Evolution* 1969; 23: 252–63.

7 Simberloff DS. Using island biogeographic distributions to determine if colonization is stochastic. *Am Naturalist* 1978; 112: 713–26.

8 MacArthur RH, Wilson EO. *The Theory of Island Biogeography*. Princeton: Princeton University Press, 1967.

9 Vuilleumier F. Insular biogeography in continental regions. I. The Northern Andes of South America. *Am Naturalist* 1970; 104: 373–88.

10 Vuilleumier F. Insular biogeography in continental regions. II. Cave faunas from Tessin, southern Switzerland. *Syst Zool* 1973; 22: 64–76.

11 Brown JH, Kodric-Brown A. Turnover rates in insular biogeography: an effect of migration on extinction. *Ecology* 1977; 58: 445–9.

12 Janzen DH. Host plants as islands in evolutionary and contemporary time. *Am Naturalist* 1968; 102: 592–5.

13 Janzen DH. Host plants as islands. II. Competition in evolutionary and contemporary time. *Am Naturalist* 1973; 107: 786–90.

14 Rosenzweig ML. *Species Diversity in Space and Time* Cambridge: Cambridge University Press 1995.

15 Simberloff DS. Species turnover and equilibrium island biogeography. *Science* 1976; 194: 572–8.

16 Gilbert FS. The equilibrium theory of island biogeography: fact or fiction? *J Biogeogr* 1980; 7: 209–35.

17 Connor EF, McCoy ED. The statistics and biology of the species–area relationship. *Am Naturalist* 1979; 113: 791–833.

18 Colwell RK, Winkler DW. A null model for null models. In: Strong DR *et al.*, eds. *Ecological Communities and the Evidence*. Princeton: Princeton University Press, 1984: 344–59.

19 Diamond JM. Avifaunal equilibria and species turnover on the Channel Islands of California. *Proc Natl Acad Sci USA* 1969; 64: 57–63.

20 Lynch JF, Johnson NK. Turnover and equilibria in insular avifaunas, with special reference to the California Channel Islands. *Condor* 1974; 76: 370–84.

21 Diamond JM. Comparison of faunal equilibrium turnover rates on a tropical island and a temperate island. *Proc Natl Acad Sci USA* 1971; 68: 2742–5.

22 Terborgh J, Faåborg J. Turnover and ecological release in the avifauna of Mona Island, Puerto Rico. *Auk* 1973; 90: 759–79.

23 Simberloff DS. When is an island community in equilibrium? *Science* 1983; 220: 1275–7.

24 Simpson BB. Glacial migrations of plants: island biogeographical evidence. *Science* 1974; 185: 698–700.

25 Diamond JM. Biogeographic kinetics: estimation of relaxation times for avifaunas of Southwest Pacific Islands. *Proc Natl Acad Sci USA* 1982; 69: 3199–203.

26 Case TJ, Cody ML. Testing theories of island biogeography. *Am Scientist* 1987; 75: 402–11.

27 Shrader-Frechette KS, McCoy ED. *Method in Ecology: Strategies for Conservation*. Cambridge: Cambridge University Press, 1993.

28 Shafer CL. *Nature Reserves. Island Theory and Conservation Practice.* Washington: Smithsonian Institution Press, 1990.

29 Newmark WD. A land-bridge island perspective on mammalian extinctions in western North American parks. *Nature* 1987; 325: 430–2.

30 Western D, Ssemakula S. The future of the Savannah ecosystems: ecological islands or faunal enclaves? *S Afr J Ecol* 1981; 19: 7–19.

31 Gilbert F, Gonzalez A, Evans-Freke I. Corridors maintain species richness in the fragmented landscapes of a microecosystem. *Proc R Soc London B* 1998; 265: 577–82.

32 Zimmerman BL, Bierregaard RO. Relevance of the Theory of Island Biogeography and species-area relations to conservation with a case from Amazonia. *J Biogeogr* 1986; 13: 133–43.

33 Burridge AA, Williams MR, Abbott I. Mammals of Australian islands: factors influencing species richness. *J Biogeogr* 1997; 24: 703–15.

34 Alcover JA, Sans A, Palmer M. The extent of extinctions of mammals on islands. *J Biogeogr* 1998; 25: 913–18.

35 Thornton I. *Krakatau—the Destruction and Reassembly of an Island Ecosystem*. Cambridge, Ma: Harvard University Press, 1996.

36 Whittaker RJ, Bush MB, Asquith NM, Richards K. Ecological aspects of plant colonization of the Krakatau Islands. *Geo J* 1992; 28: 201–11.

37 Schmitt SF, Whittaker RJ. Disturbance and succession on the Krakatau Islands, Indonesia. *Symp Br Ecol Soc* 1998; 37: 515–48.

38 Fryer G, Iles TD. *The Cichlid Fishes of the Great Lakes of Africa: their Biology and Evolution*. Edinburgh: Oliver & Boyd, 1972.

39 Carlquist S. *Island Life*. New York: Natural History Press, 1965.

40 Case TJ. A general explanation for insular body size trends in terrestrial vertebrates. *Ecology* 1978; 59: 1–18.

41 Carlquist S. *Hawaii, a Natural History*. New York: Natural History Press, 1970.

42 Wagner WL, Funk VA (eds) *Hawaiian Biogeography: Evolution on a Hot-Spot Archipelago.* Washington: Smithsonian Institution Press, 1995.

43 Keast A, Miller SE (eds) *The Origin and Evolution of Pacific Biotas, New Guinea to Eastern Polynesia: Patterns and Processes.* Amsterdam: SPB Academic Publishing, 1996.

44 Carson HL, Clague DA. Geology and biogeography of the Hawaiian Islands. In: Wagner WL, Funk VA, eds. *Hawaiian Biogeography: Evolution on a Hot-Spot Archipelago.* Washington: Smithsonian Institution Press, 1995.

45 Stone BS. A review of the endemic genera of Hawaiian plants. *Bot Rev* 1967; 33: 216–59.

46 Carr GD *et al.* Adaptive radiation of the Hawaiian silversword alliance (Compositae–Madiinae): a comparison with Hawaiian picture-winged *Drosophila*. In: Giddings LY, Kaneshiro KY, Anderson WW, eds. *Genetics, Speciation and the Founder Principle.* New York: Oxford University Press, 1989: 79–95.

47 Raikow RJ. The origin and evolution of the Hawaiian honey-creepers (Drepanididae). *Living Bird* 1976; 15: 95–117.

48 Sibley CG, Ahlquist JE. The relationships of the Hawaiian honeycreepers (Drepaninini) as indicated by DNA–DNA hybridisation. *Auk* 1982; 99: 130–40.

49 Olson SL, James HF. Descriptions of thirty-two new species of birds from the Hawaiian Islands: Part I. Non-Passeriformes. *Ornithol Monogr* 1991; 45: 1–88.

50 James. HF, Olson SL. Description of thirty-two new species of birds from the Hawaiian Islands: Part II. Passeriformes. *Ornithol Monogr* 1991; 46: 1–88.

51 Mayr E. Die Vogelwelt Polynesians. *Mitt Zool Mus Berlin* 1933; 19: 306–23.

52 MacArthur RH, Wilson EO. An equilibrium theory of island biogeography. *Evolution* 1963; 17: 373–87.

53 Whittaker RJ, Jones SH, Partomihardjo T. The rebuilding of an isolated rain forest assemblage: how disharmonic is the flora of Krakatau? *Biodiversity Conserv* 1997; 6: 1671–96.

54 Bush MB, Whittaker RJ. Krakatau: colonization patterns and hierarchies. *J Biogeogr* 1991; 18: 341–56.

55 Pratt HD, Bruner PL, Berrett DG. *A Field Guide to the Birds of Hawaii and the Tropical Pacific.* New Jersey: Princeton University Press, 1987.

Index

Note: page numbers in *italics* refer to figures, those in **bold** refer to tables

abyssal zone 249
Acacia spp. 150
   environmental factors 58, *59*
accidentals 31, 36
Achillea lanulosa, height of 105
acid rain 229
adaptation 6
adaptive radiation 279–80
   Hawaiian Islands 284–8, *285,*
      286, 287
Aegotheles, distribution 166
aerosols, sulphuric acid 126
Africa 144–8, *145*
   bird species 24
   endemic species **138**
   fauna 145–6, *147*
   flowering plants 24
   plant species *24*
   vegetation changes *215*
   see also East Africa
Africa, southern 62, 148–9
agriculture 207–14, *208, 210, 211,*
      212, 213
air pollution 225–6
air temperature *225, 226*
Alaska, succession following glacial
      retreat 27
algae 69
   number of species **12**
algal bloom 246, *247,* 248
alias problem 11
alleles 100
Allen's rule 105
allopatric speciation 65, 107
altitude, and plant height 105
Amazon Basin
   bird fauna 171
   flowering plant species 24
   rainfall 119
   tropical rainforests 171
Amazon frog, breeding requirements
      272
Amazonia, tropical forest 19
Ambrosia dumosa 80
American grey squirrel, spread of 61
ammonoids, extinction of 126
Anax imperator 43–4
Angara flora 120

angiosperms 127
   dispersal 142
   percentage representation *22*
   see also flowering plants
animals, domestication of *211,*
      213–14
Antarctic Divergence 244
Antarctic ice cap, melting of 227
antitropical distribution 254
ants 279
   number of species *2*
antshrikes 268
aphids 22
Aphodius holdereri (dungbeetle) *47,*
      49
apomorphic state 162
apple maggot 65
Apteryx australis 34
Aquilapollenites 128
arachnids, number of species **12**
Araucaria 132
Arbutus unedo *48,* 49
archibenthal zone 249
archipelago speciation 263
arctic charr, different forms of *65*
Arctic Ocean 244
Arcto-Tertiary flora 157
Argyroxiphium 285, *286*
aridity 115, 144, 214
arks 152, 161
Armadillidium spp. 57–8
armadillo 95
arrow-worm 245, *246*
Artemisian herba-alba 80, *81*
arthropods 11–12
Asiamerica 118, 173
associations *see* communities
aurochs 214
Austral Polar Current 244
Australia
   climate 58, *59*
   continental shelf *150*
   Early Tertiary deposits 125
   endemic species **138**
   flora 149–50
   fruit-eaters/omnivores 139
   lack of carnivores in 139
   pollen diagrams *186, 187*

proportion of introduced plant
      species **62**
   separation from Antarctica 243
   wheat yield *230*
   zoogeographical region 137
Australopithecus afarensis 204
avocet 64

bacteria, number of species **12**
bamboo grass 275
barley 209
barn swallows 5
barnacles 68–9
   competitive exclusion 59–60
Barosaurus 124
barriers 36–8, 112
   East Pacific 254
   Mid-Atlantic 254
   shelf faunas 253–4
   to interbreeding 103–4, *104*
   ways of overcoming 36–8
bathyal zone 249
   fauna of 250
bathypelagic zone 241
bats 5
bays 132
beans 210
beard-tongue, pollination *104*
beech
   changing distribution of *231*
   disappearance of 31
beetles
   body size and abundance *21*
   distribution of 11
   flightless 288
   number of species 2, *3, 4,* 11–12
bell miner, predatory habits 68
Benguela Current 115
bennettitaleans 121
benthic organisms 241, 243
Bergmann's Rule 5, 105
Betula pubescens 201
biodiversity 1
   diversity in time 26–31, *27, 28, 29,*
      30
   in ecosystems 77–80, *78*
   gradients of diversity *13,* 14–24,
      14–22

biodiversity (*cont.*)
　hotspots 24–6, *24, 25*
　marine 31–2
　number of species 10–14, **12**,
　　13, 14
　in oceans 239
biomass 19, *27, 28*
biomes 81, 83
　changes in 234–6, *235*
　in changing world 92–3
　and climate 84
　terrestrial distribution *84*
　zonation of 83
biotic factors 59
bipolar distribution 254, *255*
birds
　accidental visitors 36
　breeding, Central and North
　　America *14*
　distribution 53
　Hawaiian Islands *287–8*
　island distribution **266**
　migration 36–7
　number of species 20, *21*
　Rakata Island 278
blanket mires 218, 221
blood groups 109
bloom *see* algal bloom
blue–green algae 275
bog growth 219
bog mosses, sulphur damage to
　228
Boreal plants 155–8, *156*
boreotropical flora 132
Borrelia burgdorferi 4
Bothriospondylus 124
brachiopods, extinction of 126
Brachiosaurus 124
Brazil, rainforests 82
bristle-worm 245
British Isles, proportion of introduced
　plant species **62**
brush opossum 151
Bufo spp. 103
butterflies
　analysis of taxa 168–9
　climate and population *232*
　Rakata Island 278
　species richness *16, 17*
　specific food plants 72

C₃ plants 55, 213, 230
C₄ plants *55, 56*, 213, 230
Calamagrostis lapponica 229
Calamites 120, 121
California
　endemics in 51
　Mojave Desert 80
　Quercus agrifolia 74
Camargue, spatial separation 64

camels 155
Cape flora 148
Cape floral kingdom 148
capybara 153
carbon dioxide
　atmospheric 225, 234
　bubble analysis 193
　fixation 73
　glacial period 172
　in sea water 244
　see also greenhouse effect
Carboniferous Period *117*
　global cooling in 119
Caribbean
　coral reefs 253
　hurricanes 264
carnivores 75
　lack of in Australia 139
catchfly, distribution 53, *54*
Cathaysian rainforest flora 120
caytonias 121
Cenozoic Period
　climate changes 128–31, *130*
　flora 131–3
　South American fauna 153–5,
　　154
　zoogeographical distribution
　　138
Centre of Origin hypothesis 258
centres of dispersal 172
ceratopsians 124
chamaephytes 82
Chenopodium album 229
Chicxulub crater 126
chimneys 251
chimpanzees 34, 203, 204
China 121
chironomid midge, cladogram *163*
chlorofluorocarbons 225
chlorophyll 246
chromosomes 100
²cicadas, dispersal patterns 169
cichlid fish, speciation in 26, 65–6,
　103, 172
cinnamon 132
circumboreal distribution 11
cladism 162
cladistic biography 163, *164*
cladograms *162*
　podonominine chironomid midge
　　163
　vertebrate groups *164*
Clements model of vegetation 73,
　74
climate
　biome modelling 88–9, 91
　and biomes *84*
　changes in 224–7, *225, 226*
　deterioration of 218–20, *219*
　diagrams 88, *90*

and fire 218
glaciation *192*
Late Cretaceous/Cenozoic Period
　128–31, *130*
patterns of 84–8, *85, 87, 88*
seasonality 133
see also Ice Ages
climate envelopes 232
climatic relicts 45–9, *46, 47, 48, 49*
climatic wiggles 176–7
climax forest 31
clines 105–6
Clovis Man *207*
cockle 64
coconut palms 264
Cocos Islands *255*
collared dove 37
common shrew, distribution *216*
communities 72–5, *73, 75*, 183
　changes in 234–6, *235*
　global 80–4, *81, 83, 84*
competition 106
　reduction of 63–7, *64, 65, 66, 67*
competitive exclusion 59
coniferous forests (taiga) 81, 84
　Pacific Islands *265*
continental drift 114–15
continental rise 240
continents
　changing patterns of 118 *127*
　movement of 119–20
continuous vegetation 74
convergences 244
Cook Islands, bird genera **266**
coral reefs 251, 255–9, *257, 258*
　as balanced environment 69–70
　Caribbean 253
　locality cladogram *170*
coral trees 283
corals
　biogeography 256
　generic diversity *257*
　hermatypic 256
　see also coral reefs
cord-grass, polyploidy in 105
Cordaites 120
Coriolis effect 86
Cornish heath, distribution 217
Corophium volutator 64
corridors 37
creosote bush 80
Cretaceous Period 115, 254
　climate changes 128–31, *130*
　dinosaurs in 124
　extinction event 126–7
　flora 121, 131–3
　plant groups in *127*
　vertebrates of 118
　world geography 118, *127*
crop species, spread of *212*

crustaceans
   distribution 56, 57, 253
   number of species **12**
Cyanea, Hawaiian Islands 286
cyanobacteria 275
cycads 121

Dacrydium 132
daisies 38–41, *39, 40, 41*
   dispersal 264
damselflies 41–4, *42, 44*
Daphne Major, finch population 98,
   99
Darwin's finches 96
date palm *36*, 132
decomposers 75
deep-sea trenches 113, 241
deep-sea vents, as ecosystems 77
desert pupfish 190, *191*
deserts 112, 115
detritus feeders 75
Devonian Period 119
dichotomies 162
Dicroidium 121
diet, and diversity gradient 17
dingo 125
dinosaurs 124
   extinction of 126
   North American 153
dispersal 165
   centres of 172
dispersalist theory of distribution 7,
   161
displaced terranes 173
disruption of habitats 75
distribution 6–7, 34–71
   antitropical 254
   bipolar 254, 255
   circumboreal 11
   disjunct 7
   dispersalist theory 7
   limits of 34–6, *35, 36*
   patterns of 34–71
   vicariance theory 7
disturbance of habitats 75
diversity *see* biodiversity
dodo 280
dogwood 132
domestication 207–14, *208, 210, 211,*
   212, 213
dragonflies 41–4, *42, 44*
Drosophila 284
   speciation 103
drought, effects on finch population
   98
dry lands 214, *215*
Dry Tortugas, ant species 279
Dryosaurus 124
Dubautia 285–6
dungbeetles 47, 49

dwarf palm *35*
Dysoxylum 277

Early Cenozoic Period, world
   geography 118–19
Early Devonian Period *117*
earthquakes 26
East Africa
   cichlid fish species in 26, 65–6
   savanna grasslands 63, 79
   species coexistence in 63
East Pacific Barrier 254
eastern phoebe, distribution *53*
ecological biogeography 7
economy of hypothesis 162
ecoregions 93
ecosystems 75–7, *76*
   and biodiversity 77–80, *78*
ecotrons 78
Edith's checkerspot butterfly,
   climate and population *232*
einkorn *208*, 209
El Niño effect 92, 98, 227
elephant birds 280
elephants, origins of 147
Ellesmere Island 132
Emiliani curve 177
emmer wheat 209
Enallagma spp., distribution 41–4,
   42, 44
endemic organisms 49–51, *50*
endemic species 49–51, *50*
   degree of endemicity **138**
endemicity and history 169–70,
   170
environmental gradients 52–6, *52,*
   53, 54, 55
Eocene Period
   climate 132
   world geography 118–19
Ephedra 80
epicontinental seas 115, 129, 240
equatorial belt 119
equatorial submergence 249
Erythina 283
estuaries 56
eucalypt tree, distribution 164,
   165
Eucalyptus 149
euphausids 245
euphotic zone 241, 247
Euramerica 118, 121, 173
Eurasia 133
   steppes 82
Europe
   distribution of mammals in
      216
   potential vegetation *235*
   proportion of introduced plant
      species **62**

eurytropic species 52
eutrophication 229
evapotranspiration, and plant
   diversity 18
evolution 96
   human 109–10
evolutionary theory 107–9, *108*
extinctions 13
   Cretaceous 125–7, *127*
   island 267, *268*
   of megafauna 206, *207*
   of plants 13–14

Fabaceae 78
facilitation 74
fantail 34
faunal provinces 112, *137*
Fertile Crescent *210, 213*
fig trees 277
Fiji Islands, bird genera **266**
filter 37
finches 95
   beak types 96, *97*
   Darwin's 96
   effect of drought on 98
   interspecies mating 99–100
fire 218
fish, species richness *258*
flamingo 64
flora
   Angara 120
   Arcto-Tertiary 157
   Australia 149–50
   boreotropical 132
   Cathaysian rainforest 120
   Cenozoic Period 131–3
   Cretaceous Period 121,
      131–3
   Gondwana 121
   Jurassic Period 121
   Madro-Tertiary 157
   New Zealand 152
   sclerophyll 82, 151
   Southern Africa 148–9
   Southern Hemisphere 144
   see also flowering plants
floral kingdoms *140*
floral regions 112
flowering plants 127–8
   Africa 24
   Amazon basin 24
   current distribution 140–2, *140,*
      141
   dispersal of 127–8
   distribution *22*
   geography of 142–4, *143*
   Pacific Islands *265*
   South-East Asia 24
fogging 11
food webs 75, *76, 79*

forests
  birch and pine 177
  clearance 220
  coniferous 81, 84
  deciduous 177, *235*
  movement of 200–3, *202*, *203*
  Rakata Island 277
  spread of 218
  subtropical monsoon 83
  temperate 19–20, 23
  see also trees; tropical rainforests
fossil record 5, 6
fossils 95, 172–3
founder principle 266
frogs, Central and South America
  15
fruit-eaters 139
fruitflies 103
  Hawaiian Islands 284, *285*
fungi, number of species **12**
fynbos vegetation 149

Galápagos Islands 95, 96
Gammarus spp., distribution 56, *57*
gas bubble analysis 193, *194*
genes 100
genetic drift 101
genotype 101
geological time scale *114*
geophytes 82
Geospiza spp. 96, 267
  beak types 96, *97*
giant butterwort, distribution 217
Gigantopteris 120
ginkgos 13, 121
glacial drift deposit 175
glacial relicts *46*, 171
glaciations 23, 119, 175
  causes of 192–5, *194*
  climatic conditions *192*
  nomenclature 176
  vegetation *182*
Gleason model of vegetation 73, 74,
  91, 234
global cooling 127, 199
global warming 227
  see also greenhouse effect
Glossopteris 120, 121
Gobi Desert 115
Gondwana 118, 163
  dinosaurs in 124
  flora 121, 144
  flowering plants 128
  movement of 119
gorillas 203
  disjunct distribution 49, *50*
grapes 221
grasses
  competitive exclusion 60, *61*
  stimulation of growth 229

grasslands
  plant species 82
  Rakata Island 275
  vegetation *21*
grazers 75
green plants 75
greenhouse effect 92, 119, 127,
  193, 225
greenhouse gases 226
Greenland
  ice sheet *199*
  oxygen isotope curve *219*
Grenzhorizont 219
grey hair grass, distribution *52*, 53
guanaco 95
guinea-pig 153
Gulf Stream 115
Gunnera 131

habitats
  disruption of *75*
  disturbance of *75*
  see also microhabitats
hadal zone 249
  fauna of 250
hadopelagic faunas 241
halophytes 80, *81*
Haloxylon persicum 80, *81*
Hawaii
  amphipods 170
  proportion of introduced plant
    species *62*
Hawaiian damselfish 69
Hawaiian honey-creepers 190
Hawaiian Islands 280–8, *282*
  evolutionary radiations 284–8,
    285, *286*, *287*
  mechanisms of arrival 283–4
hazel
  distribution 220
  interglacial distribution 217
heather 229
helophytes 82
hemicryptophytes 82
hemlock 183
hemp 221
herb paris, absence of in Ireland 217
herbivores 75
  and increasing numbers of plant
    species 69
hermatypic corals 256
Himalayas, vegetation *83*
Hipparion 146
Holarctic region
  endemic species **138**
  mammals 155–8, *156*
Holocene Period 177
hominids 203–4, *204*
hominoids 203
Homo erectus 205

Homo habilis 205
Homo sapiens 8, 203, 205
honey-creepers, Hawaiian *287*
Horse Latitudes *85*, 86
hotspots 262
Hudson's Bay 115
humans
  emergence of 203–6, *204*, *206*
  environmental impact 220–1
  modern 206, *207*
hummingbirds, pollination by *104*
hydrological cycle 77
hydrophytes 82
hydrothermal vent faunas 251
hydrozoans 256

Ice Ages 7, 49, 136, 144, 157, 166
  changes induced by 175–96
  climatic changes 255
  confinement of species 172
  Little 219, 221
  see also glacial relicts; glaciations
ice caps *183*
  bubble analysis 193
  melting of 227
icehouse effect 119, 127
iguana lizard 264
Imperata cylindrica 275
India 144–8, *145*
Indian Ocean
  coral reefs 255
  winds and currents 246
individualistic model of vegetation
  73, 74, 91, 234
insects
  diversity 15, *16*, 17
  number of species **12**
insects, Hawaii 288
interacting factors 56–9, *57*, *59*
interactions between organisms *4*
interbreeding 102–3
  barriers to 103–4, *104*
interglacials 177–82, *179*, *180*, *181*
  current 197–200, *198*, *199*
  present 217–18
  vegetation *182*
internal predators 68
interstadials 177–82, *179*, *180*, *181*
intertropical convergence zone 86
Ipswichian interglacial *179*
Ireland
  climate 49
  post-glacial flora and fauna *216*,
    217
island arcs 262–3
islands 9, 262–90
  adaptive radiation 279–80
  biogeography 267–71, *268*
  Hawaiian Islands 280–8, *282*, *285*,
    286, *287*

islands (*cont.*)
    nature reserves 271–2
    problems of access 263–6,
        265, **266**
    problems of survival 266–7
    Rakata 272–9, *273, 274, 278*
    types of 262–3
isopolls *202, 203*

jellyfish 253
Jurassic Period 115
    flora 121
    plant groups in *127*

K/T boundary
    clay layer 126
    marine plankton record 127
    reduction of flowering pollen
        127
    soot deposits 126
Kentish plover 64
keystone species 70, 78, 272
kiwis 34, 152, 280
Komodo dragon 280
krill 245
Kure Island 282
Kuroshio current 129

lacewings, speciation 107
lactase deficiency 110
Lake Victoria 26
last glacial 183–92, *185, 186, 187,
    188, 189, 190, 191, 192*
Late Permian Period *117*
Laurasia 118
    floral provinces 121
    flowering plants 128
laurel 132
Laurentide glacier
    discharge of meltwater 198
    extent of *199*
lentils 209
Lepidodendron 120
Liliiflorae 170
    distribution *141*, 142
lime 201
    absence of in Ireland 217
limiting factors 51
linkage diagrams *169*
Liriodendron 264
Little Ice Age 219, 221
lobeliads, Hawaiian Islands 286–7
Lucy (early hominid) 204
Lycopodium 120
Lyme disease 4
lynx, and prey switching 68

Madagascar 148
    rainforest distribution *8*
Madro-Tertiary flora 157

magnolias 132
    as evolutionary relics 44, *45*
maize 210, 213, 230
major nodes 165
mammals
    Australia 150
    distribution 137–40, *137*, **138**,
        140, 216
    early spread of 124–5
    geography of 142–4, *143*
maple, dispersal 264
marine biodiversity 31–2
marine environments, salinity 56
marsupials 124, 125, 150
masting 4, 5
meadow saffron, polyploidy in 105
Mediterranean area *145*
Megaegotheles, distribution 166
megafauna 206, 235
    extinction of 206, *207*
Meiji Seamount 281
meteoric impact, and Cretaceous
        extinction event 126
methane
    atmospheric 225
    bubble analysis 193
    production of 194
Metrosideros 283
mice, white-footed, interactions
        involving *4*
microbes, diversity 12
microclimate 19, 20
microhabitats 64
Mid-Atlantic Barrier 254
Milankovich cycles 193
milk, and adult digestion 110
milkweed 72
moas 152, 280
moisture index 91
Mojave Desert 80
molluscs 253
    number of species **12**
montane pine forest 184
    limited range 51
mosaic pattern of vegetation 74
moufflon 214
mountain avens 197, *198*, 218, 229
mountain ranges 112
    Cenozoic Period 129
mountain (varying) hare 46–7
mutations 101
myxomatosis 69

natural selection 95–6
nature reserves 271–2
Neanderthals 205
Nearctic region 155–6
    endemic species **138**
nektonic organisms 241, 243
nematodes, number of species **12**

Neocalanus 248
neoendemism 51
Neonauclea 277
Neotropical region, endemic species
        138
neritic realm 241
Neuropteris 120
New Caledonia, bird genera **266**
New Hebrides, bird genera **266**
New Zealand
    flora 152
    proportion of introduced plant
        species **62**
Newfoundland, prey switching 68
nitrogen fixation 78, 79
nitrogen overload 227–9, *228*
nitrogen oxides 225
nitrogen-fixing bacteria 78
Noah's ark 152, 161
Nominostomias 249
Normapolles 128
North America
    agriculture 221
    butterfly population *232*
    dinosaurs 153
    prairies 82
North Pacific Polar Front
        Convergence 244
Norwegian mugwort, glacial relict
        46, *47*
Nothofagus 128, 132, 142, 143,
        144
    Australia 151
    New Zealand 152
novelty **95–111**

Obik Sea 118, 128
ocean baselines 166
ocean basins 241–3, *242*
    dynamics of 243–51, *245, 247*
ocean domains 246–9, *247*
ocean floor 239, *240*
    life on 249–51
oceanic conveyor belt 86, *88*, 194,
        227, 243, 244
oceanic gyres 243, 244, 246
oceans 112, 238–61
    adverse effects of food gathering
        from 8–9
    biodiversity 239
    biogeography 241
    life in 245–51
        nutrient distribution in *245*
        zones in 239, *240*
old field succession 28, *30*
Older Dryas event 197
Oligocene Period, climate 132
omnivores 139
open-sea realm 241
orang-utans 203

Oriental region
  endemic species **138**
  zoogeography 137
osmundas 121
overgrazing 147
owls, coexistence of species 63
oxygen 56
  heavy 176–7
  isotope curves *178, 183, 219*
  normal 176–7
  in sea water 244
oystercatchers
  feeding behaviour *64*
  sympatric speciation 108
ozone 225

Pacific Equatorial Convergence
    244
Palaearctic region 155–6
  endemic species **138**
  fauna 158
palaeobiogeography 173
palaeoendemism 51
palaeomagnetism 115
Palaeocene Period *114, 130*
  plant groups in *127*
palms
  date *36*, 132
  global distribution *35, 36*
pampas 82
Panama Isthmus 254
  completion of 243
  coral reef biota 259
  formation *154*
  fossil marine faunas 118
  nodes of 166
panbiogeography 164–7, *167*
Pangaea 118, 121, 243, 255
  land animals of 124
Panthalassa 243
parasites 72
parsimony 162
pattern cladism 163
peatlands 177, 219
  erosion of 228
Peganum harmala 80
pelagic organisms 65, 241, 258
Permian Period 120
Permo-Carboniferous plants 118
phalangers 152
phanerophytes 82
phenotype 101
photoperiodism 49
photosynthesis 55–6, 75
photosynthetic algae 251
phyletic tracks 165, 166, 167–9, *168,*
    169
phylogenetic biography 162–3
physical limitations 51, *52*
physiognomy 80, 88

phytoplankton 246, 248
picoplankton 248
pitfall traps 2
placental mammals 124, 125
planktonic organisms 60, 241
plant community 72
plant formations 82
plants
  adaptation to carbon dioxide
      changes 194
  diversity gradient *16, 17, 18*
  domestication of *211*
  green 75
  height and altitude 195
  'life form' 82
  mountain *50, 51*
  Permo-Carboniferous distribution
      118
  proportion of introduced species
      62
  salt-tolerant 89
  stomata 194
  world distribution of productivity
      19
plate tectonics 113–15, *113, 114,*
    161, 173
Pleistocene Period
  biogeography 170–2, *171*
  biological changes 182–3, *182,*
      183
  glacial/interglacial stages *181*
  South American fauna 153–5,
      154
plesiomorphic state 162
pluvials 190
Podocarpus 132
Polar Domain 246, 248
pollen diagrams 179, *180*, 220
  Australia *186, 187*
  British Isles *200*
  United States *185*
pollen grains, importance of 179
pollen records 128, 179
  K/T boundary 127
pollination *104*
pollution 226
polychaete worms 253
polygons *176*
polyploids 104–5
pond terrapin 220
prairies 82
predators 67–70, 76
  internal 68
  islands 280
Pretiglian 175
prey 67–70
  islands 280
prey-switching 79
primary consumers 76
protozoans, number of species **12**

Psaronius 120
psyllid bugs 68
punctuated equilibrium model of
    evolutionary change 10
purslane 221
pygmy shrew, distribution *216*

Quercus agrifolia 74

ragweed 221
rainforests
  Brazil 82
  canopy profiles *29*
  destruction of *8*
  humid *188*
  Madagascar *8*
  see also tropical rainforests
Rakata Island 272–9, *273,*
    274, 278
  coastal environment 274–5
  life on 275–9, *278*
Rapoport Effect 23, 24
Rapoport's Rule 23
Rattus tiomanicus 278
red deer 220
  distribution 217
Red Sea evaporate 253
red spruce, changing distribution of
    231
red squirrel, decline of 61
redshanks 64
redundant species hypothesis 78
refugia *206*
relict distributions
  climatic relicts 45–9, *46, 47, 49*
  evolutionary relicts 44, *45*
Reuverian 175
rhesus blood group 109
Rhipidura fuliginosa 34
ring species 102
rivet hypothesis 77, *78*
rules 105–6

sagebrush steppe 184
Sahara Desert, rainfall trends 233
St Helena 280, *281*
salamanders, interbreeding of
    102–3
salinity 56
  changes in *220*
Salsola 80
salt-marsh community, competitive
    exclusion in 60, *61*
salt-tolerant plants 89
Samoa, bird genera **266**
sand deserts, distribution *189*
sapropels 214
sauropods 124
savanna grasslands 63, 79, 133, 147,
    155

Scaevola toccata 283
scaly anteater 147
Scaptomyza 284
sclerophylls 82
  Australia 151
Scottish red grouse, territorial
    behaviour 106
sea level, changes in 7–8, 215–17,
    216
sea mounts 263
sea plantain 218
sea temperature *225*, 253
sea urchins 69
secondary consumers 76
seeds, dispersal to islands 264
Sequoia 132
shallow-sea realm 241, 251, *252*
shelduck 64
shelf break 240
shelf faunas
  barriers between 253–4
  faunal breaks within 252–3
  latitudinal patterns in 254–5
Siberia 120, 121
sickle-cell anaemia 109
Silurian Period 118
Sitka alder 27
sloth 95
snails, evolution of *108*
snow 219
snowshoe hare 68
Soay sheep 214
Society Islands, bird genera **266**
solar energy 76
  trapping 73
solar radiation *225*
Solomon Islands, bird genera **266**
South Africa, veld 82
South America
  fauna 151–5, *154*
    Late Cenozoic 153
    Late Cenozoic/Pleistocene
      153–5, *154*
    Late Cretaceous/Early Cenozoic
      152–3
  forests in *171*
  fossils 121
  pampas 82
South-East Asia, flowering plants 24
South Polar ice sheet 119
South-East Asia 144–8, *145*
  continental shelf *150*
Southern Africa
  flora 148–9
  proportion of introduced plant
    species **62**
Southern Subtropical Convergence
    244
sparrows, individual characteristics
    96

spatial separation 64
speciation
  archipelago 263
  cichlid fish 26, 65–6, 103
  finches 99–100
  salamanders 102–3
species
  coexistence of 63
  definition of 101–2
  number of 10–14, **12**, *13, 14*
species interaction 59–63, *61*,
    62
species richness 28
Sphenodon 152
sphenopsid 120
spreading ridges 113, 115
springtails, glacial relicts *46*
squash 210
stability 79
starfish 68
starlings, displacement of native
    species 60–1, *62*
stenotopic 52
steppes 82, *235*
sticklebacks, competition in 106
stoat, distribution *216*
strawberry tree, distribution 217
successions
  climax 31
  following glacial retreat 27
  old field 28, *30*
sugar cane 275
sulphur dioxide 226
sulphur overload 227–9, *228*
sunspots 225
sweepstakes routes 37, 263
sympatric speciation 65, 107

taiga 81, 84, *235*, 265
Tamarix spp. 80, *81*
tanager, specialization in *66, 67*
tapirs 155
taxa 6
taxonomy 80
tectonic plates *113*, 114, 263
teosinte grains 213
Terminalia forest 274–5
tertiary consumers 76
Tethys Sea 243, 259
thermal blanket effect 225
thermohaline circulation *see* oceanic
    conveyor belt
therophytes 82
Thomson's gazelle 63
thorn maggot 65
Tibetan Plateau 129
ticks, interactions involving *4–5*
tills 177
Timonius 277
toads, interbreeding 103

Tonga, bird genera **266**
top predators 76
topi 63
tracks *see* phyletic tracks
Trade Winds 86
Trades Domain 246, 247
transformed cladism 163
tree-groundsels *39, 40, 41, 72*
trees
  drought deciduous 83
  North America *18*
  St Helena *281*
  species richness *16, 17, 18*
  spread through North America
    202, 203
  see also forests
trenches *see* deep-sea trenches
Triassic Period
  fauna of 124
  mammals in 124
Tribulus cistoides 98
Triticum aestivum 208
Triticum monococcum 208
Triticum tauschii 208, 209
Triticum turgidum 208
Tropic of Cancer 85
Tropic of Capricorn 85
tropical rainforests 23
  Amazon Basin 171
  Carboniferous Period 119
  filling of gaps in 27, *28*
  food webs 75, 76
  leaf cover *20*
tropics
  animal life in 22
  productivity 18
tulip trees *45*, 183
tundra 133, 177
  plant species 82
  temperature effects on 234
  temperatures 88

United States, proportion of
    introduced plant species
    62

vascular plants, number of species
    12
vegetation science 73
veld 82
vertebrates, number of species **12**
vicariance 7, 161, 251
Viking funeral ships 161
vines 132
viruses, number of species **12**
vitamin D, and skin coloration
    109
volcanic activity 26, 266
volcanic islands 262–3
volcanoes 263

Wallacea 151–2
wanderers *138*, 142, *143*
weasel, distribution *216*
weathering 77
weevils 68
Westerlies Domain 246, 247, 248
wheat
    evolution of *208*
    yield *230*

whitethroat, population density *233*
wiggle matching 177
Wilkesia 285–6
wing nut tree 183
woodlice, interacting environmental
        factors 57–8
woodrats 5–6
    change in body size 5, *6*
    faecal pellet size 5, *6*

Younger Dryas event 197, 198, 199,
        214, 227

Zea mays 213
zebra 63
zoogeographical regions 112, *137*
zooplankton 248
zooxanthellae 256
Zygophyllum eurypterum 80, *81*